Lines of Power

Driver and guard proudly pose with their electric train on the LBSCR 'Elevated Electric' system (so called because of the overhead 6.6kV AC power lines) in pre-Grouping days. This innovative system, opened on 1st December 1909, operated between Victoria, London Bridge and Croydon, being extended by the Southern Railway to both Sutton and Coulsdon. The system was later converted to a low-voltage DC arrangement, the last high-voltage train operating on 29th September 1929. *K Robertson archive*

Contents

Foreword

I am delighted to be able to provide the foreword for this book on the history and application of the use of electricity as a source of power for moving trains on the UK network. This book provides what I believe to be the most complete record of railway electrification projects carried out in the UK, from the earliest schemes up to the present day.

The book also provides revealing examples of the impact of politics and economics on the prospects for electrification projects. It lays bare the effects of the lack of commitment to a structured and resourced plan for electrifying Britain's railways, from both private companies and governments. Fixed infrastructure costs for railways, such as earthworks, tunnels, bridges, track and signalling, are high, but these assets were constructed in the main without much thought being given to the possibility of accommodating the additional infrastructure requirements for electrification. As a result, the cost of electrifying established infrastructure was and is significant, especially within the confines of the UK structure gauge, yet, as we see from this book, the possibility of using electricity for traction purposes has been around for some considerable time. It could have been anticipated that the era of cheap and plentiful coal and oil would not last and that accommodating and carrying around power-generating equipment, initially in steam then diesel trains, is both costly and inefficient. The quest for railways to operate at higher and higher speeds, have modern passenger comforts such as air conditioning and wi-fi, while providing a means for trains to return energy to the supply system when braking, can only be met by electric operation.

One consequence of the lack of a long-term plan and a rolling programme of electrification was the lack of control over system specification and design standards. This has resulted in the present mix of alternating current (AC) and direct current (DC) systems and several attempts to choose both a system and operating voltage. This led directly to the initial construction and then costly conversion of the original DC overhead line systems to AC and the subsequent further conversion of the AC system voltage from 6.25kV to 25kV. As a junior engineer I can recall the Electrification Department of the British Railways Board (BRB) producing a proposal to convert the Southern Region DC lines to AC overhead, justified by avoiding the renewal costs of the DC track-side power distribution equipment. It was not surprising that the project was not approved by the then Board due to the disruption to train services and the rolling stock implications that would have been involved. As a further example of the lack of vision, an electrification project usually must bear the cost of re-signalling when a long-term plan could have specified traction immune signalling when it was initially installed.

Finally, the need for railways to be more efficient and safer and now to decarbonise its operations were not perhaps as important to the decision makers of the past as they are today. The 25kV AC system is 20 per cent more efficient than the DC system in delivering power to a train by having reduced transmission losses. It is safer by removing exposed electrical conductors from the track, and perhaps now the drive to decarbonisation will finally provide the incentive to engage in a rolling programme of electrification that this nation has so sadly lacked. If so then it must be driven by people with vision and not repeat the mistakes of the past.

Noel Broadbent CEng MIET

Introduction

Climate change demands that the industrial scale combustion of fossil fuels is stopped. To stabilise the climate we must halt the increase in atmospheric carbon dioxide (CO_2) by 2050. Annual CO_2 emissions into the atmosphere must reduce to the level of annual CO_2 removal from the atmosphere by natural processes and/or technology; that is net zero. The UK has set a net zero target binding the government to deliver against the 2050 obligation.

Transport is the biggest generator of greenhouse gas emissions. This presents the railway with an opportunity to leverage the inherent energy efficiency of rail transport and become a major part of the overall climate change solution. Immediate modal shift from road to rail offers considerable opportunity – a diesel freight train produces emissions of approximately 10 per cent of those from the same freight tonnage moved by multiple diesel lorries on the road. Trains hauled by electric traction supplied from green electricity generation are completely carbon free. The railway is the only land transport system for heavy freight that can meet that ultimate zero greenhouse gas target. That should set a strategic trajectory for modal shift to and decarbonisation of rail across the next five decades.

Since practical electric traction motors were pioneered in the 1860s and finessed through the following two decades it has been clear that trains powered by electricity are, when viewed from a railway operating perspective, superior to any other available traction type. The urgent demand to tackle CO_2 emissions provides an extra dimension to that advantage. The transcontinental networks of high-speed passenger and heavy haul freight, which are the arteries of mainland Europe, offer a clear model for what is achievable. Why then is the UK so far behind in electrifying our railways?

In the UK railway development from its very start was unplanned. Whereas the role of governments in Europe has generally been to steer and drive national integrated transport plans, in the UK the role quickly became one of seeking to contain and regulate competing private companies making increasingly speculative investment motivated largely by the prospect of profit. That same model was also evident from the start of the adoption of electric traction by the independent rail companies prior to the First World War. Market forces and profit-motivated competition may be fertile ground for innovation, but the consequence is diversity and incompatibility of systems. Ultimately, this leads to waste and abortive cost.

In this book John Buxton and Donald Heath start by providing an insight into the early history of electric traction, from the pioneers to the early adopters.

From 1890 onwards railways in the UK began to experiment with implementation of various forms of electric traction.

The building of deep tube underground railways in London commenced against the backdrop of competition from electric tramways built by many municipal authorities. In America, tram technology had been adapted and advanced by the ground-breaking work on traction motor control by Frank Sprague and the emerging interest in rail traction from the electrical engineering industrial giants Westinghouse and Edison. In 1890 when the City and South London Railway (C&SLR) needed an electric traction solution it adopted a third-rail system based on American technology. That third-rail system was looked at with admiration by the London & South Western Railway (LSWR). This company adopted a conductor rail for the Waterloo and City Line and so by 1917 that railway was embarked on a development that today gives the UK the only main line conductor rail system in the world.

Other railways adopted the overhead system of electrification, resulting in both alternating current and direct current systems being in operation prior to the First World War. Early chapters of the book present a chronology of each of those systems with the early experience of electric railway operation before and between the world wars being clearly explained.

After the Second World War the railways needed repair, replacement and modernisation. The nationalised British Railways (BR) published a plan to modernise the railways in 1951. That plan included a move from steam traction to electrification, a long-term rolling programme of electrification with completion in about the year 2000. The severe post-war economic austerity and seductive rival plans for adoption of diesel trains frustrated the electrification element of that modernisation plan. The book describes how the UK approach to electrification matured through this period and explains why 25kV AC emerged as the favoured system replacing 1500V DC. Two further rolling programmes for electrification were agreed with the government, published in 1977 by BR and 2010 by National Rail (NR). Each has failed. The book sets out in detail reasons for those failures. Along the timeline a stop–start approach to electrification schemes for individual routes has progressed. Early BR schemes, particularly the Euston, Birmingham, Manchester and Liverpool (EBML) project, fell victim to overspending and late delivery.

In 1963 the EBML works were stopped by the Minister of Transport (then Ernest Marples) who was more sympathetic to the road building programme, and intolerant of the cost escalation in the electrification investment plans. BR reacted to the demand to reduce cost and improve delivery of electrification. By 1967 a remarkable transformation in technology and design and in project management was the result. By the late BR period delivery of electrification was at its zenith, the London (Hitchin) to Edinburgh ECML works undertaken between 1984 and 1991 being the culmination of a huge effort to maximise those improvements. Donald Heath was Project Director for the ECML works, and, as an engineer on that project, I am proud to have been part of the team that allowed him to headline the project completion certificate for ECML: 'One Year Early and Under Budget'. The insight in the section covering the ECML programme is a rare first-hand account from the unique perspective of the Project Director. This book offers an assessment of that BR success, and explains how the lessons from that success might be interpreted to inform the vital work in support of decarbonisation.

Sections in the book cover the technical aspects of traction fixed equipment and affected railway systems. However, the book is not meant primarily to be an engineering textbook; rather, it is a detailed guide to the highs and lows across the history of railway electrification in the UK. Understanding of that history is impossible without understanding the influence of national politics and national economics, the structure and power balances in railway management and the tide of public opinion.

John Buxton and Donald Heath set out the history of electrification in the UK, but much more significantly they present in-depth analysis of that history. They pick up on a wide spectrum of issues that individually hinder or help in electrification programmes. Without an understanding of history mistakes are always likely to be repeated. Network Rail has repeated in the Great Western Electrification Project (GWEP) the mistakes made by BR on the West Coast Route in the 1960s. This book is rich with examples that provide such opportunities to learn.

With the need to decarbonise the railway comes ever more need for electrification to be delivered at the lowest possible cost and as reliably as possible. There is no margin to accommodate poor financial management and/or long delays in completion. Political, economic and public support can only be secured by affordable and dependable performance.

This book will be required reading to understand what the past can teach for all who would seek to propose, develop, or implement railway electrification projects.

Peter Dearman FREng CEng FPWI FIET FIMechE

Preface

This book is a subjective review of the progress of electric power and its direct application, by means of conductive infrastructure, to UK rail traction over two centuries.

The primary objective of this discourse is to give an insight into the history, politics and economics of railway electrification and raise awareness of the consequences that past dictates have had on UK electrification progress. Most importantly of all, our intent is that the lessons from the past will better inform industry leaders, politicians and stakeholders, enabling more informed decisions to be made going forward.

To put the progress of UK railway electrification in context, an outline of the history of developments from the nineteenth century to the present day is given in Chapters 1–15. This account takes cognisance of the national, railway company and industry policies of the time, coloured by our own first-hand knowledge and experience. The reader should be aware that it is difficult to make comparisons between the period prior to the Second World War and post-war phases because both the industry and the outside world has changed so significantly.

Chapter 16 reviews the economics and politics that have influenced and framed the decision – making over the decades. This part of the book brings into focus the reasons for the decisions made together with the arising direct and/or indirect consequences.

Chapter 17 indicates how rail electrification might contribute to minimising carbon emissions to help meet the UK's net zero target, revealing a new driving force for the adoption of a rolling programme of electrification.

Chapter 18 summarises the lessons learned from past experiences and advises how electrification might be undertaken in a more efficient and effective manner in the future.

Lastly, Appendices 1–8 contain interesting facts, figures and specific project information that supports the preceding chapters, but that would be too cumbersome to include in the main text.

Upon reading this history of railway electrification in the UK, the reader will quickly come to understand that, from the earliest days, implementation has progressed on a 'start–stop' ad hoc basis, lacking both vision and coordination from a national point of view.

Calls for a rolling programme of electrification have regularly been made for more than a century, but, apart from the Southern Railway's third-rail schemes south of the Thames prior to the Second World War, no coordinated development has been realised. Somewhat surprisingly, the strongest voice for electrification following the Second World War came from a railwayman more usually credited as being one of Great Britain's best steam locomotive engineers, Robert 'Robin' Riddles, CBE.

As a young London & North Western Railway (LNWR) Crewe Works Premium Apprentice in 1909, Riddles was so forcibly struck by the advantages of electric traction that he was determined to study electrical engineering. Unfortunately, to his disappointment, he was moved to Rugby and could not further his interest there and so became a mechanical engineer. Returning from France after the First World War having served as a sapper in the Royal Engineers, Riddles could see that the prospects for electrification on the London, Midland & Scottish Railway (LMS) were limited due to the company's weak financial position. However, he kept abreast of modern traction developments in both Europe and America during the 1920s and 1930s.

Following distinguished war service as Deputy Director General, Royal Engineers Equipment at the Ministry of Supply and Vice-Chairman of the LMS, Riddles became the member of the Railway Executive for Mechanical and Electrical Engineering at nationalisation in 1948. He played a key role in developing a strategic plan for future modern traction on Britain's railways, which was published in 1951. This led to, first, a significant increase in the building of diesel shunters and, secondly, construction of a fleet of railcars – diesel multiple units (DMUs) – for secondary and branch line services. In addition, while Riddles was extremely keen to initiate a rolling programme of main line electrification, it was widely recognised that the implementation would take many years and needed guaranteed power supplies from the National Grid.

Riddles's view was that the fleet of modern steam locomotives that he was building would 'fill the gap'. Other voices within the Executive, notably J.L. Harrison, Chief Officer (Administration), were calling for a modest start to be made on dieselisation. However, it was recognised that the technology at that point in time had not yet yielded a solution with sufficient tractive effort to take the place of main line steam locomotives.

Riddles was strongly opposed to wholesale dieselisation as either an interim step, or as a final traction solution. Privately, he was concerned that dieselisation would abstract funds from his electrification plans. In his endeavours to pursue a comprehensive electrification policy, together with his Chief Officer (Electrical Engineering), S.B. Warder, he initially succeeded in establishing 50 cycles per second AC as the new standard for future overhead electrification schemes. This forms the basis of the industry's standard today.

Riddles's untimely departure from railway employment was primarily due to the change of government in 1951. The new administration sought to devolve far more authority to the Regions, thus undermining the potential for a national strategic approach to electrification. Politically, the idea of modernising the railway by dieselisation was also gaining ground. Harrison led a team that drafted an unpublished plan, 'A Development Programme for British Railways', which, while recommending electrification as the main way forward, also included a pilot build of main line diesel-electric locomotives. On top of this, Riddles was told that, due to the impending reorganisation, his post would be abolished, and his position might be downgraded.

Unsurprisingly Riddles decided to take early retirement. On 30 September 1953, the day the Railway Executive was folded, he left the railway industry. The vision of a rolling programme of electrification faded with his departure.

It was a wretched end to a great man's railway career.

Had the reorganisation not taken place and had Riddles not been quite so single-minded in opposing the dieselisation elements of the Harrison plan, the roll out of electrification in the UK would undoubtedly have been more efficiently and effectively implemented. The programme would have been taken forward on a continuous production basis rather than the actual disjointed developments recorded in this volume to date.

A quarter of a century after his departure, Riddles mused that, if his rolling electrification policy had been followed, the West Coast, East Coast and former Great Western Main Lines would have been electrified by that time. Furthermore, had the same rate of progress been maintained thereafter, all the routes that Network Rail currently propose to electrify would have been completed by the end of the twentieth century. There is no doubt that the cost per single track kilometre would have been far lower than that resulting from the 'start–stop' nature of the ad hoc electrification schemes that have been progressed to date and, today, we would have a cleaner, more efficient and faster carbon-neutral railway.

An opportunity lost indeed!

So, let's not see history repeated regarding the industry's current proposals. To decarbonise the railway and improve efficiency, the politicians, the Department for Transport (DfT), the Infrastructure Commission (NIC), the leaders of the industry and the trades unions need to come together to agree a sound, long-term programme, and just as importantly, have the iron determination to see it through.

On a final note, we should like to emphasise that the views expressed in this account are purely our own, based on our research together with our knowledge and extensive experience gained from long careers in the industry. We have not sought, nor been given, any sponsorship from any rail company or organisation. Our intent is simply to proffer advice to assist the industry in developing a robust electrification and decarbonisation strategy together with improved and more cost-effective means of implementation.

A Note about Track Diagrams

Track diagrams relevant to the text are included where appropriate. Note however that the diagrams are schematics and not detailed plans, and, where space is limited, it has not always been possible to include all stations, particularly for longer or more complex track layouts.

Readers wishing to find more details are advised to view www.trackmaps.co.uk for a list of books detailing current track layouts or, alternatively, refer to the many various historical publications available.

John Buxton BSc (Hons) C Eng MICE FWPI
Donald Heath OBE BEng (Hons) FICE FPWI COMPIRSE FIM
2025

Glossary

(including references to specific types of railway rolling stock)

Wi-fi   Wireless Fidelity – a wireless technology standard for wireless Internet access

outturn   actual expenditure details, often as compared to planned or budgeted

sectorisation   the dividing of the railway network into profit centres (sectors), which directly add to the railway company's overall bottom line, where each is responsible for managing their respective operations, maintenance, enhancement and safety requirements

blockade   a formalised closure and full possession of a section railway route infrastructure, usually taken to permit physical works to be carried out on the system

hybrid   relating to rolling stock, having two or more sources of propulsion – diesel, electric, battery and hydrogen; trains may also be described as bi-mode or tri-mode

balise   electronic beacon or transponder placed between the rails of a railway as part of a train protection or control system

SCADA   Supervisory Control and Data Acquisition, a computerised real time data and control system

PDMX   Primary Digital Multiple System – a system developed by STC Communications to break into a 30 channel pulse code modulation system (PCM) intermediately so as to extract individual channels from the data stream without the need for a full termination of the system. Sometimes called 'drop and insert'. It was very useful for the distribution of the signalling SSI channels to local spots where points, signals and so forth were sited

turnouts   points (also referred to as switches and crossings)

Types of 750V third rail DC EMU (electric multiple unit), indicating by initial numeral the number of cars within the set, with subsequent alpha codes distinguishing the type of set, thus:

2/4EPB   later designated Class 415 (4EPB) or Class 416 (2EPB)

4-CEP   later designated Class 411

4-CIG   later designated Class 421

4REP   later designated Class 432

4VEP   later designated Class 423

5-WES   later designated Class 442

Types of locomotives, indicating arrangement of powered and unpowered wheels

Bo-Bo   an electric or diesel-electric locomotive having 2 bogies each with 2 powered axles

Co-Co   an electric or diesel-electric locomotive having 2 bogies each with 3 powered axles

2-Co-2   axle arrangement of locomotive, identifying powered axles ('Co' = 3 powered axles, '2' = 2 unpowered axles)

Chapter 1
Pioneers

Towards the end of the 1800s, the practical development of electrical technologies discovered and developed earlier in the century by Alessandro Volta (1745–1827), André-Marie Ampère (1775–1836), Georg Simon Ohm (1789–1854) and Michael Faraday (1791–1867) began to be applied to rail traction applications. A short resume of these developments follows.

In 1799, Alessandro Volta constructed the first generator of electricity, the voltaic pile, comprising a number of pairs of copper and zinc discs placed one above the other. Later, he developed the much more efficient voltaic cell, immersing copper and zinc plates in acidulated water, thus producing the first battery.

In 1820, Oersted discovered that an electrical current deflected a compass needle, and Ampere demonstrated that neighbouring current carrying conductors exert a force on one another. In the same year, Arago and Davy discovered that bars of iron could be magnetised by the passage of a current through insulated wires wound around them.

In 1821, Faraday demonstrated that a conductor carrying an electric current could be made to rotate around the pole of a magnet. This was followed in 1823 when Barlow constructed a 'wheel and axle' motor, comprising a pivoted copper disc, located between two poles of a magnet, which rotated when a current passed from the axle to the circumference.

Faraday constructed the first electrical generator in 1831, on similar apparatus to Barlow's, by rotating the copper disc to produce a current in the wire connecting the axle to the circumference. Henry constructed an oscillating motor in the same year and, subsequently, more refined motors were developed by Jacobi, Elias and Froment.

From 1832 to 1846, Pixii, Wheatstone, Cooke, Scoresby and Joule undertook research into magneto generators and thus progressed the further development of electric motors.

These early experiments demonstrated that electrical energy could be generated and converted into mechanical energy and vice versa. However, it was to be many years before practical use of these concepts could be realised. Nevertheless, step by step, over the next fifty years or so, progress was made in the development of both motors and generators as follows:

1834–8	Jacobi constructed a small boat driven by a battery-powered motor, which he demonstrated on the River Niva in North-West Russia.
1835	Davenport constructed a model electric vehicle, which ran around a small electric railway, current being supplied externally to the vehicle by three cells.
1839	Davidson built a small standard gauge electric locomotive powered by forty cells contained within the unit's body. The locomotive is said to have achieved a speed of 4mph on the Edinburgh to Glasgow railway but was subsequently ignored by the railway's mechanics, who feared that such motive power might supersede the steam locomotive.
1840	Pinkus obtained a patent for conveying electrical current through rails.
1847	Farmer demonstrated a small battery-powered locomotive in Dover, USA.
1847	Lilley and Cotton built a small locomotive at Pittsburgh, which ran on a circular track, the current from a battery being transmitted to the car through the rails.
1850	Hall, a scientific instrument maker, exhibited a small battery-powered vehicle on rails at Boston, USA.
1850	Page built a larger locomotive powered by 100 cells, which was tried on the Baltimore and Ohio Railroad near Washington, USA. The development of this vehicle was partly financed by means of a grant from Congress.
1851	Hall constructed a model to demonstrate that power could be transmitted through the rails to a moving rail car running on them.
1850–67	Development of the dynamo by Pacinotti, Wilde, Gramme, Siemens, Varley and others.
1867–9	Werner Siemens suggested the possibility of elevated electric railways and went on to demonstrate the transmission of power from a generator to a motor.
1879	Siemens's Pioneer Railway – the first successful demonstration of electric traction at the Berlin Exhibition
1880–2	Edison built a third of a mile-long demonstration electric railway from the front of his laboratory yard at Menlo Park, New Jersey, USA. The line curved around a hill and returned, forming a U-shape. He later built a three-mile long experimental line in Menlo Park. Both lines fed power to the traction units through the running rails.
1881	Siemens demonstrated the first use of overhead line electrification (OLE).

1881   Crystal Palace Railway, Sydenham, UK – the Siemens train ran for two years at this location.

1883   The American DAFT company constructed the outstanding 'Ampere' locomotive, which was tested on the Saratoga Mount McGregor and Lake George Railroad. The loco picked up current from a central conductor rail and was probably the first electric traction unit to pull a full-sized standard gauge passenger car. The company went on to produce a successful electric rack railway traction unit in 1888.

1883   Volks Electric Railway, Brighton – probably the world's first commercial electric railway, as opposed to a tramway. The line was developed and built in the UK in 1883 at Brighton by Magnus Volk, an inventor and engineer. When opened, the 0.4km long narrow-gauge railway was originally electrified at 50V DC, which was subsequently raised to 160V DC when the line was extended to 0.8km (0.5 miles) in 1884. The line, now 1.6km (1 mile) in length, is still operational today (2025) as a tourist attraction and currently operates on 110V DC.

1883   The Portrush Railway, Northern Ireland rivals the Volk's Electric Railway as the world's first commercial electric railway. The line, promoted by brothers William Acheson Traill and Dr Anthony Traill, was a 11km (7 miles) long, three-foot gauge line linking Portrush to Bushmills. Initially the line was constructed to carry both passengers and freight (primarily granite, basalt and iron ore) to Portrush, the nearest seaport. William Thomson (later Lord Kelvin), who was a close friend of William Traill, suggested the use of electric traction and introduced the Traills to his friends, the Siemens brothers. The railway opened in 1883, and was extended to the Giant's Causeway, a popular tourist site, in 1888. The passenger cars were of two types, low-powered ones, which could run solo, and more powerful types, which could pull up to three trailer cars. The railway became a popular tourist line, surviving until 1949. Power for the third-rail DC system was generated by means of two 52hp Alcott water turbines, located at Walkmill Falls, driving a dynamo. This equipment, itself a pioneering achievement, was developed by Werner Siemens and produced 100 amps at 250 volts. Later, overhead line partially replaced the third-rail system.

The line was known as the Giant's Causeway Railway and, today, two miles at the Giant's Causeway end have been revived, albeit utilising diesel traction.

1885	The three-foot gauge Bessbrook and Newry Tramway linked the Bessbrook Spinning Company's flax mills with the main line railway at Newry. Dr Edward Hopkinson was responsible for equipping the line, including the two cars, which each had one motor. Power was generated by two water-powered Edison-Hopkinson dynamos producing 72 amps at 250 volts. The third rail was located centrally between the rails. The line carried passengers and 28,000 tons of freight per annum at its peak, finally closing in 1948. Hopkinson's success resulted in him being invited to assist in the construction of the City and South London Railway, which used Edison-Hopkinson dynamos, and Hopkinson designed electric locomotives.
1886	Between 1886 and 1893 the Ryde Pier, Manx, Blackpool and Snaefell electric tramways were constructed. However, overall, during this period, tramway developments in the UK fell well behind the extensive networks developed in the USA, Germany and later France.

Towards the end of the nineteenth century, it was becoming clear that electric rail traction had potentially many advantages over steam power, particularly in respect of power to weight ratio. The potential for rapid acceleration of passenger trains, especially suburban services, and the power to haul heavy trains on steeply graded routes was soon recognised.

Two early pioneers of this new rail traction technology warrant special attention – Werner von Siemens and Frank Sprague.

Werner von Siemens – Demonstration Line

Werner von Siemens developed and built the first passenger hauling electrically powered rail vehicle, which he demonstrated at the Berlin Industry Exhibition in 1879. The track was only of a temporary nature for the exhibition, although Siemens did go on to produce the first electric 'streetcar' system at Gross-Lichterfelde in 1881. This 2.5km (1.5 miles) long, 180V DC system was the world's first electric tramway. From that time, Siemens developed many tramways in Germany. The construction of the first German electric railway system commenced in 1891, opening in 1902.

Frank Sprague – The Sprague System

Frank Sprague, who established the Sprague Company, worked with Edison. Initially, he took a leading role in the building of Street Tramways in the USA for which he developed a successful overhead distribution and collection system. He also pioneered the mounting of electric motors on vehicle bogies that enabled electric-powered cars to run on uneven track. His axle-hung, nose-suspended motor became a world standard. Sprague went on to develop solutions to many of the shortcomings of early streetcar operation. He combined his innovations into a successful system in 1888, dubbed the 'Sprague System', which became a maxim for 'reliable system' and was adopted all over the world, including in the UK. The 'Sprague System' was further developed for heavy rail systems for both overhead and conductor rail systems. By 1891, more than half of the 200 or more electric systems in the USA were supplied by Sprague. In 1894 Sprague developed the system for multiple-unit control, which became an essential feature of electric passenger railways. The Sprague System first came to the UK in 1890 when the City and South London Railway was constructed using electricity power generation and low voltage DC traction based primarily on USA practice. Much of the UK systems developed thereafter have been based on Sprague's principles.

An Enigma – Swansea and Mumbles Railway

A special mention is made here of this fascinating line for two reasons:

- opening in March 1807, the railway was the first in the world to regularly convey fare-paying passengers;
- it was the first system to attempt powering rail vehicles in UK commercial service from on-board battery accumulators.

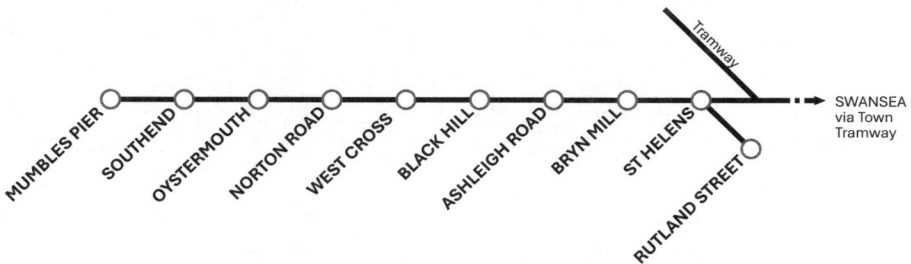

Swansea and Mumbles Railway

Initially, its horse-drawn rail vehicles, converted from road carriages, conveyed people between 'the dunes' at Swansea westwards along the perimeter of Swansea Bay to Mumbles, a fishing and oyster harvesting village on the west side of the bay. Steam power began to supersede horses in the 1860s but an innovative attempt was made in 1902 to provide traction power using battery accumulators. However, the stored electrical power in the early batteries drained too quickly so the idea was abandoned. This is the first recorded example of a battery-powered rail vehicle in UK commercial passenger service.

The railway, which was expanded over the years and converted to standard gauge, remained independent all of its life and was eventually electrified using a 650V DC overhead system in 1929. Sadly, the line closed in 1960 in a sea of local controversy.

Missed Opportunities

Notwithstanding the UK developments outlined above, Britain by and large failed to grasp the opportunities afforded by electric technology. This is particularly poignant because the great British railway pioneer, George Stephenson, is said to have prophesised as early as 1847 that electricity would be the future form of motive power. By contrast, in the USA, engineers and entrepreneurs stepped up to take the lead, not only in electric traction, but in all aspects of electrical engineering. This undoubtedly enabled that country to become the world's premier industrial nation. From the 1880s, the USA emulated what Britain had done in the steam-driven Industrial Revolution, by rapidly building up expertise in a new technology, leaving other nations, including Britain, in its wake. Unfortunately, Britain was firmly wedded to developing steam power, and thus was slow to make the transition to electricity. The UK development of an electric industrial base and supply network, based upon the new technology, therefore lagged well behind the USA. In Germany, too, the new technology was being firmly embraced and developed at pace from the late nineteenth century.

A further cause of Britain's sluggish progress was the late development of electric street tramway systems in the UK. With little expertise and experience being acquired in the new technology powering these systems, the potential development of the country's electrical engineering industry was still further disadvantaged.

Furthermore, early British electric railway developments were not inter-related, nor was any attempt made to develop a common standard. The politicians of the day showed little interest in the technology, and there was no coordination between entrepreneurs, the engineering industry, financial institutions and operators. With the focus still mainly on developing steam power, Britain lost the opportunity to take a lead in the electric revolution and thus the ability to develop a new global enterprise. This lack of vision indisputably set in train the decline of the UK's industrial power base.

America and Germany were thus left to drive forward this new strategic technology, and, as a result, early British railway electrification from the 1890s was initially based on these nations' practices.

The City and South London Railway and its Impact on the Development of London Underground Electrification

Although the London Underground system is outside the remit of this book, the electric traction developments of this system are relevant to the UK's main line railways, and so a brief inclusion of its early history is presented here.

Subsurface railway development began in 1863 with steam-hauled trains of the Metropolitan Railway running in 'cut and cover' tunnels just below street level. However, the highly sulphurous smoke emitted by the locomotives made travelling conditions in the tunnels a very unpleasant experience. The development of this railway's clean electric trains and new deep-level tunnelling developed later, brought about London's public transport revolution.

The City and South London Railway, built initially to link the City of London with the Elephant & Castle, was the first major railway system in the world to adopt electricity as a means of traction power. Based on USA electric innovation and practice, this first deep-level tube railway, powered by a three-wire 500V DC system, was opened in 1890 by the Prince of Wales and used electric trains from the outset. Each electric train comprised a four-wheeled traction unit, built by Mather & Platt of Manchester, hauling a number of windowless carriages with power picked up from an off-set third rail within the running rails. At Stockwell, a depot and generating station were constructed. Subsequently,

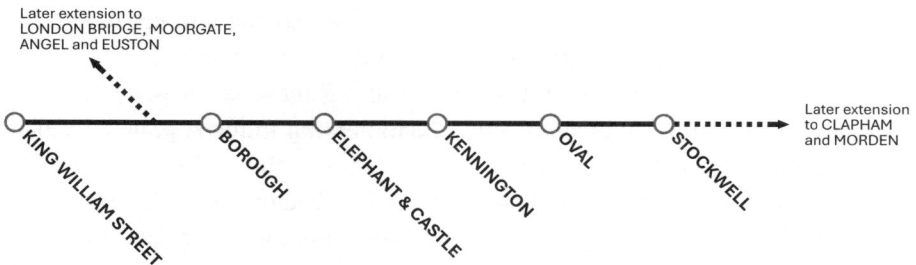

The City and South London Railway

following arbitration by the Board of Trade, the standard DC fourth-rail system was adopted.

The American electric traction technologies, pioneered and proved on the City and South London Railway, shaped the development of subsequent underground railways in London and profoundly influenced the introduction of conductor rail surface systems in the UK. Prior to the Great War, expansion of the underground network was rapid. The City and South London subsequently became part of the Northern Line.

Developments on the Underground railway system had a profound impact on early main line electrification, influencing the adoption of low voltage DC systems, particularly in London, Liverpool, Manchester and Tyneside. The early electrified suburban lines developed by the London & North Western Railway (LNWR) were actually made fully compatible with the Underground system. This enabled both main line and Underground trains to share the same infrastructure and power supply systems to mutual benefit, particularly with regard to their respective network extensions. Where the track is shared with third-rail over-ground stock, the central rail is bonded to the running rails and the outside rail electrified at 660V. This allows both types of train to operate satisfactorily.

The London Underground network was primarily developed north of the Thames where the geology is more favourable; the above-ground main line railway network, much of which was to be electrified, dominated the area to the south.

Waterloo and City Line (London & South Western Railway (LSWR))

Following on from the City and South London Railway, the next London tube line to open in 1898 was uniquely an underground line that was operated for nearly a century by main line railway organisations. It was the Waterloo and City Line, which was built by the Waterloo and City Railway Company and operated by the London & South Western Railway from the outset. This company absorbed the railway into its own system in 1907. As its name suggests, the line linked the City of London with Waterloo Station.

Although the railway was to use American rolling stock, it was a German company with significant business interests in the UK, which supplied the electrical generating and distribution equipment. This was the well-known Siemens Company. Power generation was by means of five boilers, providing the steam for five high-speed steam engines coupled to dynamos. The two-pole compound-wound dynamos delivered 500V at no load and 530V under full load. A sixth high-speed engine and dynamo was subsequently installed in about 1899.

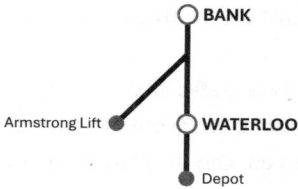

BANK

Armstrong Lift   WATERLOO

Depot

Waterloo and City Line

The standard gauge permanent way was supported on longitudinal timbers and the 530V DC traction current was fed to the trains through an inverted steel channel 'side contact' conductor rail arrangement.

Under the contract, Siemens also provided a single cab four-wheel electric shunter. This little traction unit, delivered in 1898, was provided to shunt the coal trucks and deliver them to the generating plant. It was equipped with two 60hp (45kW) traction motors, which, although modest, were sufficient for it to perform its basic duties, but it was found to be too small to rescue failed trains in the tunnel. This little electric shunter was, however, quite reliable and remained in use until 1969. It can now be viewed at the National Railway Museum, York.

In view of the urgent need to acquire a more powerful rescue vehicle, the LSWR Chief Mechanical Engineer, Dugald Drummond, had a larger traction unit with two four-wheel bogies built, which entered service in 1901.

In 1915 the original Siemens dynamo equipment became redundant as the LSWR commissioned a large generating station at Durnsford Road, Wimbledon, which also supplied its expanding third-rail surface network. Power for train operation on the Waterloo and City Line also switched to this source from December 1915, and the original Siemens 'dynamo' plant thereafter served only ancillary equipment on the line together with the heating and lighting of the main LSWR Waterloo offices. The traction voltage on the Waterloo and City Line was upgraded to the standard 600V DC 'top contact' system at this time.

The original passenger vehicles manufactured by the American Jackson and Sharp Company were imported through Southampton Docks in completely knocked-down kit form. The kits were assembled into eleven 29m long two-car sets (comprising one motor coach and one trailer car) by the LSWR at Eastleigh Works.

Siemens provided the electrical equipment and traction motors for the units. The trains usually ran as four-car formations. In 1899, five additional single motorcars were procured from Dick, Kerr and Company, primarily to operate the off-peak and weekend services when passenger demand was low.

From 1908, the underground railways in London began to market themselves through common branding as '*Underground*' but the Waterloo and City Line was the only underground passenger line in London not to sign up to this arrangement. In 1923, the line passed to the Southern Railway at the Grouping, subsequently transferring to British Railways ownership at nationalisation in 1948.

In the 1980s the line became the responsibility of the Network SouthEast (NSE) sector, and this organisation undertook a major project to refurbish the infrastucture and replace the 1937-built rolling stock with new Class 482 trains. These works were completed in 1993 with the operations and ownership of the line being transferred to London Underground in 1994 connected with the rail privatisation process.

Metropolitan Line

The Metropolitan Railway was originally built as a conventional railway to connect the main line rail terminals at Paddington, Euston and King's Cross to the City of London. Opening in 1863, significant lengths of the route were constructed by means of 'cut and cover', much of it beneath what at the time was known as the 'New Road' between Edgware Road and King's Cross. The line then proceeded by means of tunnels and cuttings to Smithfield via Farringdon Road. It thus formed, almost by default, the world's first underground railway and was initially worked by steam locomotives hauling gas-lit carriages. Unlike most other underground lines in the capital, the tunnels are only just below the surface and are of similar profile to those on other British main lines. It is referred to as the 'sub-surface' part of the London Underground system.

The line was extended in three directions, reaching Hammersmith in 1864, forming in cooperation with the District Railway, the 'Inner Circle' in 1884, and a branch from Baker Street northwards through Harrow to Aylesbury (later extended to Quainton Road), into what was to become known as 'Metroland'. Alongside the route, the Metropolitan Railway purchased and developed large swathes of land into middle-class housing estates. The company's shrewd investment in land and housing paid dividends, bringing further patronage to the railway in addition to the profit made from property development.

By 1900, patronage was falling primarily due to the opening of the new electric Central London Railway from Shepherd's Bush to the City, but passengers were also becoming increasingly intolerant of the polluted atmosphere in the Metropolitan line's tunnels. Conversion to electric traction was seen as essential to halt the decline in passenger numbers.

The Metropolitan Railway had considered electrification from the 1880s, but the technology was in its infancy, and agreement was not immediately forthcoming from the District Railway, which shared the ownership of the Inner

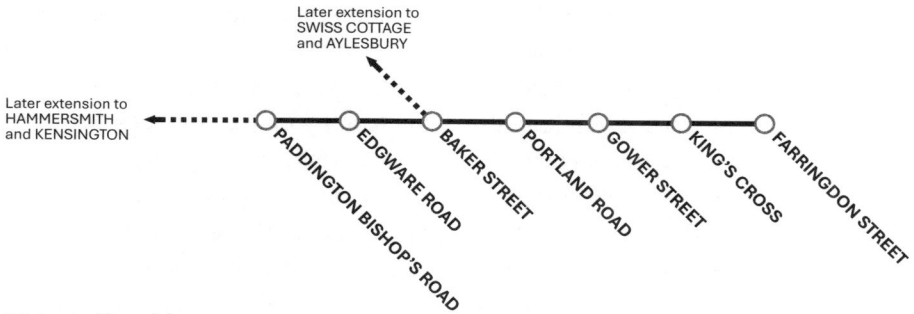

Later extension to SWISS COTTAGE and AYLESBURY

Later extension to HAMMERSMITH and KENSINGTON

PADDINGTON BISHOP'S ROAD — EDGWARE ROAD — BAKER STREET — PORTLAND ROAD — GOWER STREET — KING'S CROSS — FARRINGDON STREET

Metropolitan Line

Circle. However, in 1900, a jointly owned train of six passenger vehicles ran an experimental service from Earl's Court to High Street Kensington for six months. This was a success, and so the Metropolitan and District joint committee looking into the matter recommended an overhead three-phase AC system developed by the Hungarian company Ganz. However, the District Railway was taken over by the Underground Electric Railways Company of London in 1901. This company was led by the American Charles Yerkes, who favoured the Sprague third-rail DC system similar to the City and South London Railway and Central London Railway. The Ganz system was therefore subsequently rejected in favour of the standard DC fourth-rail system. This system was deemed necessary in metal-lined tunnels where stray electric currents had to be kept low to avoid corrosion and interference with regard to the linings and utility systems.

The Metropolitan Railway, together with the District Railway, began to electrify their underground lines in 1902. In addition, the surface line from Baker Street to Harrow was electrified by 1905, as was a new route built between Harrow and Uxbridge which opened in the same year.

Initially, all services were operated by seven-car electric multiple unit (EMU) trains built by the Brush Electrical Engineering Company and Metropolitan Amalgamated Carriage and Wagon Company to American-inspired designs. However, subsequently a small fleet of Sprague-type electric traction units (A-class) were built, which, among other things, hauled morning and evening business trains to and from the City, comprising carriage sets including Pullman cars. It is interesting to note that these A-class traction units encompass components from all the great pioneers of American electric traction, viz. Brush, Edison, Sprague, Thomson-Houston and Westinghouse.

To serve the newly electrified lines, the Metropolitan Railway constructed a 10.5MW coal-fired power station at Neasden, supplying 11kV 33.3Hz current to five substations that converted this to 600V DC.

On 1 July 1933, the Metropolitan Railway was amalgamated with the Underground Electric Railways Company of London to form the railway element of the London Passenger Transport Board. However, the physical nature of the Metropolitan Line meant that it retained much of its unique character for thirty years or so after this date, particularly in respect of the surface lines, where the ambiance was that of a main line railway rather than an underground line.

Mention should be made of the Metropolitan's relationship with the Great Central Railway Company's 1899 London Extension. This new main line railway joined the Metropolitan Railway at Quainton Road, and shared the tracks to West Hampstead Junction, before branching off towards Marylebone Station. Interestingly, the Great Central Railway never promoted electrification itself at the London end of its route.

Midland Railway

Another pioneering scheme was the Midland Railway's electrification of the 15km (9.5 miles) Lancaster–Morecambe–Heysham line that came into operation in 1908. It was not really an enhancement justified by traffic requirements; its raison d'être was to form a test bed to explore the potential for further mainline electrification for the company. This scheme was the first high-voltage overhead electrification in the United Kingdom being energised at 6,600V 25Hz, AC. The equipment was similar to that used to power the London, Brighton and South Coast (LBSC) South London scheme.

Power for the scheme was generated at the Midland Railway's own power station at Heysham, which had originally been built to supply the electrical equipment at the harbour. Additional machinery was added to both increase its capacity and to supply single phase current, as it had originally been a direct current station.

Electric working between Morecambe and Heysham began on 13 April 1908, with services commencing on the Lancaster to Morecambe section on 8 June of that year. Goods trains remained steam-hauled.

The design and implementation of the line's electrification was carried out under the direction of Mr W.B. Worthington, the Chief Engineer of the Midland Railway, by Mr J. Sayers, the Telegraph Engineer of the company, and Mr Argyle, the Northern Divisional Engineer. The scheme was very similar to the Hamburg – Altona Railway, the chief difference being the new design for the overhead line electrification (OLE), which was Mr Sayer's work.

The scheme comprised the electrification of the double track sections between Heysham and Morecambe and Morecambe and Lancaster Green Ayre,

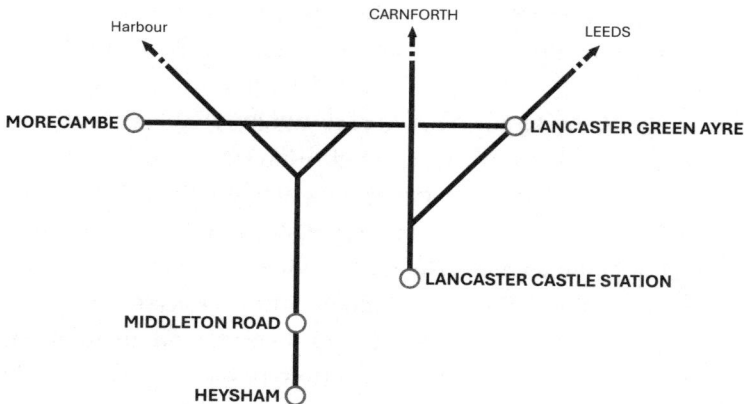

Lancaster–Morecambe–Heysham

together with the single line from Lancaster Green Ayre to Lancaster Castle Station. This was equivalent to 34 single track kilometres (21 miles).

As the first overhead line single phase AC electrification in the UK, the design and implementation successfully overcame many difficult technical challenges.

The contact wire was suspended by short (100mm) loops from an auxiliary wire, which, in turn, were held by two auxiliary cables, with each span broken up into six short auxiliary spans. The gantries, generally supported on Norwegian fir creosoted poles with thin, solid crossbeams to form simple portals, were connected by a separate overhead earthing cable, which made connections to earth every 800m (0.5 miles). More substantial steel poles and lattice gantries were provided at Morecambe Station and one or two other locations to allow for long spans.

The cars collected power by means of bow 'trollies' measuring 2.16m (7ft 1in) wide. The bow was designed to be used for travelling in either direction. The contact wire was initially staggered by 611mm from the centre line and each 915m (3000ft) section was strained by means of 365kg (800lbs) weights through pulleys at a terminal gantry. However, the stagger was found to be too great and the contact wire strain too low for reliable operation so was modified before commissioning, the sharply curved Lune Viaduct presenting a particular problem in this respect.

The rolling stock initially comprised three motor cars and four trailers, built by the Midland Railway Company at Derby to the design of the company's Carriage and Wagon Engineer, Mr David Bain. Electrical equipment was supplied by Siemens for two of the cars and Westinghouse for the other. The Siemens control equipment was all electric, which was preferred by the

Midland, whereas the third car had Westinghouse equipment, with electro-pneumatic control, thus providing a practical comparison of the two systems for this pilot scheme.

Each end of each of the trailer and motor cars was equipped with driving apparatus and, in addition, old coaches were provided to strengthen the trains, particularly for use by workmen between Morecambe and Heysham. The motor cars were 18.3m (60ft) long and 2.7m (9ft) wide and comprised three large saloons with seating for 72 passengers. The trailer cars were 13.1m (43ft) long and 2.7m (9ft) wide and comprised one open saloon with transverse seating for 56 passengers. Electric lighting was provided for all the cars, and, in addition, the motor cars were provided with electric heaters from new (and subsequently fitted to the trailer cars, which were initially only used in the summer season).

Both the infrastructure and the rolling stock were designed to deliver a 15-minute service frequency between Morecambe and Lancaster and a 20-minute service between Morecambe and Heysham, utilising just one train on each route. Normal service levels were less frequent, however.

Technically, the scheme proved to be a success but, unfortunately, the Midland Railway did not progress any further schemes.

In 1952 the original rolling stock was withdrawn as it was life expired. Services reverted to steam power while the power supply was upgraded to 25kV 50Hz, and some alternative electric stock provided. However, the main reason for this was not for normal service reasons. It had been decided that the line was to be used again as a test track to trial British Railways' proposed new high voltage electrification standard. As the original electrification was a high voltage overhead line system, converting the line could be undertaken with few alterations to the infrastructure and at minimal cost (see Chapters 6 and 16). Trial running under 25kV started in November 1952, with the commercial service starting on 17 August 1953.

The 1950s trials, which were a success, were key to British Railways adopting its new 25kV 50Hz electrification standard. On completion of the trials, the line remained electrified until it closed under the Beeching cuts in 1966.

Chapter 2
The First Rapid Transit Schemes

The early days of railway electrification marked a period when electricity was being experimented with for traction purposes. Schemes that were promoted were primarily driven by competition, particularly from new electric tramway systems. Capacity enhancements, the promise of operating cost reductions and the elimination of pollution also played a part. As a result, this haphazard process spawned many schemes and proposals utilising different voltages, AC/DC and power supply 'pick-up' arrangements on the trains.

The UK thus developed a considerable number of diverse railway electrification systems, which are briefly described in this chapter.

The Development of Liverpool's Electrified Network

Liverpool Overhead Railway (LOR)

The UK's first electric railway built as a conventional standard gauge 'open-air' line was the standard gauge Liverpool Overhead Railway (LOR). An elevated railway had been proposed in 1852 and again in 1877, but both proposals came to nothing, although powers were secured by the Mersey Docks and Harbour Board for the latter scheme. In 1888, the Liverpool Overhead Railway Company was formed by a group of prominent businessmen who obtained the Docks Board powers by means of an Act of Transfer. Two prominent engineers, Sir Douglas Fox and James Greathead, were appointed to design and manage the construction, which commenced in 1889, the first section of which opened on 4 February 1893 by the Marquis of Salisbury. Public services commenced on 6 March.

Although steam traction was first considered, it was rejected due to the risk of sparks emitted by the locomotives, igniting the many flammable cargoes stored in close proximity to the railway, on both the ships at the quays and on the dockside.

The LOR was not only the world's first elevated electric railway, but also the first to be protected by an automatic electric signalling system. It was a third-rail line operating at 525V DC, initially running between Alexandra Dock and Herculaneum Dock, with twelve intermediate stations.

The line was built on an elevated iron structure and was thus a very prominent feature of the area, colloquially known as 'The Dockers' Umbrella'.

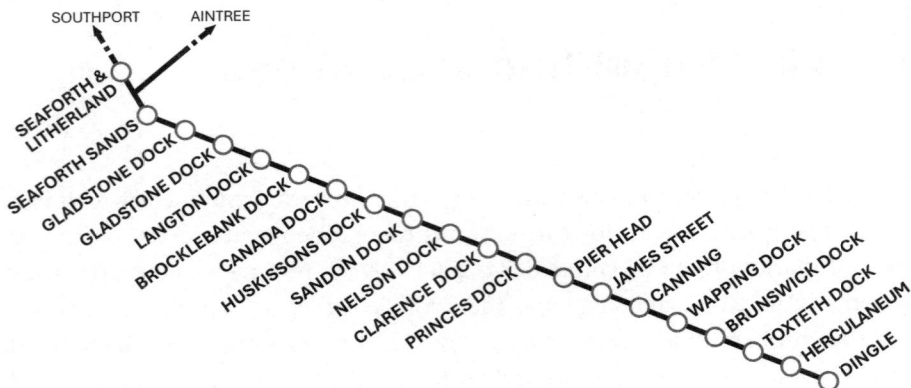

Liverpool Overhead Railway

A rapid, frequent service was offered with electric two-car passenger units (motor driving car and trailer driving car) serving the closely spaced stations. The line was subsequently expanded to the full length of the Liverpool Docks' estate (10km, 6.5 miles) with extensions to Seaforth Sands on 30 April 1894 and Dingle on 21 December 1896. Ironically, for an overhead railway, the latter section required the boring of a half-mile long tunnel!

The LOR was expanded further on 5 July 1905 to the Lancashire and Yorkshire Railway's Seaforth and Litherland station, enabling a connection to be made with the L&Y's Liverpool to Southport Line.

Regular through services were then operated initially to Southport, utilising twelve specially built lightweight cars constructed by the Lancashire and Yorkshire Railway. These services were, however, withdrawn by 1914, and passengers for Southport had latterly to change trains at Seaforth and Litherland railway station instead. The lightweight cars were transferred to the Southport and Cossens's services but returned as required until the outbreak of the Second World War for the Dingle–Aintree race traffic.

The extension to Seaforth and Litherland had brought the LOR trains parallel with the LYR's North Mersey Branch with a connection between the two being installed in 1906. The new junction, Rimrose Road, enabled the LOR services to access Aintree Sefton Arms LYR Station. Regular services were introduced between there and Dingle but only operated until 1908, although special LOR services were always put on during Aintree race days.

An additional station was added at Gladstone on 30 June 1930 serving the dock of the same name, but Princes Dock Station permanently closed following extensive bombing damage on 13 March 1941. Custom House Station was renamed Canning in 1945 as the nearby Custom House had been totally destroyed during the war.

Latterly, the sets were lengthened to three-car units, and some were modernised after the war.

The line escaped nationalisation in 1948 but, unfortunately, by 1956 the overhead structure, particularly the deck plates, were extensively corroded and considered to be beyond economic repair. The repair costs, estimated at £2M, were beyond the financial resources of the company and, as no support was forthcoming from the City Council or the Mersey Docks and Harbour Board, the line was closed on 30 December 1956, despite considerable public protest.

Mersey Railway

The Mersey Railway, opened in 1886, was a 7.5km (4.75 miles) line between Liverpool and Birkenhead via the Mersey Railway Tunnel. Subsequently, it was extended to link Birkenhead Park, Rock Ferry and Liverpool Central by 1892. With gradients as steep as 1 in 27, it was initially operated by heavy steam locomotives. Steam power was also used to operate the lifts, hoists, ventilation and drainage pumps. With low thermal efficiency, the cost of operating the railway and its associated systems was high. Patronage was falling due in part to the polluted, choking atmosphere caused by the steam locomotives in the tunnel under the River Mersey. With passengers reverting to use of the river ferries, the railway was in financial difficulties by 1900.

The successful introduction of electrification on the City and South London Railway and the Liverpool Overhead Railway convinced the Mersey Railway directors that electrification was the way forward but, with the company one step away from bankruptcy, there were no funds to finance it.

A lifeline was fortuitously provided by the American George Westinghouse Jr, who had set up companies in the USA and Europe manufacturing electrical equipment, power plant and braking systems. He established the Westinghouse Electric and Manufacturing Company at Trafford Park, Manchester in 1899, anticipating a British electrification boom. Westinghouse considered that he needed to demonstrate his company's wares by electrifying a short length of railway and, with its challenging infrastructure and the existing use of heavy locomotives, he considered that this was an ideal line to show that an electric railway could take over all steam-operated services.

His offer to the Mersey Railway Company that the line be electrified at his company's expense was, unsurprisingly, accepted!

As there was no power supply, Westinghouse built a 7.7MW railway power station at Shore Road, which not only supplied traction power, but electricity for the afore-mentioned infrastructure systems. The 600V DC railway electrification arrangement utilised a four-rail system with a positive outer rail, with

WEST KIRBY
HOYLAKE
MANOR ROAD
MEOLS
NEW BRIGHTON
MORETON
WALLASEY GROVE ROAD
LEASOWE
WALLASEY VILLAGE
BIDSTON
BIRKENHEAD NORTH
BIRKENHEAD PARK
CONWAY PARK
HAMILTON SQUARE
BIRKENHEAD CENTRAL
GREEN LANE
ROCK FERRY
BEBINGTON
PORT SUNLIGHT
SPITAL
BROMBOROUGH RAKE
BROMBOROUGH
EASTHAM RAKE
HOOTON
CAPENHURST
BACHE
CHESTER
LITTLE SUTTON
OVERPOOL
ELLESMERE PORT

SOUTHPORT
BIRKDALE
HILLSIDE
AINSDALE
FRESHFIELD
FORMBY
HIGHTOWN
HALL ROAD
BLUNDELLSANDS & CROSBY
WATERLOO
SEAFORTH & LITHERLAND
BOOTLE NEW STRAND
BOOTLE ORIEL ROAD
BANK HALL
SANDHILLS
MOORFIELDS
Mersey Tunnel
JAMES STREET
LIME STREET
LIVERPOOL CENTRAL
BRUNSWICK
ST MICHAELS
AIGBURTH
CRESSINGTON
LIVERPOOL SOUTH PARKWAY
HUNTS CROSS
WIGAN

PRESTON
ORMSKIRK
AUGHTON PARK
TOWN GREEN
MAGHULL NORTH
MAGHULL
OLD ROAN
AINTREE
ORRELL PARK
WALTON
KIRKDALE
RICE LANE
FAZAKERLEY
KIRKBY
HEADBOLT LANE
WIGAN

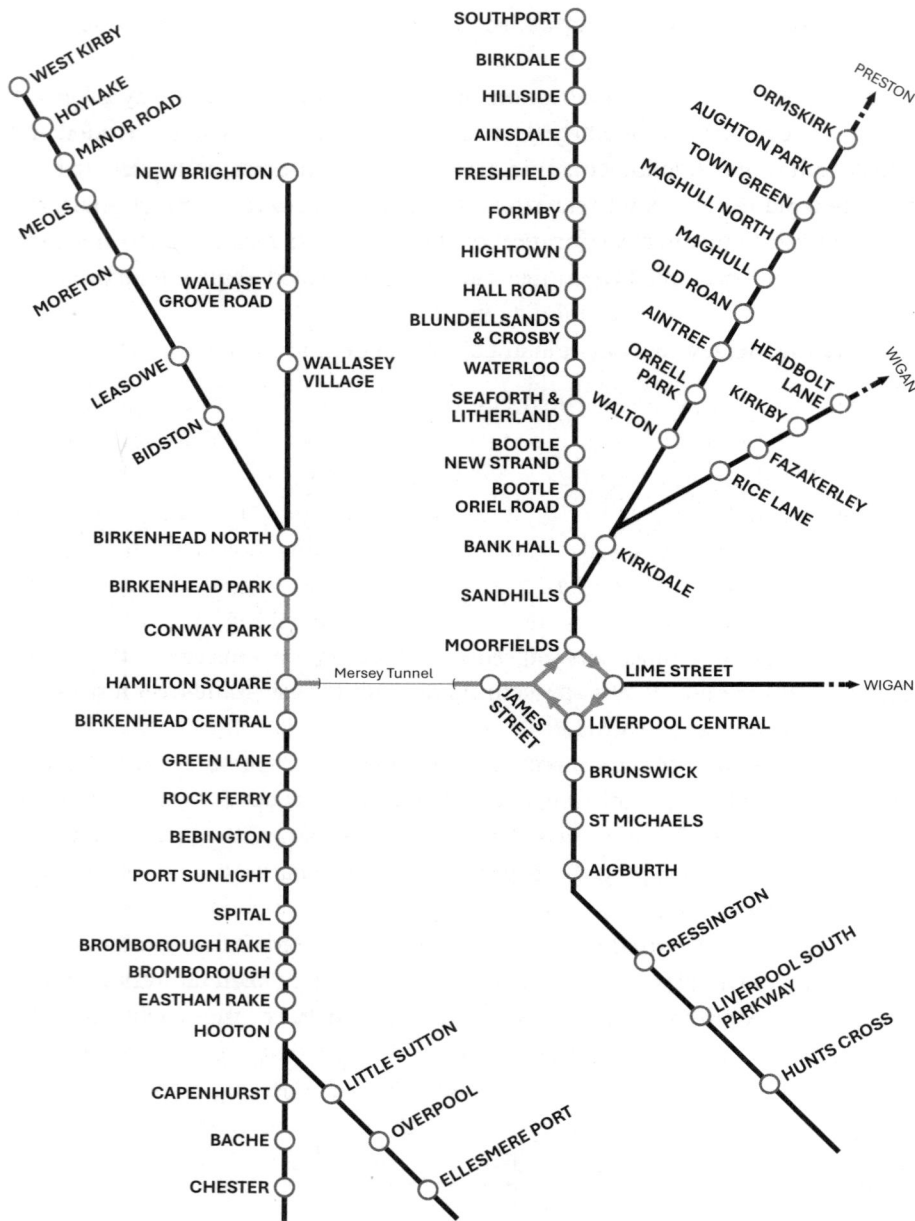

Mersey Electrified Railway System

storage batteries to improve the load factor so avoiding the need for transformers and rotary converters. The batteries also provided a back-up in case of power failure to enable trains to be rescued from the tunnels. The work was carried out without interfering with the steam service at a cost of circa £400,000 (£41M at 2025 prices). The Mersey Railway was the first in Britain to convert as a complete entity from steam to electric operation in 1903.

The new electric rolling stock comprised twenty-four motor cars and thirty-three trailer cars, which were formed into two-car or four-car sets. The stock had British-built timber bodies, with electrical equipment and bogies imported from the United States. The power station at Shore Road included passive provision to accommodate future electrification of the Wirral Railway, although this did not come to fruition until 1938. The line was opened for electric operation in April 1903 following an inspection by the Board of Trade.

While the Mersey Railway remained independent at the Grouping of 1923, it became closely aligned with the London, Midland and Scottish Railway's Wirral electric train services from 1938. The link between the Wirral and Mersey Railway system enabled fast through journeys by electric services to be made directly into the City of Liverpool. The Mersey Railway became part of the nationalised British Railways in 1948.

Lancashire and Yorkshire Railway (L&Y)

From 1903, the Lancashire and Yorkshire Railway had also been busy electrifying their busy commuter routes centred on Liverpool Exchange Station. The company was the first in the UK to electrify a mainline route. The 37km (22 miles) line to Southport and Crossens opened in March 1904. Initially, 600V DC was provided to a fourth-rail system, although this was later converted to 625V DC third-rail. A power station was built at Formby, supplemented by substations and battery stations located along the route. Sheds, stores and workshops were built at Meols Cop.

The rolling stock provided at the opening comprised 58ft 6in (18m) long open saloons. The carriages were electrically lit, initially formed into four-car sets. The two driving motor cars were third class, powered by four 150hp motors.

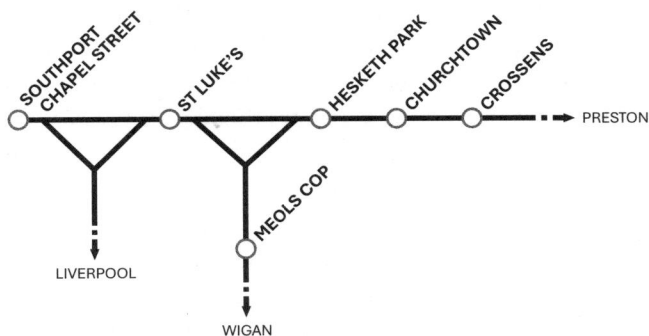

Southport and Crossens Railway

The traction current was controlled from driving cabs at both ends of the train. A vacuum braking system enabled the trains to haul non-electrified stock. First-class and third-class trailers completed the consist. All the stock was wooden bodied on steel underframes. Later build stock was 63ft 6in (19.5m) long and could seat up to 103 passengers. Subsequently, the section between Southport and Crossens was operated by the lightweight trains, originally provided by the Lancashire and Yorkshire Railway for the through LOR services, until these were withdrawn in 1945.

The Southport line was followed by electrification of the two routes to Aintree in July and December 1906, later extended to Ormskirk in 1913. For these services, the Lancashire and Yorkshire Railway provided different passenger cars that were incompatible with the Southport trains. Initially, twelve third-class motor coaches and six third-class trailers were built. From 1910 to 1914 a further seven first-class and twenty-three third-class trailers, plus four driving trailer thirds and eight motor coaches (with 250hp motors), were added to the fleet.

The L&Y built two electric traction units. The first, No. 1, was built on the frames of an Aspinall 2-4-2T and had a jack shaft drive. It was designed to collect current from top contact third rail and in sidings from overhead wires, for safety reasons. It probably wasn't very successful as it had been withdrawn by 1919. The second, No. 2, was a four-wheel battery-powered steeple cab design for shunting wagons at Clifton Junction power station. It was rather more successful than the first, lasting until 1947.

By 1915, the Lancashire and Yorkshire Railway had electrified 60km (37.5 miles) of route. Taken together with the Liverpool Overhead Railway and the Mersey Railway, Merseyside then had an electrified suburban network that was second only to London in terms of route length.

Following the grouping, the London, Midland and Scottish Railway built eleven electric units for the Ormskirk route in 1926–7, which operated services until 1964. The LMSR also replaced the Southport units in the early 1940s. The pre-Second World War years also saw the electrification of the Wirral lines from Birkenhead to West Kirby and New Brighton in 1938 and this company also modernised the stations and signalling on this route and the former Lancashire and Yorkshire routes from Liverpool Exchange, including replacement of the original rolling stock.

The Liverpool lines survived the Beeching era intact, and, with the subsequent extensions and the opening of the underground tunnels, the network has developed into a highly regarded suburban transportation system. In 1974 the route under the Mersey from the Wirral, which terminated below Liverpool Lime Street Station, was converted into a terminal loop servicing all three of the main line terminal stations. Three years later the Link Line was opened, which enabled through running from Exchange to Central and, following electrification of the Cheshire Line Committee route, as far as Hunt's Cross.

The Development of Manchester's Electrified Network

Bury to Holcombe Brook (L&Y)

Electrification of the first railway in the Manchester area came about as a result of an experimental scheme promoted by a Preston firm, the Dick, Kerr and Company in 1912. The 6km (4 miles) long route was electrified, utilising overhead distribution, at the company's expense, as Dick, Kerr and Co needed a line to demonstrate their electric traction supply and motive power equipment in connection with a tender for a large railway contract in Brazil.

Electrification of the route between Bury Bolton Street and Holcombe Brook was completed in July 1913 and two trains were built to operate the passenger service at the Lancashire and Yorkshire Works at Newton Heath. The sets, constructed with steel underframes and aluminium body panels, comprised one driving motor coach having four Dick, Kerr and Co traction motors, plus one driving trailer car. The motor coach had a pantograph for current collection from the overhead catenary.

On the successful completion of the Dick, Kerr and Co's trials, the Lancashire and Yorkshire Railway purchased the equipment and stock in 1916. Initially, a ten-minute interval service was maintained, which it was possible to do with just one set in use at a time.

Dick, Kerr and Co did not win the Brazilian contract but benefited when the Lancashire and Yorkshire Railway decided to use their equipment for the subsequent electrification of their commuter lines.

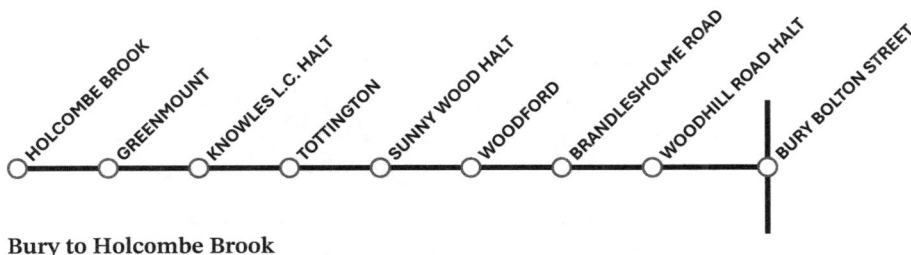

Bury to Holcombe Brook

Manchester Victoria to Bury (L&Y)

In 1916 the 16km (10 miles) route between Manchester Victoria and Bury was electrified at 1.2kV DC. The voltage chosen was the maximum permitted by the Board of Trade for outside third conductor rail electrification and because of the high voltage, the third rail was protected. The third rail was encased with a timber guard assembled in a channel form with a large surface area available for the shoes of the electric multiple units to make contact with the side of the specially profiled conductor rail.

Initially, a fourth rail was laid between the running rails for electrical return purposes, there being no contact between this rail and the trains. Subsequently, this rail was considered to be obsolete and was gradually removed over time.

The new electric motor car sets were notably the first passenger vehicles in the UK to be of all metal construction, excluding wood completely in favour of steel and aluminium. The interiors were arranged in saloons and all cars, whether motored or trailing, had driving compartments at each end. Distinctive features, in addition to their massive appearance and riveted construction, were the recessed passenger swing doors and roller-shutter doors giving access to the guards' compartments on each motor coach. A further interesting feature was the use of buck-eye couplings, there being no conventional buffers, merely nominal blocks acting as 'dumb' buffers.

All the high-tension electrical equipment was mounted inside a compartment, the door of which was interlocked, requiring the isolating switch connecting the control leads to the supply from the shoes to be opened before entrance could be gained. The motor coaches weighed 54 tons and the trailers 29 tons.

Electric operations between Manchester Victoria and Bury commenced in April 1916. However, the Lancashire and Yorkshire Railway's Works at Horwich was at that time heavily involved with war work, so deliveries of the

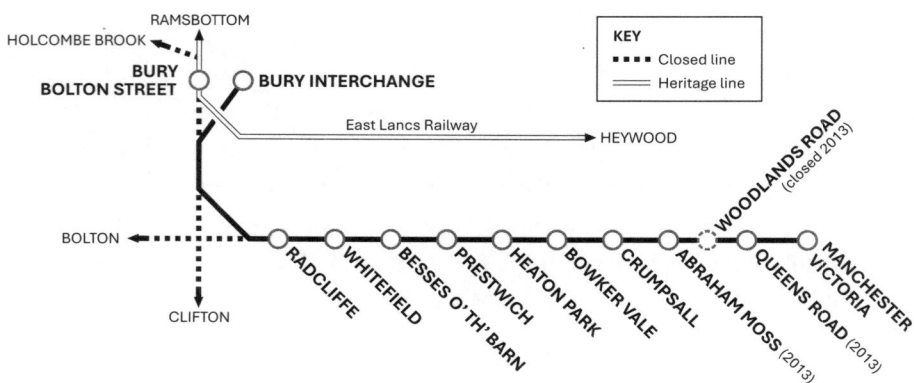

Manchester Victoria to Bury

new electric stock was delayed. It was, therefore, not until August 1916 that all remaining steam-hauled services were withdrawn.

Commencing in 1917, the Bury to Holcombe Brook system was converted to the same 1.2kV DC third-rail system and subsequently integrated into the Manchester to Bury electric services in March 1918. Before conversion, the 3.5kV DC supply had actually failed, so, as a temporary power source, 1.2kV DC was fed from the Bury line to the overhead wire and thus via the pantograph through the vehicle to the motors of the temporarily modified 1.2kV DC train sets.[*]

Generally, the trains ran on the network as five-car units. All vehicles were gangwayed throughout. A five-car set could seat 389 passengers.

The original Manchester to Bury cars were the last pre-grouping electric stock to remain at work on British Railways. They were replaced in 1959/60 by BR built Class 504 two-car units, which continued to serve the Victoria–Bury route until it was converted to a light rail system.

The Lancashire and Yorkshire Railway envisaged the Manchester to Bury electrification as the start of a widespread electrification of suburban lines from Manchester Victoria but the Great War halted progress. The railway's Board tried to resurrect their embryonic electrification projects in 1919, but it was not to be. The post-war economic difficulties and the uncertainties leading up to the Railway Grouping caused their plan to be put on hold. The London, Midland and Scottish Railway's immediate focus after 1923 did not include electrification schemes and so little was done until the 1930s, when even these projects were cut short by the Second World War.

The line to Holcombe Brook was de-electrified 1951, as it was not considered to be economic to renew the outdated electrical equipment. Push-pull steam working took over for a short time until May 1952 when this little used line closed.

Following the publication of the 1956 System of Electrification for British Railways Plan, consideration was given to converting the side-contact DC conductor-rail arrangement to a 6.6kV AC OLE system. On this short line, there were numerous overbridges together with a tunnel, so this proposal was significantly cheaper than providing the necessary electrical clearances for 25kV AC. However, in the event, this scheme was not progressed further.

Subsequent development of the network had to wait until 1989/90, when, as part of the development of the Manchester Metrolink, the Manchester to Bury line was converted to a 750V DC Light Rail system. Over the last thirty years, the electric tram system has been extended to Altrincham, Ashton, Oldham, Rochdale, Manchester Airport and Manchester Docks.

---

[*] It is interesting to note that the two original sets were not converted for further use on the Manchester to Bury system. After a considerable period in store, they were converted experimentally to diesel railcars by the LMS.

Manchester South Junction and Altrincham Railway (MSJ&AR)

With the introduction of an electric tramway along the road that paralleled the 14.5km (8.5 miles) long Manchester to Altrincham Railway in the early part of the twentieth century, patronage of the steam rail service was severely affected. While a number of railway electrification options were considered to counter this competition, no action was taken until after the Grouping when the Manchester, South Junction and Altrincham Railway Committee was formed. The Committee represented the joint interests of the London, Midland and Scottish Railway and the London and North Eastern Railway. The latter company was keen to develop the 1500V DC overhead system for mainline electrification, which was recommended as the national standard by the Kennedy Committee in 1920 and the Pringle report in 1927 (see Chapter 5).

The two railway companies jointly developed the electrification project, with the LMS taking the lead in installing the new infrastructure and designing and procuring the twenty-two new three-car non-corridor, wooden-framed electric multiple units from the Metropolitan Cammell Company. Both first- and third-class accommodation was provided.

A new depot was built on the site of the original MSJ&AR Bowdon terminus, south of Altrincham and Bowdon station. New suburban stations were opened at Navigation Road and Dane Road. The station, which had previously been called the Old Trafford Cricket Ground, was renamed Warwick Road and opened daily on a permanent basis.

Electric services between Manchester and Altrincham commenced in May 1931, providing a faster and more frequent service than the steam trains they replaced. Passenger numbers increased substantially, and the fast trains also encouraged builders and property developers to site new housing developments along the route bringing further patronage – an early example of what became known in railway circles as the 'sparks effect'.

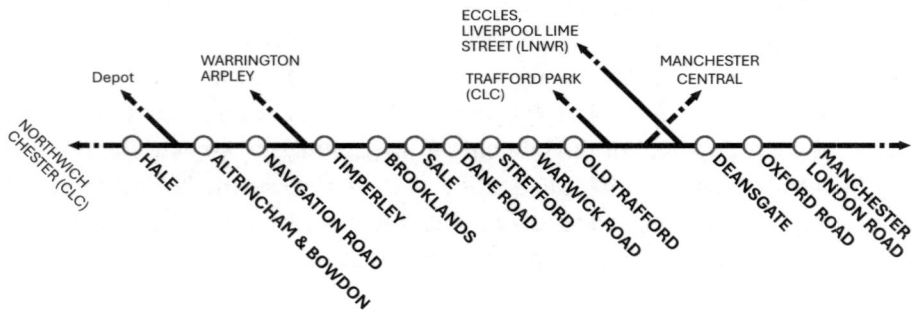

Manchester and Altrincham Railway

The line was converted to 25kV AC in 1971 by British Railways. However, the section between Altrincham, Trafford Park and Cornbrook Viaduct later reverted to DC power at 750V DC and became incorporated into the Manchester Metrolink.

Early Electrification in South London (LBSCR)

Following the award of powers to the London, Brighton and South Coast Railway in 1903 to electrify their suburban lines, the company decided to use

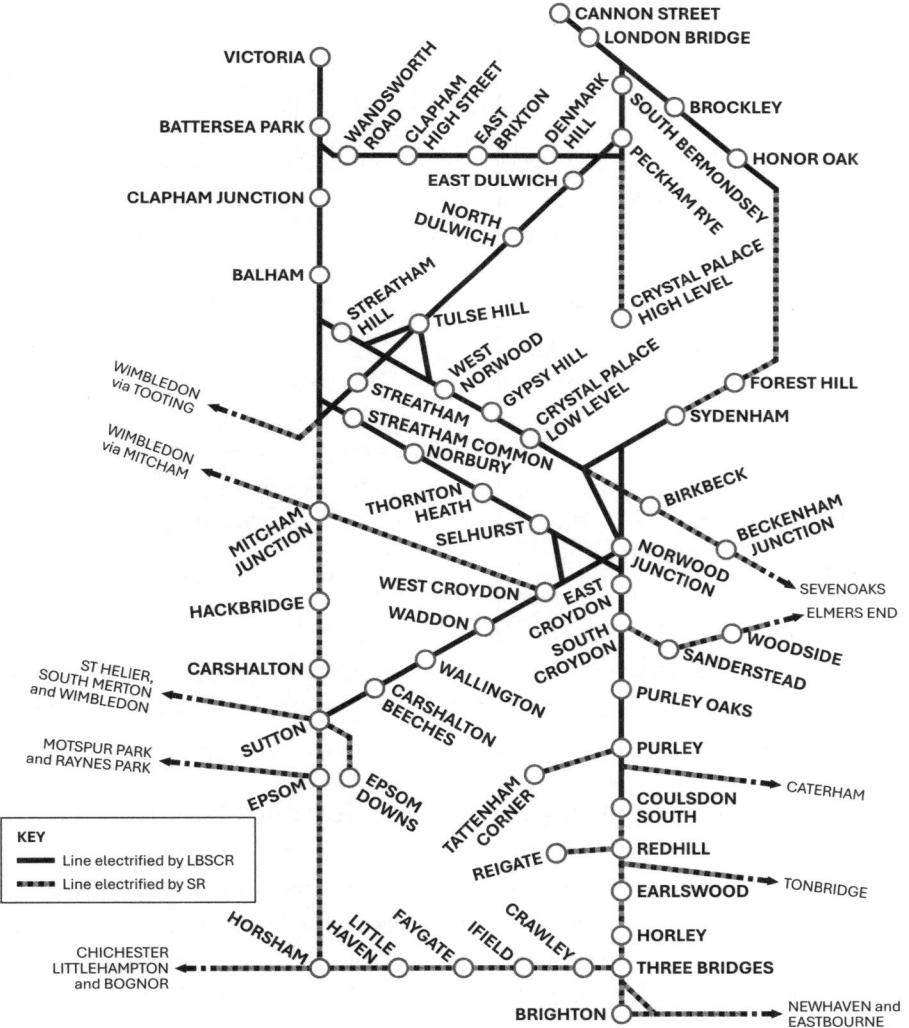

LBSCR Electrification and Further Extensions by SR

a 6.6kV 25Hz overhead wire system as recommended by the company's consulting engineer, Sir Philip Dawson. Tramway engineering contractors, Robert W. Blackwell, won the main contract with Allgemeine Electricitats Gesellschatt (AEG) of Berlin supplying the electrical equipment. The first line converted was the South London Line from London Bridge to Victoria via Peckham Rye, totalling 16km (10 miles). New servicing sheds for the electric trains were built at Peckham Rye. The London Electric Supply Corporation fed power from Deptford generating station to the railway's Peckham distribution room and the Queens Road switch cabin.

The full public electrified service commenced in December 1909. To operate the services, eight three-car sets were supplied by the Metropolitan Amalgamated Carriage and Wagon Company. Each set was formed of a third-class driving motor brake on each side of a first-class trailer vehicle. The vehicles were just over 62ft (19m) long.

Current collection was by means of a bow collector mounted above the brake compartment behind each cab with the rear one used for each direction of travel. Two traction motors were provided per motor car fitted to both axles of the leading bogie.

The electrification was later extended to Crystal Palace and West Croydon, doubling the electrified mileage, and a new depot was opened at Selhurst. Unfortunately, plans for further extensions were put on hold by the outbreak of the First World War.

In 1921 the company drew up plans to extend the overhead electrification to Brighton, but the post-war economic situation and the impending Grouping of the railways intervened. However, one final extension to both Coulsdon and Sutton was completed, after the Grouping by the Southern Railway in 1925. The Southern had, however, previously decided to standardise future electrification schemes on the London and South Western Railway third-rail DC system, Gradually, the LBSCR AC overhead system was replaced, the last section between Coulsdon North and Sutton being converted to third-rail DC in September 1929.

The Development of the Tyneside Electrified Network (NER)

At the very beginning of the twentieth century, the North Eastern Railway (NER) suburban routes around Newcastle were facing stiff competition from the newly developed electric tramways. George Gibb, the NER General Manager, submitted a memorandum to the NER board members in 1902, bringing to their attention the fact that, since the arrival of the new electric tramway systems, the railway's fare box receipts had fallen by over 50 per cent. He suggested that something needed to be done to minimise the continuing diversion of traffic, which at the time seemed inevitable. Gibb proposed that 'steps be taken to run the passenger trains by electricity' and estimated that the Tynemouth lines and the Ponteland lines could be electrified including the provision of new electric stock, for just under £200,000 (£20M at 2025 prices).

The board agreed that the company develop a scheme to electrify the 40km (25 miles) circular route* from Newcastle Central Station via Percy Main to Tynemouth, and back to Newcastle New Bridge Street Station via Whitley Bay, Monkseaton, Benton and Jesmond. The scheme also included electrification of the East Coast main line between Heaton Junction and Benton with the curves electrified from Benton Quarry Junction to the Gosforth to Monkseaton route thus allowing services to operate via the south-east curve to Monkseaton.

Approval for the scheme was given on 15 December 1902. The British Thompson-Houston company were given the contract to supply the electrical equipment for this 'third-rail' project together with the coaches. Siemens Brothers and Co were contracted to supply the high tension (HT) cables and the responsibility for the equipment for the substations given to the Westinghouse and Electric Manufacturing Co. The value of these contracts was just over £300,000 (£31M at 2025 prices). In addition, the North Eastern Railway built the carriages itself at York Carriage Works.

The NER thus developed the UK's first electric suburban network, the first stage of which opened in March 1904, starting just one week after Lancashire and Yorkshire Railway's first electric Liverpool service. The remainder of the NER network opened in August 1904. The consulting engineer engaged to oversee the project was Charles H. Merz, who was later a driving force behind the development of the UK's National Grid.

Electrification was by means of 600V DC outside third-rail with the trains' collector shoes making contact with the top surface of the live rail. Power was supplied from the Newcastle-upon-Tyne Electric Supply Co. New servicing sheds for the electric units were built at Heaton.

Of special interest is the fact that experiments were conducted with both a conventional outside third rail and an inverted third rail (a U-shaped rail laid on its side facing the track) in 1901. For the latter system, the pick-up shoes bear upwards on the lower surface of the upper arm of the conductor rail, thus providing protection from 'icing-up' in cold weather.

---

* Travel between New Bridge Street and Newcastle Stations was not immediately possible, the city centre stations having been built by different railway companies. It was not until 1909, following extensive infrastructure alterations, that it became possible to run a circular service from Central to Central. However, despite this, it was not until 1917 that such a regular circular service was introduced.

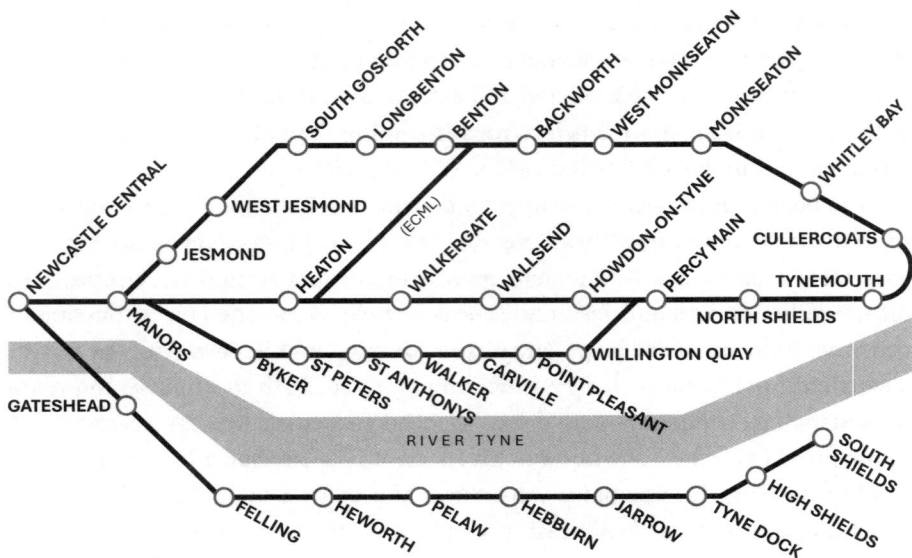

Tyneside Electrified Network (NER)

An experimental length of inverted third rail was put in place between South Gosford and Gosforth West Junction in 1901. It is not known if the experiment was intended to test the pick-up arrangements for the Tyneside network or if it was a trial of one of Charles Merz's ideas. Interestingly, in 1919, Merz and his business partner McLellan proposed the same inverted rail system for the York to Newcastle main line electrification scheme.

The decision was made to proceed with the conventional third-rail pick-up (that is, from top of rail) despite the knowledge that ice forming on the head of the conductor rail would cause pick-up problems during the cold north-eastern winters. To counter this, Charles Merz devised a third-rail head ice scraper that could be fitted to the electric cars. This was found not to be robust enough and a strengthened version that the driver could adjust was developed and was, apparently, reasonably effective.

From 1938, however, three of the original cars were converted into de-icing units, equipped to apply Kilfrost de-icing fluid to the head of the conductor rail and deployed when the air temperature was forecast to fall below freezing.

The third rail was originally set 1ft 7¼" (0.49m) from the running face of the nearest rail, with the surface of the conductor rail being 3¼" (0.08m) above the running rail surface. The conductor rails were altered to meet the 'standard' dimensions recommended by the Pringle Committee in 1938, when the South Shields Line was electrified to the same standard. The remarkable feature of the original construction works was the speed at which electrification was accomplished. The circular route was electrified by the end of 1903, with the Ponteland branch following at the end of March 1904.

Immediately following the introduction of the electrified services, not only were many passengers regained from the trams, but much new traffic was attracted, particularly from stations outside the city. Operating costs were said to have been reduced by 50 per cent and traffic nearly doubled to more than 10 million passengers carried over the ten years to 1913. Perhaps, unknowingly, the North Eastern Railway management were experiencing their first 'sparks effect', although of course this term did not come into use for another half century! Unfortunately, many UK railway managements, immersed in a cost-saving culture, failed to recognise the value of this, considering that the main benefits of electrification were either a means of winning back customers lost to other means of transport or as a way to reduce operating costs. Thus, bereft of this additional 'new business' benefit, a number of potential electrification schemes contemplated thereafter failed to get off the starting blocks.

The new rolling stock, which had distinctive vertical match-boarded side and end panels, was built at York Carriage Works. The 88-seater cars had clerestory roofs and were fitted with 'Cowhead' couplers but had no buffers. Buffers were, however, provided on the parcels vans, which could then haul other rolling stock. Motors of 125hp were supplied by British Thomson-Houston Co and fitted at York Works. Later, cars built after 1909, were fitted with more powerful motors. Further rolling stock was built over the years to cope with both growing passenger demand and to provide replacements for rolling stock lost in a serious fire at the Heaton Shed in 1918.

The dockside branch connecting the Newcastle Quayside to Trafalgar Yard near Manors East Station was also electrified in 1905. This mile-long line descended in a semi-circle with gradients as steep as 1:27 passing through three tunnels and was thus difficult to work with steam locomotives. Two centre-cab traction units (Class 'Electric 1', later designated 'ES1'), NER No. 1 and 2 were introduced in 1905 to operate the line. The frames and body were constructed by Brush Engineering with the electrical systems and four 160hp traction motors provided by British Thomson-Houston. The units picked up power from the third rail in the tunnel section but, in the sidings at both ends, overhead catenary was provided to improve staff safety.

In 1937 the NER's successor, the London North Eastern Railway, electrified the 16km (10 miles) South Tyneside route between Newcastle Central and South Shields via Gateshead and Pelaw. A new fleet of electric trains was built to operate the original North Tyneside routes, while the NER 'elliptical-roof' stock was transferred to operate the new South Tyneside electrified lines. With the exception of the three Motor Parcels Vans, which were converted into de-icing units, all of the original 1904–05 stock was withdrawn between August and December 1937. These were replaced with a new fleet of electric trains constructed by the Metropolitan Cammell Carriage and Wagon Company built to Sir Nigel Gresley's specification. Electrical control equipment was supplied by Westinghouse, and Crompton Parkinson provided the electric motors.

The passenger units comprised sixty-four twin-coach articulated units. Two motor parcels vans and two motor luggage motor thirds were also procured. Fifteen of the original NER passenger cars were retained to operate workmen's trains. Along with the introduction of the new stock, the train service was improved with trains from 'Central to Central' every ten minutes. There were also 'express' trains to and from the coast every hour between 9am and 11pm.

After the Second World War, the system was not extended any further. In 1955, some of the rolling stock was in need of replacement, so British Rail brought in third-rail stock of Southern Region design (EPB units).

However, by the 1960s, with the deterioration of the area's industrial base and changing demographics, the railway was in decline. In addition, much of the electrical equipment was in need of renewal. In December 1962, British Railways took the decision to replace the electric trains with diesel units and not to renew the electrical equipment. It was estimated that the annual loss of £300,000 per annum (£6M at 2025 prices) would be reduced by £80,000 (£1.5M at 2025 prices), although the costing methodology looks rather flimsy, to say the least, compared to present-day analyses.

The South Tyneside line was de-electrified in 1963 and the last electric trains ran on the north side of the river on 17 June 1967. However, following the 1968 Transport Act and the subsequent Local Government Reorganisation, which heralded the new County of Tyne and Wear, an electrified metropolitan railway system was to be resurrected. Parliamentary authority was granted in July 1973 for the Tyneside Metropolitan Railway Act. Construction of the Tyne and Wear Metro, as it became known, commenced in 1974 and involved taking over much of the former North Eastern electrified routes, both north and south of the Tyne. A new link by means of a bridge across the Tyne was constructed and new tunnels were bored under both Newcastle and Gateshead to improve city centre access. The new Metro was electrified by means of overhead line energised at 1500V DC.

The Development of the London and North Western Railway Electrified Network

The London and North Western Railway (LNWR) gained parliamentary powers in 1907 to expand and electrify their Euston suburban services. The Watford

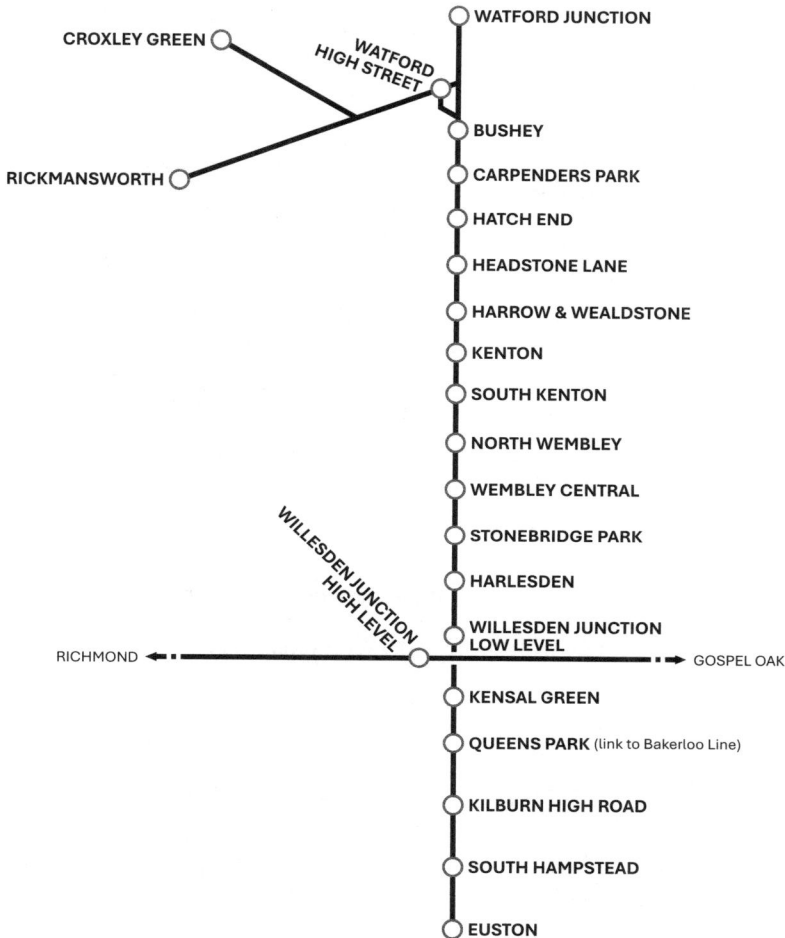

WATFORD JUNCTION

CROXLEY GREEN

WATFORD HIGH STREET

BUSHEY

RICKMANSWORTH

CARPENDERS PARK

HATCH END

HEADSTONE LANE

HARROW & WEALDSTONE

KENTON

SOUTH KENTON

NORTH WEMBLEY

WEMBLEY CENTRAL

STONEBRIDGE PARK

HARLESDEN

WILLESDEN JUNCTION HIGH LEVEL

WILLESDEN JUNCTION LOW LEVEL

RICHMOND

GOSPEL OAK

KENSAL GREEN

QUEENS PARK (link to Bakerloo Line)

KILBURN HIGH ROAD

SOUTH HAMPSTEAD

EUSTON

London and North Western Electrified Network (London DC Lines)

line project involved the provision of two additional tracks running parallel to the main lines between Camden and Watford Junction together with the electrification of 32km (20 route miles) of track. A new branch line was planned between Watford Junction and Croxley Green, branching off from the Watford to Rickmansworth line. A new link was also constructed between the LNWR and the Underground Bakerloo Line at Queens Park. New suburban stations were provided on the existing routes.

To permit through working of Underground trains, the new lines were electrified at 630V DC by means of an outer conductor rail and negative return via a fourth rail between the running lines. The LNWR provided its own power station at Stonebridge Park that provided power to the track.

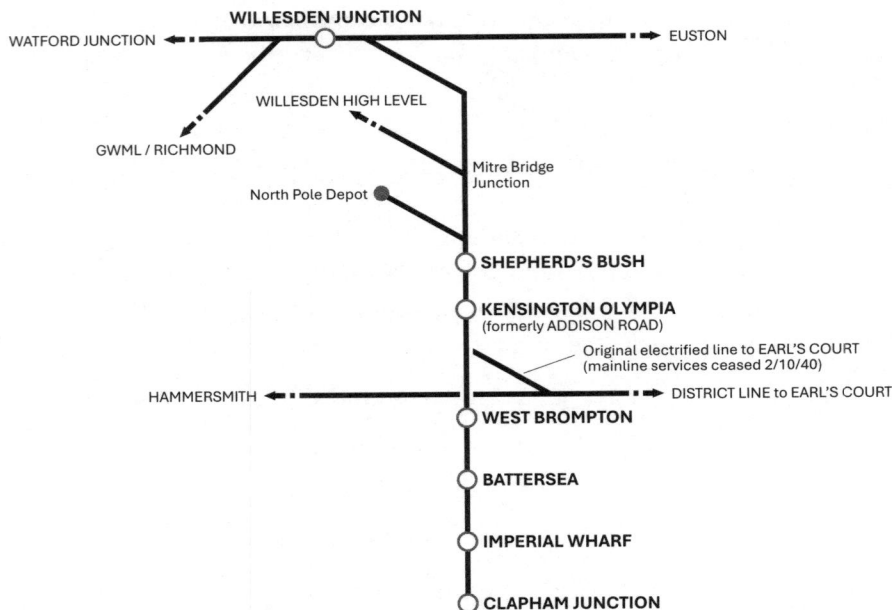

London and North Western Electrified Network (West London Line)

Work commenced in 1909, and in the same year the LNWR took control of the North London Railway (NLR).

The first LNWR electric trains ran from Willesden Junction to the District Railway station at Earls Court via Addison Road (now Kensington Olympia) in 1914. These were four three-car sets supplied by the Metropolitan Carriage and Wagon Company. Thereafter, the outbreak of the First World War considerably impeded the progress of the project. Although the link to the Bakerloo at Queens Park was completed in 1915, as the line north of Willesden to Watford Junction was not electrified until 1917, Underground trains could not use the link to provide a service northward for another two years.

The operation of services by a number of different companies on this 'joint line' brought about the need to standardise on the third-rail 630V DC system, which became known as the 'London Standard'.

The North London Railway was an amalgamation of a number of separate lines constructed between 1853 to 1860. The principal ones were:

- the North and South Western Junction Railway linking Willesden Junction to the Hounslow Loop near Kew Bridge, which commenced operations in 1853;
- the Hampstead Junction Railway between Camden Road and Willesden via Hampstead Heath, which opened in 1860.

The company therefore also decided to electrify the 23km (15 miles) long former NLR route from Broad Street Station, via Dalston Junction and South Acton, to Richmond. In addition, links were made with the Bakerloo and District Railways. The same fourth rail system was utilised as before in both instances, this being the same arrangement, used at that time, on the underground lines in London.

Thus, during the last year of the LNWR existence in 1922, with the introduction of the full electric service between Euston and Watford Junction, together with the Croxley Green branch and the North London line services, the company operated the largest electrified suburban network in the world.

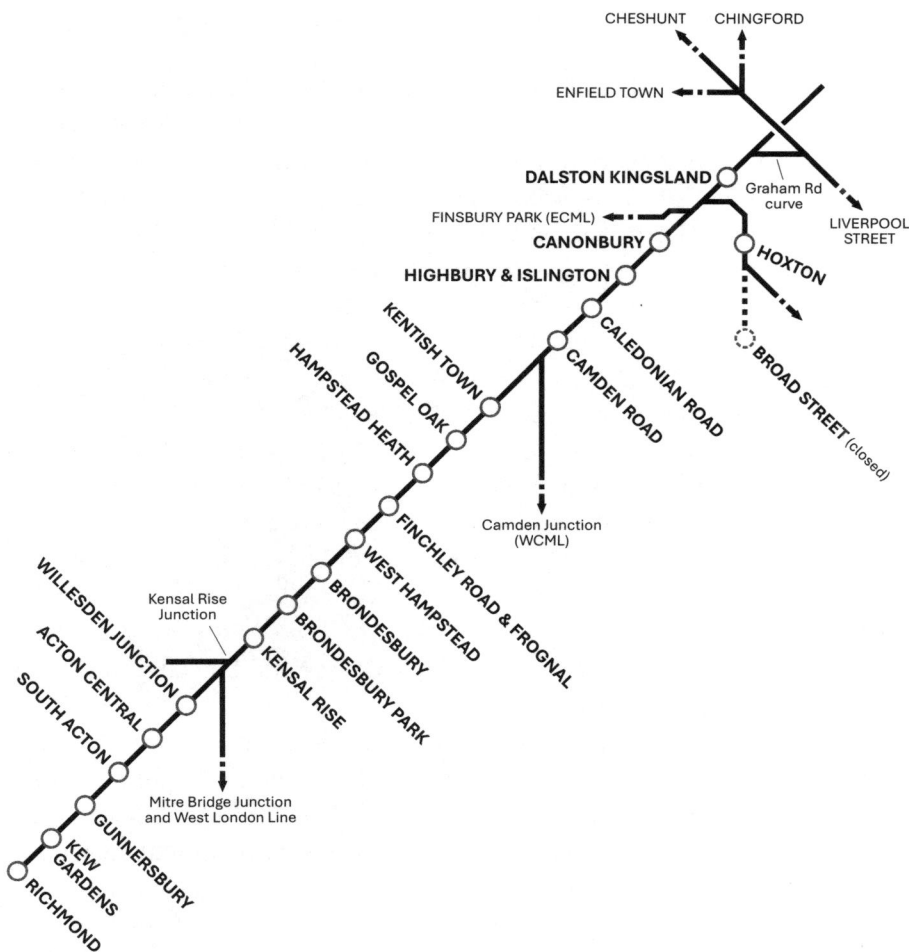

Richmond, Willesden to Dalston Junction

Something of an anachronism to be noted was that the 7.5km (4.5 miles) long Watford to Rickmansworth branch, which had been electrified on 16 April 1917, was initially operated by London Electric Railway (LER) tube trains. In this original form, power for the branch was only fed from Watford and thus the line suffered from a quite severe voltage drop at the Rickmansworth end. The LER services were augmented at peak times by steam trains to Euston and electric services to Broad Street. The newly formed London, Midland and Scottish Railway took over and re-energised the line in 1927, thus completing its electrified network in the London area. The Rickmansworth branch closed in 1952, and the Croxley Green branch was 'mothballed' in 1996. Sections of route may, however, be reopened in the future as part of the London Underground Croxley Rail Link scheme. From 1957, the original rolling stock was replaced with services operated by three-car Class 501 EMU trains, which worked the line until 1988.

The power supply and rolling stock used on this line and the North London Line was changed in the 1970s to +630V/0V. Subsequently, the electrification system was modified again to a hybrid BR DC system between Euston and Watford (with the fourth rail bonded to the running rail for the return traction current*). On the North London line, the standard BR DC third-rail traction supply arrangements were implemented, enabling the centre/negative current rail to be removed between Dalston Junction and South Acton.

The section beyond Acton South southwards retained the 630V DC fourth rail system to permit dual operations with the London Underground trains that share the route through Gunnersbury to Richmond.

Between 1983 and 1985 the DC system was extended from Dalston Junction to North Woolwich and through passenger services started running in 1985 from Richmond to North Woolwich. Thereafter, between 1985 and 1987, the core part of the North London Line from Stratford to Camden Road was electrified with the 25kV AC system to permit electric haulage between the Great Eastern/ London Tilbury and Southend routes and the West Coast Main Line. These works also included a link to the East Coast Main Line at Copenhagen Junction just north of King's Cross.

Over time, the DC third-rail equipment was removed, and the Class 501 units were replaced by dual-voltage Class 313 EMUs in 1988. From this year, traction current changes were necessary at Acton Central, Camden Road, Dalston Kingsland and Hackney Wick. As can be imagined, drivers sometimes

---

* The fourth rail was left supported by the insulator 'pots' on the sections of the route shared with London Underground Bakerloo line trains. However, beyond the London Underground limit of operations to the north of Harrow and Wealdstone, the bonded fourth rail was 'dropped' directly onto the sleepers, acting, with the running rails, as the conduit for the return current.

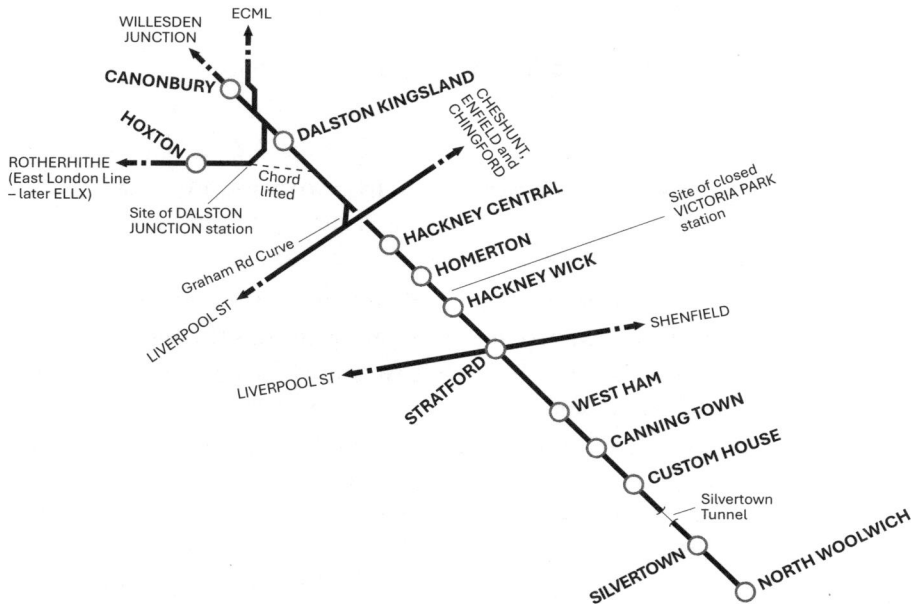

WILLESDEN JUNCTION

ECML

CANONBURY

HOXTON

DALSTON KINGSLAND

CHESHUNT:
ENFIELD and
CHINGFORD

ROTHERHITHE
(East London Line
– later ELLX)

Chord
lifted

Site of DALSTON
JUNCTION station

Graham Rd Curve

HACKNEY CENTRAL

HOMERTON

HACKNEY WICK

Site of closed
VICTORIA PARK
station

LIVERPOOL ST

SHENFIELD

LIVERPOOL ST

STRATFORD

WEST HAM

CANNING TOWN

CUSTOM HOUSE

Silvertown
Tunnel

SILVERTOWN

NORTH WOOLWICH

Dalston to North Woolwich

forgot to drop the pantograph, so there were many instances of bridge strikes, causing severe impacts to service performance.

From September 1995 through to August 1996, Railtrack electrified the section between Camden Road and Willesden at 25kV AC. The Class 313 EMUs operated the route until 2010, under the Silverlink brand, when they were replaced by Class 378 Capitalstar four-car units operated by London Overground (see also Chapters 9 and 14).

Chapter 3
Early Main Line Schemes

Great Central Railway – Grimsby & Immingham Electric Railway

The 11km (7 miles) long Grimsby & Immingham Electric railway was conceived primarily as a passenger railway by the Great Central Railway in order to provide transport, mainly for workers, from Great Grimsby to the company's vast new port development at Immingham.

The 500V DC overhead line, opened in several stages between 1912 and 1915, was based in part on tramway principles but it ran mainly on reserved track – today, it would be referred to as a tram–train inter-urban system. Electric power was the obvious choice for two reasons. First, the Great Central Railway planned not only for on-street operation in Grimsby but also to extend the line to connect with the existing Grimsby and Cleethorpes tramways. Secondly, as the company had built a large generating station at the Docks to power the cranes, ancillary equipment and lock gates, this plant could also supply the new line, which it did until 1957.

Two substations, built by Siemens Brothers, fed traction power to the line, which was transformed and converted from a 6,600V AC by means of Westinghouse 250kW rotary converters to produce the 500V DC line voltage with line feeders every half mile. When the generating station was closed in November 1957, high tension power was drawn instead from the National Grid.

The Great Central Railway's application for a light railway order in 1906 specified a 2½km (1.5 miles) street running section between Corporation Bridge

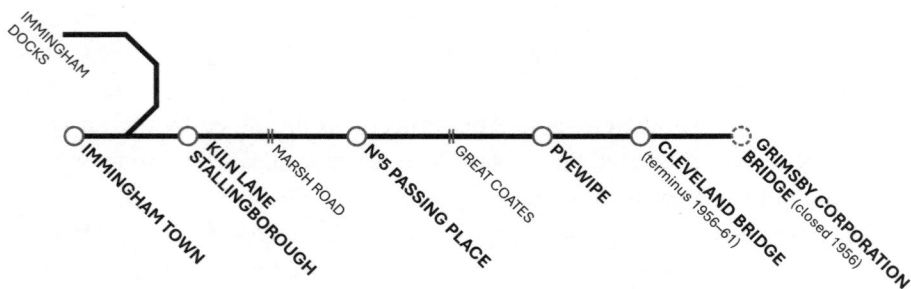

Grimsby and Immingham Electric Railway

and Cleveland Bridge in Grimsby, together with a reserved right of way between Cleveland Bridge and Immingham Dock.

The line closed in 1961.

London, Tilbury & Southend Rail Electrification Proposals

The London, Tilbury & Southend Railway (LT&S), jointly promoted by the London & Blackwall Railway (LBR) and the Eastern Counties Railway (ECR), was authorised by Parliament in June 1852. The railway was extended to Southend in March 1856 and running powers into Fenchurch Street were granted by the Great Eastern Railway in July 1875.

The railway was extended from Southend to Shoeburyness in February 1884 and the line from Barking to Pitsea opened in June 1888. Despite a close relationship with the GER, the LT&S company and its assets were sold to the Midland Railway in 1912; the GER didn't apparently want (or couldn't afford) the Tilbury and the Midland made a good offer. As part of the Parliamentary proceedings, the Midland told Parliament it would electrify the LT&S within 7 years, and subsequently commissioned Merz and McLellan to undertake a study to assess how this might be undertaken. Concurrently, the GER was putting together a series of its own proposals to enhance Fenchurch Street Station's capacity, and so the two companies entered into an agreement to work together. However, the onset of the First World War effectively thwarted all the Merz McLellan recommendations, and it was thus to be more than 40 years before electrification of the LT&S was again revived (see Chapter 8).

Shildon to Newport

Following on from the electrification of Tyneside routes by the North Eastern Railway, the company decided to electrify a route operating heavy coal and mineral trains as a precursor to the electrification of the busy East Coast Main Line between York and Newcastle. The project was promoted by the railway's Chief Mechanical Engineer, Sir Vincent Raven, who was a keen protagonist of main line railway electrification.

The line chosen for electrification was the relatively self-contained 30km (19 mile) route from Shildon, in County Durham, to Newport, near Middlesborough, and included the famous Stockton to Darlington Railway. Shildon Yard was the collection point for coal from local coalfields, and Newport Erimus Yard was the location from which coal was distributed to the docks, blast furnaces,

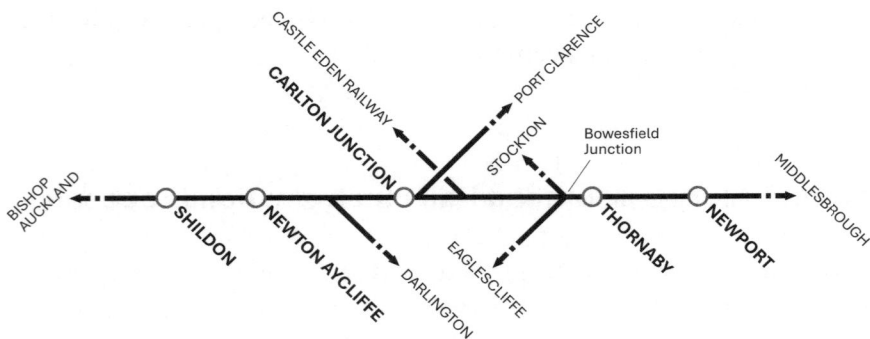

NER Shildon to Newport Railway

and iron works in the Stockton and Newport areas. The heavy mineral traffic prior to the Great War was very profitable and demand was such that, for part of the route, the number of tracks had been increased to four.

Electrification of the route commenced in 1914 utilising, for the first time in the UK, the 1500V DC overhead system. The electrification infrastructure was of a simple form, clearly designed with economy in mind. The Newcastle-upon-Tyne Electric Supply Company were contracted to not only supply the power to the system but also build the substations. The electrification was commissioned in stages from July 1915 to January 1916.

Ten freight Bo-Bo (0-4+4-0) traction units (Nos. 3–12), which owed much to American practice, were designed by Sir Vincent Raven and built at the Darlington Works between 1914 and 1919 (subsequently termed EF1 – Electric Freight Class 1). Electrical equipment supplied by Siemens Bros included the four 275hp motors per unit, connected in series, thus making the EF1 traction units considerably more powerful than the existing NER Class ES1 type provided for the Tyneside scheme. This enabled an EF1 unit to start a 1,400 ton train (seventy wagons) and haul it on the level at a minimum speed of 25mph. The No. 3 roundhouse at Shildon was adapted to stable and service the traction units.

Unfortunately, with the Great War progressing, this was just about the worst time to commission the electrification of a freight railway. With the restrictions on coal shipments during the war, mineral traffic levels were significantly reduced. After the First World War and into the 1920s, due to the war reparations foisted by the Allies on Germany, the General Strike and the Depression, coal traffic was further reduced and, by 1928, it was clear that there was insufficient traffic for all ten traction units. A proposal was even put forward to rebuild one as a diesel-electric locomotive, but this project stalled.

By the mid-1930s, against a background of declining traffic, it became necessary to replace much of the overhead equipment. Unfortunately, in these circumstances, it proved uneconomic to carry out the necessary refurbishment and renewals to maintain electric operation. The electric infrastructure was therefore taken out of use and dismantled. The Shildon yards were closed in 1935, and steam locomotives once more took over the haulage of the mineral traffic. All ten of Raven's EF1 traction units entered storage at Darlington, where they remained throughout the Second World War. Only one was recommissioned after the war and used at Ilford to perform shunting duties at the new electric carriage sheds built for the 1500V DC London Liverpool Street to Shenfield electrification.

One further aspect related to Sir Vincent Raven's electrification aspirations is worthy of note. Even during the difficult times, during and after the Great War, he continued to press the North Eastern Railway Board to consider electrifying the company's stretch of the East Coast main line from York to Newcastle. Following Raven's visit to the General Electric works in Schenectady, USA, and backed up with technical assistance from Merz and McLellan, the Board eventually authorised Raven to build a prototype 2-Co-2 (4-6-4) electric passenger traction unit in 1910. This locomotive was capable of hauling fourteen bogie coaches (450 tons) at 65mph on level track.

Numbered 13, and subsequently termed EE1 (Electric Express Class 1), the vehicle was built at Darlington Works, with electrical equipment provided by Metropolitan-Vickers. Trials were held on the Shildon to Newport route, but the 'unlucky' No. 13 never operated at high speed on a passenger line and spent most of its life in storage until it was finally withdrawn in 1950.

The London & South Western Railway (LSWR)

Tramway services that were springing up in South London in the early years of the twentieth century were having a significant impact on railway commuter services. With the appointment of a LNWR senior manager, Herbert Ashcombe Walker, as General Manager of the London and South Western Railway in January 1912, the railway's interest in electrification was heightened. With his experience of the electrification taking place on the LNWR, Walker initiated an extensive programme of third-rail suburban electrification.

A survey of the systems in use elsewhere in the UK, including a visit to the Lancashire and Yorkshire Railway's Liverpool to Southport line, reinforced his view that the low voltage (600V DC) outside conductor rail system based on American practice was the way forward.

Working closely with Alfred W. Szlumper, who was promoted from Assistant to Chief Civil Engineer of the London & South Western Railway in 1914, the two men piloted through the first programme of suburban electrification scheme between 1913 and 1916, together with the major reconstruction of Waterloo Station.

The routes programmed to be electrified were:

- Waterloo to Wimbledon (via Earlsfield and East Putney)
- Point Pleasant Junction to Wimbledon (via Twickenham and Kingston)
- Barnes to Hounslow, Twickenham
- Shacklegate Junction and Strawberry Hill to Shepperton
- New Malden to Hampton Court
- Surbiton to Guildford (via Cobham)

The programme progressed rapidly with the first section from Waterloo to Wimbledon, via East Putney, opening for electric services in October 1915. This was followed by the electrification of the routes from Clapham Junction to Twickenham, Kingston and Shepperton in January 1916, Barnes–Hounslow–Twickenham in March 1916, and New Malden to Hampton Court went live in June the same year.

This Guildford route was not to be completed as planned due to the economic difficulties brought about by the outbreak of the Great War. As a result, the electrification of this section was curtailed to Claygate, which was energised in November 1916.

The London and South Western Railway built a power station and new car sheds at Durnsford Road, Wimbledon. An extensive new depot was constructed, enabling all the 84 three-car units provided to operate the new services to be stabled there. These sets were not new, having been converted from steam-hauled carriages at the LSWR workshops at Eastleigh.

These EMUs, latterly designated 3-SUB, comprised two motor coaches and a trailer. The sets were 157ft (48m) or 160ft (49m) long, depending on the length of the converted trailer cars, and weighed 95 tons. Each unit seated between 172 and 190 people in first and third class according to length and were the first trains on the LSWR not to offer second-class accommodation. A further 24 two-coach third-class trailer units were constructed at Eastleigh in 1914 to work with the multiple units. These non-powered sets were either 105ft (32m) or 108ft (33m) long and weighed 46 tons.

Electrical and braking equipment was supplied by the UK Westinghouse Company, with motor carriages having two 275hp (205kW) electric motors, giving a total power of 1100hp per unit. Electric jumper cables were provided to enable sets to work in multiple.

Imitating the tramway competition, a head-code box was provided between the cab windows of the driving cars, where a stencil denoting a letter or number

LSWR Electrification and Further Extensions by SR and BR

could be placed indicating the route over which the train was booked. This was to become a familiar feature of Southern electric services.

Herbert Walker proved to be a shrewd railway administrator and, in recognition of his achievements, including his electrification successes, he was knighted in March 1915 and later became the acting chairman of the Railway Executive Committee in 1917.

Both the LSWR and the LBSCR therefore kicked off the development of electric suburban routes in South London. Unfortunately, the Great War put an end to electrification progress from 1916 and, with the impending Grouping of the railways in 1923, no further electrification schemes were progressed until after the formation of the Southern Railway. The new company absorbed both the LSWR and LBSCR electrified suburban routes at the Grouping on 1 January 1923.

There is much speculation about why there was very little penetration of the area south of the Thames by the underground electric railways. Until the Victoria line was built in the 1960s it was only the Northern line to Morden and the Bakerloo line to the Elephant & Castle that ran south of the Thames. It has been postulated that there was an understanding between Sir Herbert Walker and Albert Stanley (the Managing Director of London Transport), that the underground would not attempt to expand and poach traffic from the Southern Railway. There is no record of this, but it is interesting that there was no expansion into an area that clearly had great potential for underground services.

South Eastern & Chatham Railway (SECR)

Although the SECR did not actually carry through any electrification schemes during its independent existence, it is interesting to note that following World War I, the company had extensive plans to electrify its network utilising a 1500V DC system. It is also worthy of note that the company had in mind the operation of electrically powered long-distance passenger and freight traffic in addition to suburban services.

The works were not started, primarily as the company's proposal to build its own power station, at Angerstein Wharf on the Thames, was blocked by the Electricity Commissioners. This was because government policy was for power to be supplied by electric utility suppliers in the future, rather than proliferating private sources of supply.

The SECR lines were subsequently electrified in Southern Railway days using the low-voltage DC 'London Standard'.

Southern Railway

At the Grouping in 1923, the Southern Railway absorbed the suburban electrified routes of the LBSCR and the LSWR, including the underground Waterloo and City Railway, totalling 134km (85 miles). Nearly 40km (25 miles) of the total was electrified at 6700V AC (elevated or overhead collection), 92km (57 miles) of over-ground railway at 600V DC (3rd-rail) and just over 2km (1 mile) underground at the same voltage.

The new company therefore had the largest suburban electrified mileage of the newly formed 'Big Four' railway companies. The SECR had an extensive suburban network but no electrified routes and this was also brought into the group. Herbert Walker, formerly of the LSWR, became the General Manager of the Southern Railway and ex-LSWR men took the majority of the key roles in the new company.

With commuter traffic increasing, the Southern embarked upon a programme to expand electrification on all its key suburban routes. An evaluation of the installation and operating costs of the two electrification systems was made at an early stage. The third-rail system was cheaper to install and there appeared to Walker to be no significant difference in the operations and maintenance cost of either system. Walker, who had previously carried out his investigations into other schemes before deciding on the LSWR system, favoured this third-rail DC system. With the evidence presented and Walker's undoubted influence, the Southern Railway Board adopted the LSWR system as the new standard for the company. It was, however, agreed that the last LBSC style electrification to Coulsdon North and Sutton should progress, even though it would have a short life. This scheme was completed in 1925. Shortly after, on 9 August 1926, the Southern announced that the DC system was to replace the AC system.

The initial focus of the newly invigorated electrification process was on enhancements and additions to LSWR electrified lines together with schemes to energise suburban lines on the former SECR.

Extensions to the Third-rail Network

The day of 12 July 1925 was a significant point in the early history of the Southern Railway, with the first of the company's new electrification schemes going 'live'. These schemes doubled the company's electrified route mileage as follows:

- Raynes Park to Dorking (ex-LSWR)
- Claygate to Guildford (ex-LSWR)

- Victoria to Orpington via Herne Hill and Shortlands (ex-SECR)
- Victoria to Holborn Viaduct (ex-SECR)
- Crystal Palace Branch (ex-SECR).

This was the initial phase of a relentless programme of third-rail electrification by the Southern Railway, which, by the start of the Second World War, saw most of its suburban lines and some of its main line routes electrified.

The electrification between Charing Cross and Cannon Street to Orpington went live on 28 February 1926. This was followed, in July 1926, with the ex-SECR routes from Cannon Street and Charing Cross to Dartford (three routes).

Over the next three years, the following routes were either newly electrified or converted from the old LBSCR high voltage system to third-rail:

- London Bridge to Crystal Palace (low level), Caterham, Tattenham Corner
- Victoria, Coulsdon North, Epsom Downs, Norwood Junction, Selhurst and Epsom
- Victoria to Beckenham Junction
- Victoria to Elephant & Castle
- Wimbledon to Herne Hill and South Merton.

To minimise interruptions to timetabled services when replacing the LBSCR system, the overhead equipment was maintained in a functional state until the DC system was in place and ready to be brought into use. Thus, seamless changeovers to the third-rail DC system were effected.

With the conversion of the lines to Coulsdon North and Sutton to DC, the last AC train ran on 29 September 1929. By the end of that year, the Southern operated over 446km (280 miles) of third-rail electrified track and its electric services operated nearly 29 million km (18 million miles) per annum.

On the former SECR routes, the Dartford–Gravesend Central was energised on 6 July 1930, thus completing the first phase of electrification on the former SECR lines. The electrification of a number of connecting lines that had been omitted from the earlier programmes followed during 1933 and 1934. These included the two routes to Sevenoaks, one from Bickley via Otford and the other from Orpington – these being electrified in January 1935. The short spur from Nunhead to Lewisham was energised from 30 September 1935. This permitted peak hours electric workings between Dartford and Victoria and Holborn Viaduct to be initiated.

With the inner suburban routes of former LSWR, LBSCR and SECR nearing completion, attention turned to outer suburban and main line routes. By the end of the decade, the Southern would complete the electrification of nearly

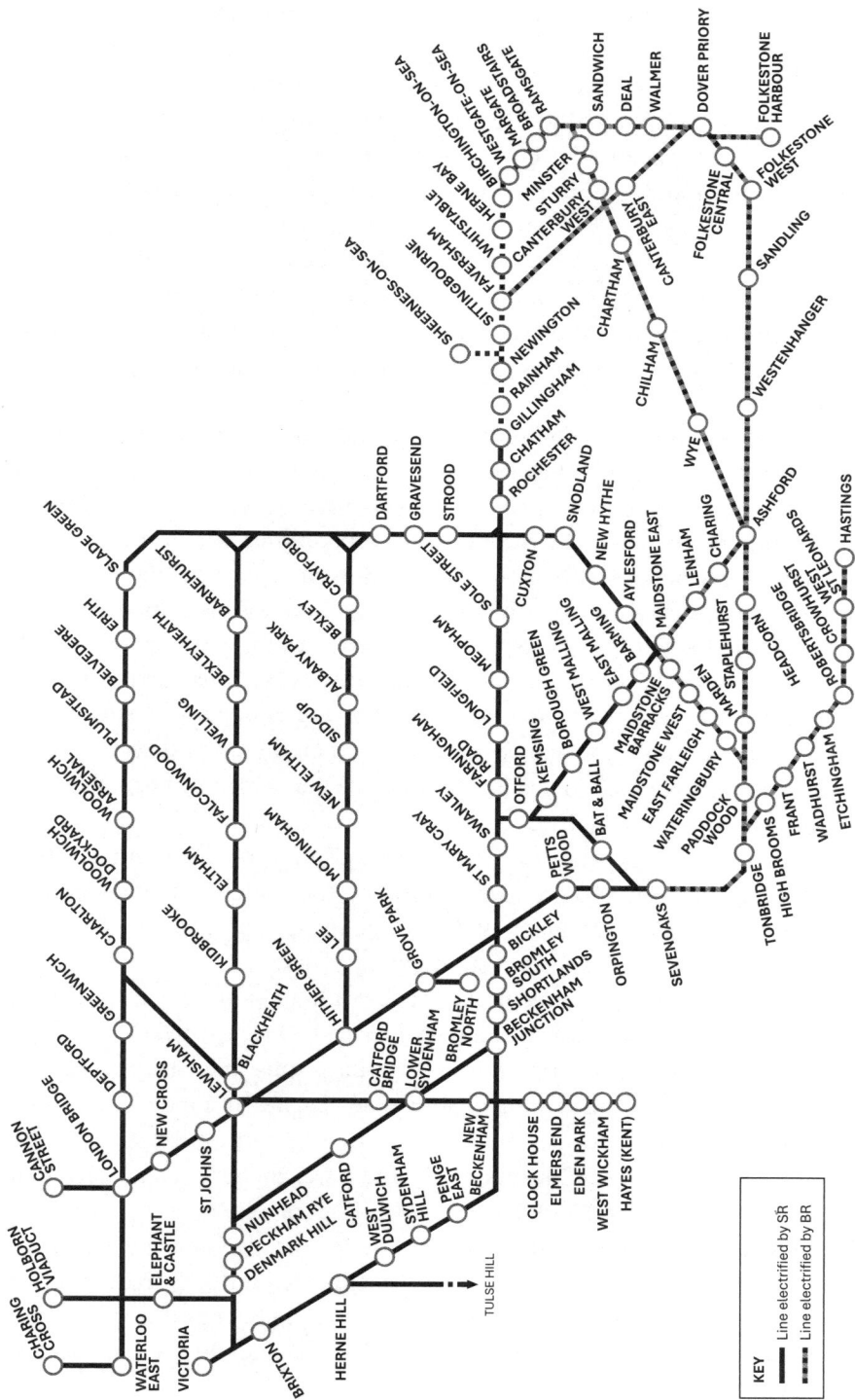

Lines Electrified by SR and Further Extensions by BR

RAMSGATE, BROADSTAIRS, MARGATE, WESTGATE-ON-SEA, BIRCHINGTON-ON-SEA, HERNE BAY, WHITSTABLE, FAVERSHAM, SHEERNESS-ON-SEA, SITTINGBOURNE, NEWINGTON, RAINHAM, GILLINGHAM, CHATHAM, ROCHESTER, STROOD, GRAVESEND, DARTFORD

SANDWICH, DEAL, WALMER, DOVER PRIORY, FOLKESTONE HARBOUR, FOLKESTONE WEST, FOLKESTONE CENTRAL, SANDLING, WESTENHANGER

MINSTER, STURRY, CANTERBURY WEST, CANTERBURY EAST, CHARTHAM, CHILHAM, WYE, ASHFORD, HASTINGS

SNODLAND, NEW HYTHE, AYLESFORD, MAIDSTONE EAST, BEARSTED, LENHAM, CHARING, HEADCORN, ROBERTSBRIDGE, ETCHINGHAM, FRANT, WADHURST, HIGH BROOMS, TONBRIDGE

CROWHURST, ST LEONARDS

CUXTON, SOLE STREET, MEOPHAM, LONGFIELD, FARNINGHAM ROAD, SWANLEY, ST MARY CRAY, PETTS WOOD, ORPINGTON, SEVENOAKS

OTFORD, KEMSING, BOROUGH GREEN, BAT & BALL, WEST MALLING, EAST MALLING, BARMING, MAIDSTONE BARRACKS, MAIDSTONE WEST, EAST FARLEIGH, WATERINGBURY, PADDOCK WOOD, MARDEN, STAPLEHURST

SLADE GREEN, BARNEHURST, CRAYFORD, BEXLEY, ALBANY PARK, SIDCUP, NEW ELTHAM, MOTTINGHAM, GROVE PARK

ERITH, BELVEDERE, PLUMSTEAD, WOOLWICH ARSENAL, WOOLWICH DOCKYARD, CHARLTON, GREENWICH, DEPTFORD, LONDON BRIDGE, CANNON STREET, CHARING CROSS, HOLBORN VIADUCT

BEXLEYHEATH, WELLING, FALCONWOOD, ELTHAM, KIDBROOKE, BLACKHEATH, LEE, HITHER GREEN, LEWISHAM, NEW CROSS, ST JOHNS, ELEPHANT & CASTLE, WATERLOO EAST, VICTORIA

BICKLEY, BROMLEY SOUTH, SHORTLANDS, BECKENHAM JUNCTION, BROMLEY NORTH, LOWER SYDENHAM, CATFORD BRIDGE

CLOCK HOUSE, ELMERS END, EDEN PARK, WEST WICKHAM, HAYES (KENT), NEW BECKENHAM, PENGE WEST, SYDENHAM HILL, WEST DULWICH, CATFORD, DENMARK HILL, PECKHAM RYE, NUNHEAD

HERNE HILL, BRIXTON, TULSE HILL

KEY
Line electrified by SR
Line electrified by BR

Early Main Line Schemes 45

all lines to destinations within 60 minutes of London and many towns beyond, including destinations on the South Coast.

On the ex-LSWR side, electric trains reached Windsor through Staines via both the Twickenham and Hounslow routes on 6 July 1930. However, it was more than six years before there were any more energisations, until the lines between Surbiton and Guildford via Woking together with the branch from Weybridge to Staines via Chertsey were electrified.

The line from Woking to Winchester was electrified as far as Alton on 4 July 1937, and this is still the limit of the third rail on this route. The route, which was later truncated, made Alton the terminus for this electrified line, although the preserved Mid-Hants Railway extends from here to Arlesford, operating as a tourist railway.

The line from Virginia Water to Wokingham, together with the former SECR section from Wokingham to Reading, along with the branches from Ascot to Aldershot and Guildford to Aldershot were energised on New Year's Day 1939.

The Southern also had a busy new works programme, which included the construction of a new suburban route from Motspur Park (on the Waterloo–Epsom line) to Leatherhead via Chessington. The line was electrified from the outset, with the section from Motspur Park Junction to Tolworth opening on 29 March 1938, and to Chessington South from 28 May 1939. Construction was cut short with the outbreak of the Second World War and sadly never resurrected.

Earlier, another new line had been constructed on the Central Section. This line was between Wimbledon and Sutton via St Helier, which was again electrified at its opening on 6 July 1930.

Elsewhere on the Central Section, the energisation of the Wimbledon to West Croydon line occurred on 6 July 1930, and the route between Woodside and Sanderstead on 30 September 1935. The completion of these routes effectively completed the Central Section suburban network.

The important decision to undertake the first full main line electrification in the UK was made by the Southern Railway Board in 1930. The 83km (51 miles) route between London and Brighton was chosen as this line was increasingly subject to intense competition from express road coaches and motor cars using the newly widened main roads. The distance was too short for the speed of the steam trains to show any significant benefit, and it was therefore considered that only a fast, intensive rail service would stand any hope of regaining the lost traffic.

This was a bold decision as much of the line ran through country, which, in the prevailing depressed economy of the time, appeared unlikely to offer much off-peak or commuter traffic to add to that which might be recaptured from the roads. As a consequence, it was also decided that all expenditure that was not absolutely essential was to be avoided.

The first 27km (17 miles) to Coulsdon had recently been converted to third-rail, leaving 56km (37 miles) of the route to be electrified. In addition, it was decided to electrify to Reigate from Redhill and also to West Worthing as part of the scheme. To save money, track alterations were kept to a minimum. The proposed new service was therefore to be divided between Victoria and London Bridge. No enhancements were to be made to or through Croydon. The route from Coulsdon to Redhill had been part of the original route to Brighton but was owned by the SECR until 1923. In the past, there had been frequent disputes between the LBSCR and the SECR, who both shared the busy section between Croydon and Redhill. Eventually, the LBSCR built an avoiding line between Coulsdon North and Earlswood, which became known as the Quarry Line. Fortuitously, this cut-off could be used by the fast trains to Brighton avoiding Redhill. Both the original line through Redhill and the Quarry Line were to be electrified.

However, to avoid conflictions with the Redhill, Tattenham and Oxted services and the cost of major track alterations, the fast Brighton services using the Quarry Line were routed on to what had been the suburban lines that restricted speed due to their sharp curvature.

As already noted, it was also decided to electrify the first 3km (2 miles) of the SECR from Redhill to Reading as far as Reigate to cater for the hoped for increase in commuter traffic from the town. Some track improvements were therefore required at Redhill.

South of Redhill, few track alterations were called for. On the two-track section between Balcombe Tunnel and Preston Park, Up and Down loops were added between Copyhold Junction and Haywards Heath. At Brighton Station itself, some remodelling and rationalisations were undertaken, although on a fairly modest scale.

From the signalling point of view, the more intense service required shorter headways between Coulsdon and Preston Park. It was therefore decided to equip the line south of Coulsdon with new colour light signalling. The cost was justified in that thirty-three signal boxes could be fully or partially closed.

The first stage of the electrified route from Coulsdon to Redhill, Reigate and Three Bridges was opened on 17 July 1932. The southern half of the line to Brighton and West Worthing was energised on 1 January 1933.

In view of the rapid progress made on the Brighton scheme together with the favourable reaction of the public, the Southern Railway decided to undertake a further 96km (60 miles) extension to Lewes, Eastbourne and Hastings. Work started at the end of 1933 and the public electric service was inaugurated on 7 July 1935. Stage 2 of the extension scheme extended the electrification from West Worthing to Havant, including the Littlehampton and Bognor

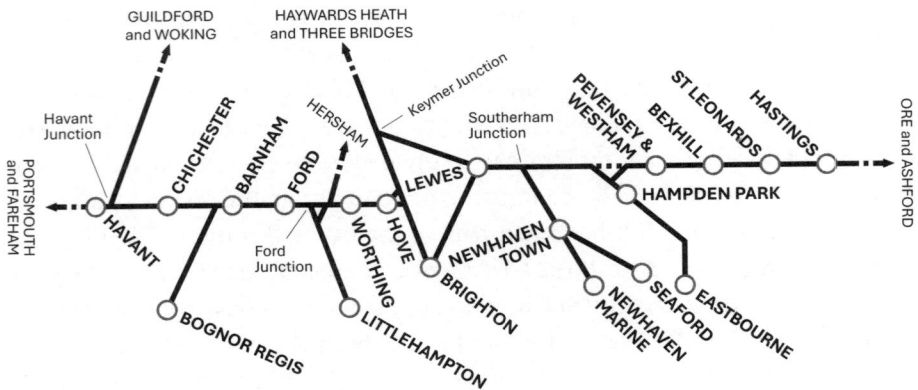

GUILDFORD and WOKING

HAYWARDS HEATH and THREE BRIDGES

Keymer Junction

PEVENSEY & WESTHAM

ST LEONARDS

HASTINGS

Havant Junction

CHICHESTER

HERSHAM

Southerham Junction

BEXHILL

ORE and ASHFORD

PORTSMOUTH and FAREHAM

BARNHAM

FORD

LEWES

HAMPDEN PARK

HAVANT

Ford Junction

WORTHING

HOVE

NEWHAVEN TOWN

BOGNOR REGIS

LITTLEHAMPTON

BRIGHTON

NEWHAVEN MARINE

SEAFORD

EASTBOURNE

SR Coastway Lines

branches, where the line would join up with the soon to be electrified Portsmouth Direct Line.

On completion of the Brighton schemes, the Southern Railway possessed a total of 718km (444 route miles) and 1,850km (1,156 track miles) of electrification.

Electrification of the Portsmouth Direct Line followed with the electric train service commencing on 4 July 1937. The scheme covered the electrification of the routes from Hampton Court Junction to Portsmouth Harbour, Weybridge to Staines and Woking to Alton. This was by far the biggest project that had been attempted by the Southern Railway equating to 152 route km (95 miles) and 388 single line km (242 miles).

Work undertaken in connection with the electrification included the complete rebuilding of Woking and Havant Stations, together with track alterations and/or platform lengthening at Guildford, Haslemere, Fratton, Portsmouth and Southsea and Portsmouth Harbour.

The scheme also included extending the carriage sheds at Wimbledon together with the erection of completely new sheds at this location as well as at Fratton and Farnham. New carriage washing plants were erected at Wimbledon and Fratton, and additional electric stabling sidings were laid in at Chertsey, Woking, Guildford and Portsmouth.

Power from the newly established National Grid was fed into the line by means of twenty-six substations, which were managed by two control rooms with, in addition, the provision of eighteen track paralleling huts. Following the implementation of this scheme, the Southern Railway possessed 860km (540 route miles) of electrification making a total of 2,248 track km (1,405 track miles).

The last two lines to be electrified before the Second World War intervened were the Staines to Reading and the Mid-Kent Line to Maidstone and Gillingham. Pertinent details are as follows: Electric trains reached Virginia Water on

3 January 1937. In that year a number of electrification schemes were proposed, which included extending the route from Virginia Water to Reading. Electric services from Waterloo to Reading commenced on 1 January 1939 with services operating every 20 minutes at peak times and every 30 minutes off-peak and on Sundays. New berthing sidings were provided at Reading as part of this project.

Further extensions authorised in 1937 were the South Eastern route from Shoreham to Maidstone East, Swanley to Gillingham and Gravesend to both Maidstone West and Rochester. These routes opened to electric traction on 2 July 1939.

The Southern Railway had planned that the routes to be next in line for electrification would be the Kent Coast line followed by the Southampton and Bournemouth route. The war and the subsequent financial and political upheavals faced by the railways caused these plans to be deferred until the late 1950s and late 1960s respectively.

Chapter 4
LNER Electrification Schemes Deferred by the Second World War

Experience with the Shildon to Newport Line and the success of the Manchester and Altrincham scheme encouraged the LNER to pursue further electrification schemes in the 1930s in the form of the Woodhead route and the Great Eastern Suburban lines out to Shenfield. In addition, and although it was not intended to be a main line electrification project, the LNER did considerable preparatory work in the Finsbury Park area to enable the route to Alexandra Palace and Edgware (GN) to be handed over to London Transport for their conversion to underground lines (see Chapter 9).

The Woodhead Route (MSW)

Electrification of the Woodhead Line (the Manchester, Sheffield and Wath route), was first mooted by the Great Central Railway to alleviate the difficulties of operating heavy coal trains with steam locomotives on the steeply graded Wath–Penistone section, also known as the Worsborough branch.

Detailed plans were drawn up by the LNER in 1936 for a 1500V DC scheme. Essentially, this was a British version of the American DC practice that had evolved in the USA for mineral lines with steep gradients during the previous twenty-five years.

Physical works commenced with the erection of many of the gantries prior to the Second World War, but work was curtailed at the outbreak of hostilities. The project included a new double track tunnel between Woodhead and Dunford Bridge.

A prototype Bo-Bo locomotive, LNER No. 6701, was built at Doncaster Works in 1941 to a design by Sir Nigel Gresley. Having been tested on the few sections of 1500 V DC lines owned by the LNER, it was mothballed but re-commissioned and renumbered 6000 with an EM1 classification (Electric Mixed-Traffic 1) after the war. The loco was then loaned to the Dutch Railways in 1947 to alleviate that railways traction shortage. While in the Netherlands, it gained the name 'Tommy', after the British soldiers who had helped liberate the country.

Tommy returned to the UK in 1952 and a further fifty-seven of these 1300hp locomotives were built at Gorton Works, Manchester from 1950 to 1953, albeit to a modified design. This batch of traction units, although slightly different,

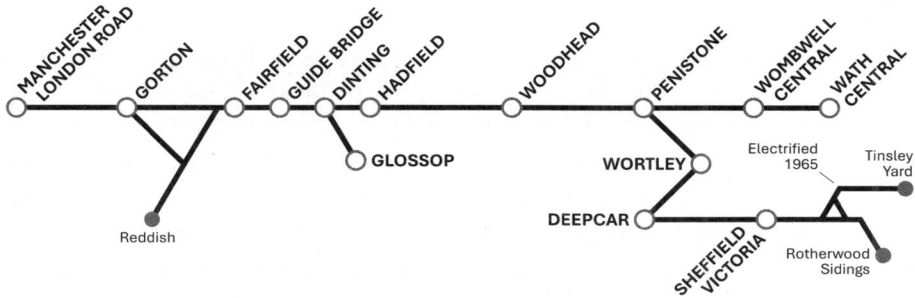

Woodhead Line (the Manchester, Sheffield and Wath route)

were also classified EM1 (later Class 76). While primarily intended for freight working, the EM1 locomotives also operated Manchester to Sheffield passenger services. Usefully, regenerative braking could be engaged between speeds of 15 and 55mph (26 and 89km/h), thus not only assisting with the braking effort on the long descents on both sides of the Woodhead Tunnel, but also feeding electric current back into the overhead wires, so reducing the overall power consumption of the line.

In addition, a further seven more powerful 2490hp EM2 (Electric Mixed-Traffic 2) Co-Co locomotives were constructed by Metropolitan Vickers during 1953 and 1954. Based on the design of the smaller EM1 locomotives, these traction units were built to primarily operate the express Manchester to Sheffield passenger services on the route.

Additionally, eight three-car EMUs with power-operated sliding doors were provided to work the Manchester, Glossop and Hadfield services. These were later classified as Class 506 units.

The infrastructure works were restarted immediately after the war, but timescales were extended in boring the new Woodhead Tunnel. The OLE was energised at 1500V DC in stages as follows:

- Wath to Penistone was energised on 2 February 1952
- Manchester to Penistone on 14 June 1954
- Sheffield Victoria to Penistone on 20 September 1954
- Sheffield Victoria to Rotherwood Sidings on 3 January 1955.

A new maintenance depot was also constructed at Reddish situated on the Fallowfield Loop line to maintain the new locomotives and EMUs.

The lines wired were from Manchester (London Road) to Sheffield (Darnall), plus Penistone to Wath as well as the branches and the new electric maintenance shed at Reddish – a total of approximately* 93km (58 miles). In 1965 the

---

* Accurate sources no longer available.

scheme was extended to Tinsley with OLE that was designed to be capable of conversion to 25kV.

However, by the time the line was nearing completion, the advantages of the 25kV AC system over the 1500V DC had become clear. The 1956 System of Electrification for British Railways Plan did not therefore include extensions to the electrification of this line. The plan stated that when it became necessary to bring the line into working contact with the London Midland and Eastern Region main lines, it would be necessary to face conversion to 25kV AC to obtain full inter-running.

Consequently, the orders for both the EM1 and EM2 locomotive were cut back, the latter type to just seven locomotives from the twenty-seven originally envisaged. Subsequently, with the decline in the number of services on the line, even these seven locomotives were to be sold to the Dutch Railways in 1968.

Extensions to the route were never even contemplated so, sadly, the Woodhead line, as a through route, became an obsolete and isolated electric system.

Furthermore, despite the major investment in the 1950s, the line was closed to passenger traffic on 5 January 1970, with the slower Hope Valley route, which had ironically been recommended for closure by Beeching, remaining open to handle all Manchester to Sheffield passenger traffic.

A decade later, with a downturn in coal traffic across the Pennines, the lack of passenger traffic and the need to upgrade or renew the obsolete electrical supply system, the writing was on the wall for the line as a through route and the decision to close the line east of Hadfield was taken on 18 July 1981.

Suburban passenger services, however, continue to operate between Manchester, Glossop and Hadfield. The remaining section of route was converted to the standard 25kVAC OLE in December 1984.

The Great Eastern Suburban Service

The LNER also decided to electrify the Liverpool Street to Shenfield section of the Great Eastern Main Line at 1500V DC. The scheme included the section from Fenchurch Street out to Bow Junction. Civil engineering works again began in the 1930s, but the Second World War once more intervened.

Prior to the outbreak of war, only a few of the clearance problems had been dealt with as well as a small number of OLE structures being erected. Plans were resurrected after the cessation of hostilities, so work recommenced in 1946 and was well in hand at the time of nationalisation.

The scheme was completed in 1949, with the first electric train running on 20 September and the full service commencing on 7 November. New electric

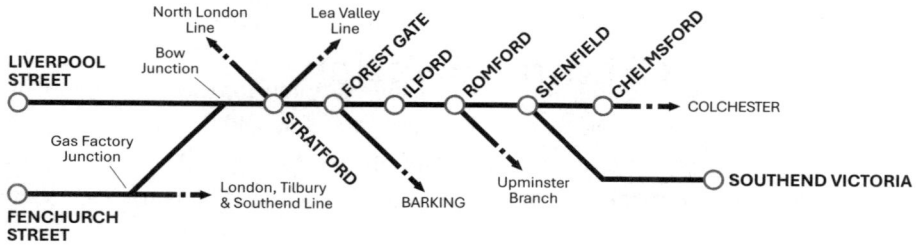

North London Line

Lea Valley Line

Bow Junction

LIVERPOOL STREET

FOREST GATE

ILFORD

ROMFORD

SHENFIELD

CHELMSFORD

COLCHESTER

STRATFORD

Gas Factory Junction

FENCHURCH STREET

London, Tilbury & Southend Line

BARKING

Upminster Branch

SOUTHEND VICTORIA

Liverpool Street to Shenfield, Chelmsford and Southend Victoria

rolling stock in the form of 92 three-car EMUs were introduced in 1949. These were later classified as Class 306 units.

The electrification was subsequently extended to Chelmsford in December 1956 and converted to 25kV (see Chapter 8).

Chapter 5
Attempts at Standardisation

The Kennedy Report

The plethora of main line electrification schemes that proliferated in the first 20 years of the twentieth century caused the government of the day to set up a committee to review the pros and cons of the various systems. The committee was chaired by Sir Alexander Kennedy and was formed with many eminent railway engineers and operators from across both the rail and electricity industry.

The committee was asked to consider:

(i) whether any regulation should be made for the purpose of ensuring that the future electrification of railways in this country is carried out to the best advantage in regard to interchange of electric traction units and rolling stock, uniformity of equipment and/or other matters;

(ii) if any such regulations are desirable, what matters should be dealt with, and what regulations should be made; and

(iii) how far it is desirable, if at all, that railways or sections of railways already electrified should be altered so that they may form parts of a unified system.

The group examined the systems currently in use in the country and by means of hearings, took cognisance of the views of domestic and foreign operators.

The Committee issued a 16 page interim report on 12 July 1920, which was published by the Ministry of Transport.

Key recommendations made in this report were that the direct current system be adopted in the future with the generation of current for such lines by 'alternating 3-phase'. Both overhead and rail conductor collection was to be permitted with the standard voltage (referred to as 'pressure' in the report) of the direct current, which should be 1500V, 'at the sub-station busbars'. The adoption of half or multiples of the standard voltage would be permitted where a proven case could be made and a suggestion was made that regulations be made to ensure power units could operate, wherever necessary, on both 600/750V and 1500V with either contact rail or overhead collection. Existing systems would be allowed to expand without change, and special consideration by the committee was given to the LBSCR's proposed extension of their single AC overhead system to Brighton. The group felt that it would be unreasonable to ask the railway company to bear the cost of changing the system to DC, so providing the proposed extension showed a substantial financial advantage it should be allowed to proceed.

The final eight-page report, published 30 June 1921, endorsed the finding of the interim report, recommending 1500V DC overhead as the future national standard. Within the three appendices, basic standards for extensions to third-rail electrified lines were agreed together with the contact wire height standard for overhead systems.

An Internet link to the technical recommendations of the interim and final reports is shown in Appendix 4.

The Pringle Report

In the light of the 1923 Grouping of the railway companies and further technical developments, the government set up another committee to re-examine the recommendations of the previous Kennedy report on 11 November 1927.

The group was chaired by Colonel Sir John Wallace Pringle, CB, FRGS who was Chief Inspecting Officer of the Railways Inspectorate of the Ministry of Transport from 1916 to 1929. Members included a number of people who had assisted with the Kennedy Report.

The terms of reference were:

To review the recommendation made by the Electrification of Railways Advisory Committee, 1921, and to report what modifications, if any, should be made in these recommendations, having regard to the developments that have taken place since that date.

The committee noted that 445km (280 miles) of electrification had been completed in the seven years to March 1928. In addition, the Southern Railway had decided to substitute the low voltage DC third-rail system for the whole of the 211km (132 miles) of the former LBSCR AC overhead electrified lines. This decision by the Southern Railway enabled the committee to simplify their recommendations in respect of the type of current for electrification, for which, 'direct current should be used for all future electrifications, or extensions of existing electrification'.

The committee's nineteen-page document was published on 23 July 1928. With the exception of the elimination of AC systems, the committee by and large endorsed the 1921 Report recommendations and made a special point of not de-barring the use of a bus-bar voltage of 750V as had become the practice on the Southern Railway.

The other main recommendations were therefore as follows:

- 1500V DC overhead with uninsulated return via the running rails and under contact current collection

- 750V DC third rail with uninsulated return via the running rails and top contact current collection
- fourth rail for tube lines and limited extensions of existing systems.

Furthermore, the committee 'left the door' open for higher voltage third-rail systems by 'under the top' collection; also, for future extensions of the Manchester–Bury line utilising side contact collection.

The committee were keen to ensure that 'inter-running' could be achieved between different networks in the future as electrification expanded across the country. Included in the report was a diagram for the proposed profile of overhead 'bow' current collectors for electric traction units and two further diagrams showing the recommended maximum loading gauges for electric traction units to permit degrees of inter-running. Two other drawings also showed the basic dimensions for third-rail top and bottom collection systems and overhead wire clearances.

These five drawings supporting the text were included in the body of the report and not in appendices as in the style of the Kennedy Report.

An Internet link to the full Pringle Report is shown in Appendix 4.

One member of the committee, Herbert Jones (Electrical Engineer of the Southern Railway), did not feel able to fully endorse the group's recommendations and so did not put his signature to the report. He considered that there should be no departure from the 1921 Report in respect of the Standardisation of Methods of 'Collection' and 'Contact', as he did not consider that any departure was necessary for inter-running. His opinions were, no doubt, influenced by his experience of the Southern Railway's extensive electrification programme. He also felt that there was no need at that time to specify the loading gauges for electric locomotives, which should be re-considered 'when the motive power of one-third of the total number of locomotives in the country is by electricity'. He was, however, in accord with the other conclusions and recommendations in the report.

1931 Main Line Electrification Report

The government appointed a further committee in September 1929 to look at main line electrification in view of the progress that was being made at that time towards the widespread availability of high-tension electrical energy.

The committee was chaired by The Right Hon. Lord Weir of Eastwood, a Glaswegian industrialist, who had previously reviewed the inefficient and fragmented electricity supply industry resulting in the Electricity (Supply) Act 1926.

This act led to the creation of the Central Electricity Board, which set up the UK's first synchronised, nationwide AC grid, running at 132kV, 50Hz and linking the 122 most efficient power stations with 6400km (4000 miles) of cables. The work commenced in July 1928 and was completed in September 1933.

Lord Weir of Eastwood was assisted by: Sir Ralph Wedgwood CB, CMG and Sir William McClintock, GBE, CVO. The Secretary to the Committee was Colonel A.C. Trench, C.I.E., R.E. (ret.)

The committee also involved Mr Kennedy, Mr Richardson and staff from the private railway companies. The group were supported in their deliberations by Messrs Merz and McLellan..

The committee were given the following terms of reference: – 'To examine into the economic and other aspects of the electrification of the railway systems in Great Britain, with particular reference to main line working, and to report their conclusions.'

A further official statement was subsequently made to the effect that the investigation was 'not to stand in the way of, or affect, the adoption of any proposed schemes of further suburban electrification by the railway groups, including the semi-suburban scheme of the Southern Railway to Brighton'.

The remit was somewhat limited in scope, there being no requirement to assess the impact on the inevitable wider impacts, not only within the industry itself, but on the country as a whole. These would include the effects on railway staff, private locomotive and rolling stock manufacturers as well as the coal and electricity industries and consumers generally. The scheme would undoubtedly impact on coal and electricity prices, and there would clearly be uncertainty about the long-term costs for electricity supplied to the railways who would then be dependent on one supplier. How the high cost of widespread main line electrification would be funded was also not to be addressed. The Weir Report, as it became known, was published on 24 April 1931.

The report endorsed the technical standards of the Pringle report together with the notion of the Central Electricity Board providing the generating stations, high-tension transmission lines and substations. It was considered that a substantial network scheme should be identified for electrification as a proving ground. Consultants Merz and McLellan favoured a pilot scheme to electrify the London King's Cross to Doncaster main line, possibly extending to Cambridge. This would enable all aspects of electrification to be considered in depth, including infrastructure form and construction, and, from operational experience, traction unit design and technical reliability. Means of assessing financial risk would then be easier to assess for future schemes. Another scheme to electrify the northern section of the West Coast Main Line between Preston and Carlisle was also considered.

While the committee estimated the cost of electrifying all key main lines at between £340M and £445M at 1931 prices (£25Bn and £32Bn) with a 7 per cent return assuming steady-state traffic levels, it was clear from the start that state aid would be essential. This was a huge sum when the national economy was still suffering from effects of the Depression. In addition, the report caused alarm by highlighting the fact that railway income and profits, as a whole, were in decline, which was further exacerbated when the committee admitted that no accurate assessment of financial risk could be predicted. The Weir Report also acknowledged the potential of the oil-electric locomotive. This was subsequently referred to as either diesel electric or diesel-electric, and has generally been simplified to the term diesel.

The Southern Railway already had its own 750V DC electrification strategy in place and most of its longer routes, with the exception of the Salisbury to the West of England route, could be described as 'semi-suburban'. The company was already electrifying the Brighton Main Line and other South Coast extensions and the Portsmouth line were in its sights.

The other three companies, the GWR, LMS and LNER, were unenthusiastic about main line electrification. All three had or were in the process of introducing new, improved express passenger steam locomotive types, so could see little benefit in electrification of main line routes unless substantial financial assistance was available from the government. This, they considered, was unlikely to be forthcoming.

With the government wrestling with the country's economic problems and three of the 'Big Four' railway companies lukewarm about electrification, the report seems to have been put 'on the shelf' and largely forgotten. As a result, the Weir Report did not stimulate any further electrification schemes, let alone a strategy. In fact, if anything, by highlighting the financial woes of the industry, the case for an electrification strategy was effectively shattered.

Consequently, only the Southern Railway progressed electrification of its network based on its pre-Weir Report proposals for its third-rail low voltage DC system. It has often been argued that the Southern Railway Board should have decided on the high-voltage overhead line system rather than the low-voltage third-rail system adopted. However, overhead lines would have been significantly more expensive to implement, involving extensive track lowering and /or bridge and tunnel works to achieve the necessary electrical clearances. Accordingly, a much smaller proportion of the Southern Network would have been electrified prior to the Second World War than was actually achieved.

As was noted in Chapters 1 and 2, some progress was made elsewhere in the Liverpool and Manchester areas with relatively small extensions to existing routes. While the LNER de-wired the Newport to Shildon route in 1935, the company did commence electrification of the Woodhead route and the London

Liverpool Street and Fenchurch Street lines to Shenfield just before the Second World War. However, these were 'piecemeal' schemes and could not be described as even the beginnings of the implementation of a route strategy. In any event, the commencement of the war curtailed the implementation of these schemes.

Some have argued that the progression of so few schemes during the 1930s was a 'blessing in disguise' by sparing the development in the UK of a multiplicity of different systems, as occurred on the European mainland. On the plus side, apart from the two aforementioned LNER DC schemes restarted after the war, this resulted in the UK being able to progress the 25kV 50Hz system north of the Thames from the 1950s. However, unfortunately, on the flip side, electrification progress in the UK has therefore always lagged behind the rest of Europe. Many of the 'non-standard' systems were subsequently converted anyway from low-voltage overhead systems to high voltage, including the Shenfield scheme.

An Internet link to the full report is given in Appendix 4.

The 'Standardisation of Electrification Order 1932'

By the 1930s, the development of both low-voltage DC and high-voltage AC systems had proceeded apace across the world. No clear advantage of one system over the other had been demonstrated and, as a result, the overall extents of both systems were similar. In the UK, while the Kennedy and Pringle Reports had made recommendations, no national system standards had been defined. Notwithstanding this, the success of the suburban schemes, and in particular the Brighton project, demonstrated the potential advantages of main line electrification.

However, there were many different views on how to proceed within the three main railway companies that had an interest in future electrification. The government of the day therefore decided to take matters in hand by endorsing the Pringle Report recommendations. The 'Standardisation of Electrification Order 1932' set the report's standards in place. The Order also permitted the use of the 3000V DC overhead line system in exceptional circumstances.

In compliance with the 1932 Order, and with the assistance of the government's low interest loans from 1935, the Southern Railway significantly expanded its third rail DC system and the LNER made a start on electrifying two lines using the 1500V DC overhead system: the Liverpool Street to Shenfield line and the MSW (Manchester to Sheffield and Wath) route. However, with the outbreak of the Second World War, all schemes were suspended, there being no resumption until 1947.

An Internet link to the full Order is given in Appendix 4.

1951 Electrification of Railways: Report of a Committee Appointed by the Railway Executive and the London Transport Executive

In April 1948, just four months after the nationalisation of the railways, the British Transport Commission invited the British Railways and London Transport Executives to appoint a committee to review all aspects of current UK electric railway operations. The review was to take cognisance of technological progress and make recommendations concerning the system or systems that should be adopted in the future.

The Executives appointed a joint committee of officers in 1950 chaired by C.M. Cock, the Chief Officer, Electrical Engineering, who had pioneered electrification on the Southern Railway. Later the distinguished consulting electrical engineer Mr F. Lydall, of Messrs. Merz and McLellan, was added to the team of railway officers. With his extensive knowledge and experience, Mr Lydall made a significant input into the committee's proceedings, but, unfortunately, he passed away before the publication of the final report.

The remit was to consider:

- whether recent progress in electrical practice has affected the conclusions of the railway electrification committee (1927) – The Pringle Committee;
- whether a single system should be adopted in all future electrification schemes, which may be contemplated;
- whether the country should be divided into separate areas earmarked for third rail and overhead wire respectively;
- in the event of two systems being approved, the position with regard to connecting services;
- the necessity for continuing to utilise the fourth rail return (standard on the London Transport Executive Railways) in the light of technical progress; and
- recent developments in the use of electrical operation of freight services and shunting.

The committee worked closely and extensively to the remit, giving much thought to standardisation while taking cognisance of technological developments that had taken place in the twenty years or so since the Pringle Committee's Report.

The report, published in 1952, by and large confirmed the main conclusions contained in the Pringle Committee's Report of 1927. Taking cognisance also of technical developments in the UK and abroad in the intervening years, including high-voltage DC and AC systems, the report recommended that the 1500 volt overhead direct current system be adopted as the standard in all future electrification schemes.

Three exceptions were noted. First, the area of the former Southern Railway where the third-rail direct current system would be continued. However, the exact boundaries on the western side would be subject to further consideration. The other two exceptions would be areas within the fourth-rail London Transport system and, possibly, in self-contained and purely urban systems in the provinces.

It should also be noted that the Report did not rule out the possibility of the 3000V DC system, but the committee's view was that the economic advantage over the 1500-volt system would only accrue on routes where traffic was relatively light. The committee recognised the advantages of the single-phase alternating current at 50Hz, noting that this system had been made more attractive by the further development of the mercury-arc rectifier. However, while not ruling it out altogether, they considered that with the information then available, any advantages would be negated by a number of technical drawbacks, particularly the greater cost of locomotives and motor coaches.

An Internet link to the full report is given in Appendix 4.

The BR 1956 Report Entitled 'Modernisation of British Railways: The System of Electrification for British Railways'

Technically, much progress had been made in the first half of the 1950s, particularly in Europe, with the development of high-voltage AC overhead systems. This was primarily due to the development of reliable mercury-arc rectifiers enabling high-voltage AC to be transformed to lower voltage DC within a locomotive or railcar to power the vehicle's traction motors. In the UK, as has already been noted in Chapter 1, the Lancaster, Morecambe & Heysham line had been, for the second time, used as a test bed and upgraded by BR from 6.6kV to trial the 25kV 50Hz electrification system. In France, rather more extensive trials were also undertaken with a similar system on the main line between Valenciennes, Thionville and Charleville. Both trials proved successful and pointed to lower 'first cost', primarily due to a significant reduction in the number of electrical substations and cabling together with lower operating costs than equivalent 1500V DC systems.

The 1956 report took cognisance of these developments and was in effect a technical addendum to the BR Modernisation Plan published in 1955.

The exercise included consultations with the UK's electrical engineering industry and other relevant authorities, including the Central Electricity Authority together with continental railway administrations, particularly

SNCF. In particular, scrutiny was focused on the experience to date of the 25kV AC 50Hz system, and the committee took cognisance of expert opinion concerning future potential.

As the work by the committee progressed and detailed results became available, an evidence-based comparison between high-voltage AC and the existing UK standards was undertaken to identify the best means of electrifying the London Midland Region's main route between Euston and Manchester/Liverpool.

On completion of a very thorough and comprehensive review, the Commission decided to take advantage of recent technological innovations and depart from the low-voltage DC recommendations of the 1951 Review and to 'adopt as a standard for future electrification the use of an overhead supply of alternating current at industrial (50-cycle) frequency, generally at a pressure of 25kV'. They excepted from this ruling only those parts of the Southern Region where a change in the existing third-rail system was not practicable.

An Internet link to the full report is given in Appendix 4.

British Railways Electrification Conference London 1960: Railway Electrification at Industrial Frequency – Proceedings

In October 1960, the British Transport Commission, in association with the British Electrical and Allied Manufacturers' Association and the Locomotive and Allied Manufacturers' Association of Great Britain, staged an Electrification Conference in London. A wide range of delegates, from both the UK and abroad, were invited to the conference to discuss and present papers of investigation findings and experience of the developing 25kV 50Hz AC electrification system.

The principal object of this conference was to showcase developments of the 25kV AC electrification in the UK. In particular, the aim was both to show how this system had been adapted to meet the peculiar needs of railways in Britain and also to share the lessons learned with overseas delegates. It was thus hoped that discussions would prove to be of mutual advantage to all who attended.

More than 40 foreign railway administrations attended, in addition to the leading experts from British Railways and British industry. While several setbacks were acknowledged, it was stated that these had only affected a very small proportion of the equipment that had gone into service. Overall, however, the conference was told that much had been learned, and that the schemes progressed at that time had proved to be a useful testing ground that would facilitate future progress. The report of the conference proceedings comprised a large number of papers presented by railway engineers and suppliers, covering rolling stock, power supply and distribution equipment, overhead line equipment, control and supervision and signalling immunisation. The report contains numerous photographs, electrical schematics, diagrams, graphs and technical analyses, showing the state of the art for rail electrification in 1960.

An Internet link to the full report is given in Appendix 4.

A second conference with the same objectives was hosted by BR at the University of York in September 1989.

Railway Construction and Operation Requirements – Structural and Electrical Clearances (1977)

This document, universally known in the railway industry as the 'Blue Book', was drafted and published by the Department of Transport in 1977. It set out the clearances to be maintained when installing new overhead line electrification systems on the national network.

Railway Electrification: 25kV AC Design on BR

A technical pamphlet was produced by the British Railways Board and published in 1988. It outlined the (then) current technologies and systems used in 25kV Overhead Line Equipment (OLE), extolling the virtues of various technologies designed to reduce the capital costs of electrifications schemes, such as structure mounted outdoor switchgear (SMOS) and overhead system design (OSD) by computer.

The booklet was given to many graduate electrification engineers (in the co-author's case, in 1991) as a simple way to introduce them to current electrification thinking; it also serves as a good record of BR's OLE design philosophy in the last few years of its existence.

Technical Specification for Interoperability (TSI)

Reference in this section is also appropriate regarding the debacle created by the Railway Industry's attempts to comply with the Technical Specifications for Interoperability (TSI) relating to the 'energy' subsystem, dated March 2008.

The Technical Specification for Interoperability applied to all new, upgraded or renewed high-speed infrastructure. This TSI lays down essential requirements for the Energy subsystem and its interfaces with other subsystems.

The Railways (Interoperability) Regulations took effect in 2011, requiring, among other things, compliance with the applicable TSI. It should be noted that no existing UK system was compliant with this standard. Nor were any multiple pantograph train operations at speeds above 100mph.

By 2013 the European Commission indicated its intention to extend the TSI requirements to the whole rail system of the EU (not just high speed) based on a recommendation from the European Railway Agency in December 2012.

Network Rail should have been aware of the requirements in 2008, as they had representatives on the drafting groups. With the UK railway's restricted loading gauge, the TSI requirements were difficult to meet, so Network Rail should have sought the practical way forward by seeking derogations through the DfT. Instead, in 2009, Network Rail decided to redesign its electrification equipment. Due to the lack of experienced design staff and poor leadership from Network Rail at a senior level, together with the questionable role of the RSSB, the design approval work proceeded slowly. Thus, there were negative consequences for all electrification projects, including those that were already underway. This had a significant effect on the largest scheme at the time, GWEP, where basic designs were incomplete in 2014, five years after authority (see Chapter 12).

ORR were also unhelpful on matters such as electrical clearances, heights of bridge parapets and wire height at level crossings, so that designers were put in the unenviable position of being unclear about key matters and so having to 'play safe' and over-engineer requirements that inevitably significantly inflated costs.

All the foregoing could have been avoided in the authors' view. Instead, this unhappy state of affairs led to vastly increased project costs, which in turn, resulted in many projects being cut back, deferred or terminated.

National Technical Specification Notices and National Technical Rules

TSIs were removed from UK law when the EU Exit Regulations were enacted in January 2021.

These were replaced by National Technical Specification Notices (NTSNs) and National Technical Rules (NTRs). RSSB issued Railway Group Standard GLRT1210 (issue 3) in December 2022 and a new Rail Industry Standard RIS-1853-ENE in December 2022 to take the place of the TSIs.

It remains to be seen if the costs of future electrification fall, as suggested at the time, with the implementation of these standards.

Chapter 6

British Railways Attempts at Developing an Electrification Strategy

The 1951 British Transport Commission Survey: 'The Electrification of Railways'

This British Transport Commission (BTC) review of electrification concentrates mainly on the technical aspects of the electrification outlined in Chapter 5.

The main strategic conclusion that the committee highlighted was that the key factor that determined whether a valid economic case could be made to electrify a line is its traffic density. The Report considered the threshold value to be 3 to 4 million ton-miles per single track mile. This is quite a high value, but it was still estimated that some 10,400 km (6,500 route miles) (43.4 per cent of the rail network) had a traffic density that would justify electrification on economic grounds. It is certainly not the only factor to be taken into account, but, to be fair, this was an engineering-led report, and probably the expectation was that the commercial aspects would be considered in more detail later.

C.M. Cock himself had developed diesel-electric traction for the Southern. With the abundance of cheap oil, then recently available from the Middle East, there was the growing impression that the senior BTC managers of the early 1950s were coming to the conclusion that a programme based on diesel-electric traction was the way forward to modernise the railway. Many considered that this option had many of the benefits of electrification but avoided the infrastructure costs, so that electrification could be restricted to a small number of core routes.

The Chief Mechanical and Electrical Engineer of the Railway Executive at the time, R.A. Riddles, was, however, pushing hard for a rolling programme of electrification of all core routes and suburban conurbations, to form, over time, a comprehensive and coherent network. Under his plan, the elimination of steam traction would be phased over a longer period, but he believed that his proposal, rather than wholesale dieselisation, was the more robust approach from a practical and economic point of view.

He disagreed strongly with the report's findings and the direction the BTC and senior managers in the Railway Executive were proposing. To a degree, history has proved him right, but the more balanced view, countenancing a modest dieselisation programme, put forward in an unpublished plan drafted by J.L. Harrison, Chief Officer (Administration), would have probably been the ideal way forward (see Chapter 16).

The 1955 Modernisation Plan Entitled 'Modernisation and Re-equipment of British Railways'

The Modernisation Plan of 1955 heralded a major change in the form of motive power on British Railways. The phasing out of some 19,000 steam locomotives was seen as a prerequisite to the improved quality of service that the plan envisaged. Equally, with the changeover it was hoped that modernisation would lead to major economies in operation.

Early in the document, the virtues of electrification were considered, noting that it could be the way forward, where traffic density permitted, to fast, clean, reliable and economic operation. Rather oddly, however, the plan took a somewhat more negative view of the associated civil, electrical and signalling engineering works that must accompany electrification schemes, rather than promoting these works as part of the modernisation process to enhance the system.

Perhaps unsurprisingly, therefore, the report went on to note how much quicker and easier it would be to adopt diesel traction, the plan stating, rather idealistically, that dieselisation 'offers many of the advantages of electricity in the shape of cleanliness, acceleration and uniform standard of performance' and 'has a further advantage that the changeover from steam to diesel working generally does not involve important civil engineering or signalling works'. Bearing in mind the backlog of infrastructure renewals resulting from the war and the uncertainties caused by the process of nationalisation, this was, to say the least, a peculiar tack to take.

As a result, although it was noted that on the lines of heaviest traffic, the potential economies are less with diesel traction than with electrification, the wording in the plan itself finally sank any aspirations for a rolling programme of electrification of the core main line and suburban railway routes.

Unsurprisingly, therefore, the plan only envisaged the electrification of three main lines (Euston to Birmingham, Manchester and Liverpool; King's Cross to Leeds and possibly York, plus Liverpool Street to Ipswich). In the event, the latter two schemes were significantly delayed.

The 25kV AC Suburban schemes were limited to those already being progressed or 'in the pipeline', namely, Liverpool Street suburban lines, London, Tilbury and Southend, King's Cross/Moorgate to Hitchin and Royston, and the Glasgow suburban lines. Similarly, for the 750V DC Southern region routes, the report only mentions the then ongoing plan for the Kent Coast electrification.

It is clear from the report that the British Transport Commission's primary objective was to eliminate steam traction as quickly as possible by replacement with diesels, without considering the wider and longer-term advantages that a rolling electrification strategy would have offered in delivering whole route enhancements.

For this the railways paid dearly. Fleets of main line diesel locomotives, often of unproven design, were delivered, usually with little or no power advantage over the steam locomotives that they were to replace, but at three times the 'first' cost. The plan stated that the diesels needed to 'achieve a very high degree of utilisation, approaching closely to their theoretical availability, to spread the capital charges'. In the event, none of the classes of diesel locomotives delivered by the Modernisation Plan achieved this goal. Frequent diesel loco failures in service resulted in a deteriorating timekeeping with additional locomotive resources kept on standby to rescue failed trains. The performance of some classes of locomotives was so poor that many locos had to be re-engined after a few years or even withdrawn from service and scrapped. In addition, infrastructure enhancements that would have improved reliability, increased linespeeds and reduced ongoing maintenance costs on core routes were also lost.

If only wiser council had prevailed.

The perceived failure of the Modernisation Plan by politicians and stakeholders, particularly the Treasury, resulted in a deep distrust of BR's financial management ability. This had a negative impact on railway investment, particularly electrification, throughout the remaining 40 year life of the organisation.

Chapter 7

British Railways Approach to Electrification Following the Modernisation Plan

As described in Chapter 5, much technical progress had been made in the first half of the 1950s with the development of high-voltage AC overhead systems.

From a technical perspective, at least, the BR Modernisation Plan took cognisance of these developments. As described in Chapter 5, the London Midland Region's main route between Euston and Manchester/Liverpool was identified for a study to compare the low-voltage DC and high-voltage AC systems. Thus, shortly after the publication of the Modernisation Plan, BR published its 1956 report entitled 'Modernisation of British Railways: The System of Electrification for British Railways'.

The committee looked at a wide range of both UK and selected European railway electrification schemes, reviewing the infrastructure, motive power, operational requirements and comparative costs.

As previously noted, the report came down firmly in support of the 25kV AC system albeit permitting further extensions of the Southern Region low-voltage system, in particular, the Kent Coast scheme. An appendix to the report included a map of the actual and proposed British Railways electrified lines.

While the Report was not broad enough in scope to be termed a national strategy, the Plan did at least propose an initial way forward for AC electrification, not only of the Euston–Birmingham, Crewe, Liverpool and Manchester main line, but also the other following routes:

- King's Cross to Leeds/York
- Liverpool Street to Ipswich, including Clacton, Harwich and Felixstowe branches
- Great Eastern, Great Northern and Glasgow suburban lines.

At the time of the report, electrification between Liverpool Street and Shenfield was almost complete, and extensions to Southend and Chelmsford were well advanced. The committee recommended converting these lines to AC and to also progress in the same manner the authorised East Anglian schemes to Enfield, Bishop's Stortford and Chingford from Liverpool Street and to Tilbury and Shoeburyness from Fenchurch Street.

With regard to electrification, the 1956 report was rather more positive than the Modernisation Report, stating that 'electrification had come to be regarded as the ultimate objective for the movement of main line traffic'.

In the late 1950s the country was still recovering from the effects of the Second World War. The economic situation was improving – characterised by Harold McMillan's 'you never have had it so good' pronouncement – but was still very fragile. The economy was very much driven by a reliance on coal, so as to minimise the importing of oil, which required overseas exchange. Despite the great achievements of the railway industry during the Second World War, the overall public view, reflected by the actions of the politicians, was that the railway was a spent industry and had only a limited and declining role to play in the future. Thus, the government of the day's aim was principally to minimise the long-term cost to the public purse. To do this, the politicians were prepared to make a significant 'one-off' investment to modernise the railway, so that, thereafter, no ongoing Treasury support would be needed.

Progress Made within the Ten Years from the Modernisation Plan

Progress over the ten years following the publication of the Modernisation Plan was slow and problematic. The 'showcase' West Coast Electrification scheme fell behind schedule and exceeded the authorised budget, necessitating re-authorisation by the British Transport Commission. Only 960km (600 track miles) of 25kV AC were completed in the period, averaging a lowly 96km (60 miles) per annum.

In parallel with the work on the southern end of the West Coast Main Line, work also started in Glasgow and in north-east London. The first lines to be electrified in Glasgow were the north side suburban lines and these are described in Chapter 8. In addition to the existing line out to Shenfield, the former Great Eastern lines in north-east London to Enfield Town, Hertford East, Bishop's Stortford and Chingford were also electrified. These were followed by the London Tilbury and Southend routes from Fenchurch Street. The north-east London and LTS projects also included the conversion of the existing 1500V DC system to 25kV AC power supply.

Following the reauthorisation of the West Coast Electrification scheme, matters improved. Management systems were overhauled and the scope reviewed to reduce costs and increase efficiency. In addition, better design and ever improving technology also helped to transform the project. By 1966, through 25kV AC electrified services were operating from London to Liverpool and London to Manchester. Twelve months later the rest of the southern end of the West Coast Main Line was completed by wiring from Rugby through to Stafford via Birmingham and Wolverhampton, including the routes through Bescot and the branch to Walsall. Further lines had also been electrified in the southern area of Glasgow.

This completion of electrification at the southern end of the West Coast Main Line delivered significant journey time savings which in turn, resulted in significant increases in patronage. Building on this success, BR set about the task of persuading the government to authorise the extension of the scheme to cover the northern end of the route. The authority to proceed was eventually received in 1970 and the project was completed in 1974.

Following this authorisation, the Conservative government asked BR to make a proposal for a significant electrification project. The BR Board submitted the southern end of the Midland Main Line as far as Bedford as being its preferred candidate as the existing commuter trains were becoming life expired and the London Midland Region had the resources and expertise to deliver the electrification. The project was formally proposed in 1977 and authorised in 1979 (see Chapter 9).

1979–1981 Joint Review of Railway Electrification

As the electrification of the southern end of the Midland Main Line (MML) progressed towards completion, the Railways Board considered how to keep together the skills that the various contractors had acquired and maintained during the delivery of this project. There were discussions with the Department of Transport (DoT) and considerable political lobbying. In May 1977, the Select Committee for Nationalised Industries recommended that consideration should be given to further electrification. The committee's view was that all relevant benefits should be taken into account and, if a sound case could be established, the government should be willing to allow BR to invest above its current investment ceiling.

BRB Appreciation Paper

In November 1977 the government agreed to review with the British Railways Board the general case for mainline electrification on the basis of an appreciation paper produced by the board. The outputs from this were produced in the form of a booklet entitled 'Railway Electrification – A British Railways Board discussion paper'. This paper made a sufficiently strong case for the then

Secretary of State for Transport to announce, in May 1978, the setting up of a Joint Review by the Department of Transport and the British Railways Board.

Joint Review

The board's Vice Chairman, David Bowick, and the Under-Secretary, Railways of the Department of Transport, John Palmer, were appointed joint chairmen. The group included representatives from the Treasury and the Department of Energy. Mr Bowick retired on 1 February 1979 and was succeeded by Michael Posner, another member of the British Railways Board.

The group's remit was *to review the case for a programme of main-line electrification, to analyse the various relevant considerations and formulate issues for decision.*

Initially, the Steering Group decided to concentrate on a financial assessment, while taking into account the possible wider implications of value to the nation, such as energy savings, environmental effects, the balance of payments and export opportunities for the rail industry, together with the possible transfer of traffic between transport modes.

A two-stage approach was also decided upon; Phase 1 would consider these factors, but a key requirement of this stage would be to determine if there was a sufficient prospect of a case for main-line electrification to justify progressing the second phase of the study. Should Phase 2 go ahead, it was determined that this exercise would involve a more detailed financial appraisal, considering in depth the implications for resources, revenue plus the wider effects of electrification. Phase 1 was completed in the winter of 1979 and the Steering Group produced an interim report. The main conclusion was that the case for further electrification looked sufficiently promising for Phase 2 of the review to proceed.

As all previous experience had shown, electrified main line railways generated substantially more revenue. This was not surprising. As noted earlier, BR management termed this phenomenon the 'sparks effect', because, as a newly electrified railway, it offered a faster, cleaner and a more intense service, attracted more passengers and boosted the local economies along the route. The Phase 2 report comprised 97 pages and was published in on 11 February 1981. Titled 'The Joint Department of Transport/British Rail Board Report on the Economic Case for a Rolling Programme of Main Line Electrification', the document recommended the electrification of a number of key routes.

The group took the advice from a large number of outside organisations, overseas administrations and knowledgeable individuals. A comprehensive financial analysis, with the aid of computer models, was undertaken, covering a number of options for main line electrification. While the report acknowledged that further refinements of the outputs could be carried out, the group stated that they were satisfied with the work they had done. The outputs were scrutinised in detail by both the Department of Transport and the Railways Board; the report considered that any further refinement would not alter the conclusions drawn.

The report conclusions were particularly strongly worded in favour:

On the assumptions made, a substantial programme of main line electrification would be financially worthwhile. All the larger electrification options examined show an internal real rate of return of about 11%; the faster options give the higher net present values.

Looking at wider effects, not taken into account in the financial evaluation, we have not identified any important disadvantage. There are two important advantages. The first is that electrification, while scarcely affecting total energy consumption, would reduce dependence on oil: the railways at present use about 3% of oil consumed by transport. The second is that a programme of electrification in the UK should assist the UK manufacturing industry to win more orders overseas, in an expanding market, and would be in keeping with the Government's policy of using public purchasing more effectively to enhance the competitiveness of British industry.

We have considered the various ways in which the financial result described might be undermined. Given the present experience of the recession, we have thought it right to consider the effect of lower forecasts of passenger and freight traffic, forecasts significantly below any the Board considers likely. We have also examined the effect of costs turning out higher than expected. Our conclusion is that it would take an unlikely combination of adverse factors to undermine entirely the prospect that a programme of main line electrification would be financially worthwhile; i.e. earn a return of at least 7% – this is in part because of the greater scope now foreseen for divergence between oil and electricity prices. Similarly, the outcome could be better than 11% if favourable chances combine.

It is worth emphasising that the group considered that only an unlikely combination of adverse factors could undermine entirely the financially worthwhile programmes outlined in the report.

An interesting insight into the physical output that may have been expected from a rolling programme of electrification was outlined in the section of the report entitled 'How Much, How Soon'. A summary of this section follows:

Noting that all the electrification programmes, except the smallest, gave an internal rate of return of 11%, it was stated that the best course of action, providing railway finances were not constrained, would be the largest and fastest programme.

This programme would take 20 years, extending electrification to Edinburgh and Aberdeen on the East Coast Main Line and from Edinburgh to Glasgow and Carstairs; to Sheffield on the Midland Main Line; across the Pennines from Liverpool to York; from York to Birmingham and Birmingham to Bristol and Reading on the North-East to South-West route; and on the Western Region to Swansea and Penzance.

It was estimated that more than 80% of passenger and 70% of freight traffic would then be electrically hauled. It was reckoned that this extensive programme would give a net present value of £305M (£1.25Bn at 2025 prices), a rate of return of 11.1%, requiring a cash flow demand between £24M to £60M (£100M to £250M at 2025 prices) a year for the first fifteen years.

The smallest programme would reach Newcastle on the East Coast Main Line, Sheffield on the Midland Main Line and York to Birmingham. Only Edinburgh to Glasgow via Carstairs would be included in Scotland. There would be no programme of electrification on the Western Region, but it was stated that this option, which would take 15 years, could form the first stage of a larger programme. Of all the options, it presented the lowest net present value of £84M (£340M in 2025 prices) and the lowest rate of return at 9.9%. On completion, 62% of passenger traffic and 38% of freight traffic would be operated by electric traction but there would be a significant amount of diesel running, mostly freight, over the electrified network.

The study intimated that, while it would seem sensible to go on beyond the smallest programme of electrification, it would not be essential to initially take a decision on the final stages of the largest programme, but to go for a 'medium-sized' programme, which could be extended at a later date as work progressed, thus enabling a rolling programme to be maintained. Such a medium scheme might, as noted in the report, include, extending electrification on the East Coast Main Line to Edinburgh; from Edinburgh to Glasgow and Carstairs; the Midland Main Line to Sheffield; from Liverpool to York; the North-East to South-West route from York to Birmingham, Bristol and Reading; and on the Western Region as far as Swansea and Plymouth. This would mean that 75 per cent of passenger and 54 per cent of freight traffic would be hauled by electric traction.

The report suggested the setting up of three or four electrification construction teams depending on whether a 'slow' or 'fast' programme was to be adopted. It was recommended that there should be a ramping-up period of three or four years before major work started to allow British Rail to complete design and private sector firms to recruit and train the necessary manpower and assemble the required resources.

It was estimated that three teams working simultaneously would complete a medium-sized programme of electrification within twenty-five years and would take about thirty years to finish the largest programme. By working on a greater number of routes at any one time, four teams would complete a medium programme in about fifteen years or a large programme in twenty years. With regard to a 'start date', the group considered that it would take about a year from the decision to proceed before any significant expenditure would be incurred and three or four years to build up a steady work rate.

Two key advantages of authorising a rolling programme were highlighted:

- It would provide for the British Railway Board a firm basis for longer-term financial planning with revised investment ceilings agreed with the government.
- The agreement of an over-arching commitment from the supply industry, the workforce and railway management should help prevent abortive expenditure and secure cost reductions resulting from the continuity of production. Such a programme should also enhance the industry's ability to compete overseas.

The group went on to note that the alternative of 'ad hoc approvals' of individual schemes could not accrue such benefits. Furthermore, it would be impossible to fully evaluate the benefits of individual electrification proposals without some consideration of the likely future extent of the electrified network. As a result, investment decisions would be distorted.

Finally, the study concluded that there were no 'wider disbenefits', e.g. land and property intrusions, and, indeed there were wider benefits e.g. increased local economic activity along each route. However, little attention was given to environmental aspects, apart from mentioning cleaner stations and trains. Issues such as clean energy and CO_2 emissions were not considered to be an issue at the time. This is evident, by implication, in the following comment contained in the report:

We have not identified any significant differences between the programmes, so far as wider effects are concerned.

Today, it is interesting to consider how much higher the environmental benefits would be rated, and the 'weight' that would now be attributed to the 'beneficial arguments'.

Summary

The 'issues for decision' were deemed to be:

- whether the key assumptions were valid;
- the effect of energy prices was very important, particularly if oil costs rose as predicted;
- the study paid little attention to environmental aspects as they were not very significant at that time, but it did indicate that electrification brought benefits;
- the study showed the added benefits coming for the larger options; the natural 'add on' effects from the opportunities for using the technical assets over wider areas and avoiding traction changes or diesel operation under the wires.

The study's key recommendations were:

- On the assumptions made, a substantial programme of main line electrification would be financially worthwhile. The greater the extent produces an internal real rate of return of about 11 per cent. Higher values come from faster implementation.
- There are no important wider disadvantages.
- Electrification would scarcely affect total energy consumption but would reduce dependency on oil.
- A programme should assist UK manufacturing to secure more overseas orders.
- Only an unlikely combination of adverse factors could undermine entirely a financially worthwhile programme – this was partly because of the foreseen difference between oil and electricity prices.
- The outcome of 11 per cent could be better if favourable changes combine.

Unfortunately, however, most of the recommendations were not taken forward. It probably didn't help that while trying to seek government support for an electrification programme, BR was at the same time trying to close the electrified Manchester to Sheffield route. No matter how good the economic case was, the perception must have been negative in many senior politicians' and stakeholders' eyes!

The BR 1980 Rail Policy Statement Document referred to the Joint Review then underway and hailed 'electrification' as the most significant decision that needed to be taken. The document emphasised that unless a programme of electrification to cover at least the main routes was authorised, BR would have to replace its fleet of DMUs with ones of a similar kind that are 'heavier, slower, less economic and significantly more expensive to maintain'.

In 1980 the country was in a deep recession and reference was made to the fact that an electrification programme would bring much needed work to 'hard hit parts of the country, in South Wales and Scotland in particular'.

Para 7.12 in the document stated, 'Electrification would improve service quality to many areas in Wales, Scotland, and Cornwall, would reduce oil consumption by about ½ million tons per annum, and strategically would be a sound investment. It would give the work force an optimism and sense of hope for the future; it must go ahead.'

It was not to be. Electrification schemes continued to be authorised slowly on an 'ad hoc' basis.

An Internet link to the report is given in Appendix 4.

Schemes Authorised Immediately Following the 1979–1981 Joint Review

ECML Stage 2

While authority for the 'hoped-for' rolling programme of electrification was not forthcoming from the government, the ECML Stage 2 scheme, to extend electrification from Hitchin to Leeds and Edinburgh, was authorised in July 1984. These works, extending over nearly 640km (400 miles), formed what was claimed in the late 1980s to be 'the longest construction site in the world'.

Utilising the Mark 3B range of electrification OLE, the section between Hitchin and Peterborough was completed in 1987, and the extensions to Doncaster, Wakefield and Leeds in 1988 and to York in 1989. Electrification was completed through Newcastle and on to Edinburgh by the middle of 1991 on time and within budget. A more detailed account is given in Chapter 9.

In parallel with this project the Edinburgh to Carstairs electrification scheme was undertaken – see Chapter 9. In addition, so as to exploit the opportunity to convert one of the Edinburgh suburban services to electric working, the 7.6km (4¾ miles) line from Drem to North Berwick was wired.

Ironically, other than these schemes, it was not the output from the 'Joint Review' process that drove electrification forward in the 1980s but the new, smarter commercial, financial and operating arrangements spawned under sectorisation, as described in Chapter 9.

Chapter 8

Schemes Promoted by British Railways 1950–1974

Very Early BR Schemes on the Great Eastern and the London Tilbury & Southend

A summary of the progress made on the Shenfield electrification scheme prior to the Second World War has been set out in Chapter 4. At the end of hostilities, the LNER dusted down its plans for this route and recommenced work on the project, utilising, by and large, the original design work. This included the LNER's own OLE design, which was also used later for the Manchester, Sheffield and Wath (MSW) scheme.

The section from Liverpool Street/Fenchurch Street to Shenfield was electrified and the first train, an AM6 (later Class 306) emu ran on 20 September 1949. The public service started on the following 7 November. Energisation, using a modified version of OLE, was subsequently extended to Chelmsford in December 1956 (the first train ran on 11 June) and to Southend Victoria in 1957 (the first train ran on 30 December 1956). At the time this extension was carried out, it had already been decided that future electrification would be at the AC industrial frequency, but as work was too far advanced on this scheme to permit a delay, 1500 V DC working was initially extended. The Shenfield to Chelmsford OLE was constructed in a way to subsequently permit the easy conversion to single phase AC, which was completed in March 1961.

In the main, the original clearances chosen for 1500V DC proved to be adequate for 25kV AC, although in the Inner London area it was felt that some of the clearances were too small and thus a lower tension of 6.25kV was selected. At these locations, the electrical clearances used for the original DC scheme were considered to be adequate for operation up to 6.25kV AC, therefore much of the original DC equipment was retained, it only being necessary to change the insulators and install booster transformers.

To prove the 25kV to 6.25kV changeover system, a test section was installed between Colchester and Clacton, which included a section of OLE that could be energised at 6.25kV. This was used to prove the operation of the EMU voltage detection and changeover equipment prior to their acceptance for service.

The downside was that this complication required the electric multiple units to be capable of working at both voltages. This required all the Class AM4

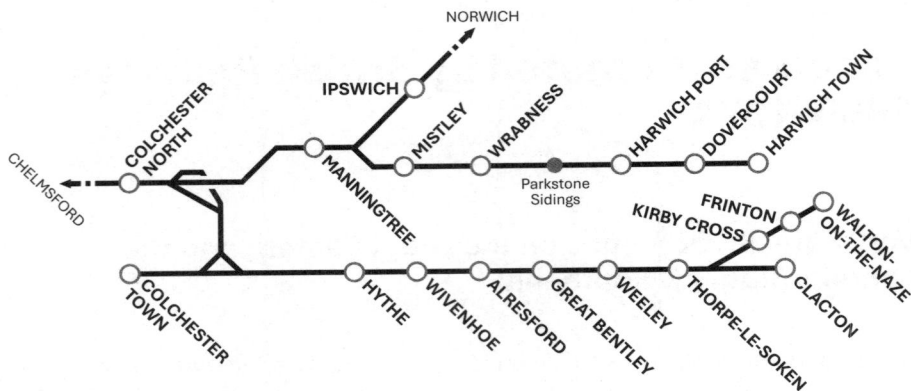

Colchester, Walton, Clacton and Harwich

(later Class 304) EMUs to be modified with the installation of mercury arc rectifiers.

However, the rectifiers proved to be unreliable, and a cause of numerous failures in traffic. This led to the tests that were carried out in the goods lines tunnels at Crewe, which found that it was possible to reduce the previously selected minimum clearance sufficient for the overhead lines into Liverpool Street and Fenchurch Street to be converted to 25kV AC throughout.

When the decision had been taken to convert the system from DC to AC it was decided to establish a small test installation utilising the new BR Mk 1 OLE system (see Appendix 1), which could be later incorporated into the main network. The 37km (23.5 miles) line from Colchester to Clacton and Walton-on-the-Naze was chosen for this and it was energised on 13 April 1959. Following the successful tests, this was joined into the main part of the network in 1962.

One of the most difficult issues associated with the conversion from low voltage overhead DC to high voltage AC concerned the safety of the overhead line maintenance staff. On the DC system, wearing very thick rubber gloves, they had been used to carrying out maintenance with the overhead lines energised. Clearly, this could not be done with the change to 25kV AC and there was, for quite a time, an apprehension that somebody might 'forget' the new regime. In the event this didn't happen, but it was, nevertheless, a very serious risk.

Following on from the Great Eastern main line, the decision was taken to electrify the Great Eastern's North East London suburban lines from Bethnal Green out to Enfield, Chingford, by reopening (to passenger traffic) the Southbury loop line to Hertford East and to Bishop's Stortford. This AC electrification again utilised the BR Mk 1 OLE system. The Southbury loop line had not had any passenger services over it since 1919. Included in this reinstatement, stations at Turkey Street, Theobalds Grove and Southbury were reopened. Surprisingly, the

CAMBRIDGE and
HERTFORD EAST

ENFIELD TOWN
TURKEY STREET
THEOBALDS GROVE
CHESHUNT
CHINGFORD
SOUTHBURY
BUSH HILL PARK
WALTHAM CROSS
ENFIELD LOCK
HIGHAMS PARK
BRIMSDOWN
EDMONTON GREEN
PONDERS END
SILVER STREET
ANGEL ROAD
WHITE HART LANE
WOOD STREET
BRUCE GROVE
NORTHUMBERLAND PARK
SEVEN SISTERS
SOUTH TOTTENHAM
GOSPEL OAK
TOTTENHAM HALE
Tottenham South Junction
STAMFORD HILL
Coppermill Junction
WALTHAMSTOW CENTRAL
Clapton Junction
ST JAMES STREET
STOKE NEWINGTON
Temple Mills Depot
CLAPTON
RECTORY ROAD
Temple Mills East Junction
HACKNEY DOWNS
STRATFORD
CAMDEN ROAD
STRATFORD
Graham Rd Curve
LONDON FIELDS
HACKNEY WICK and DALSTON
CAMBRIDGE HEATH
LIVERPOOL STREET
BETHNAL GREEN
SHENFIELD

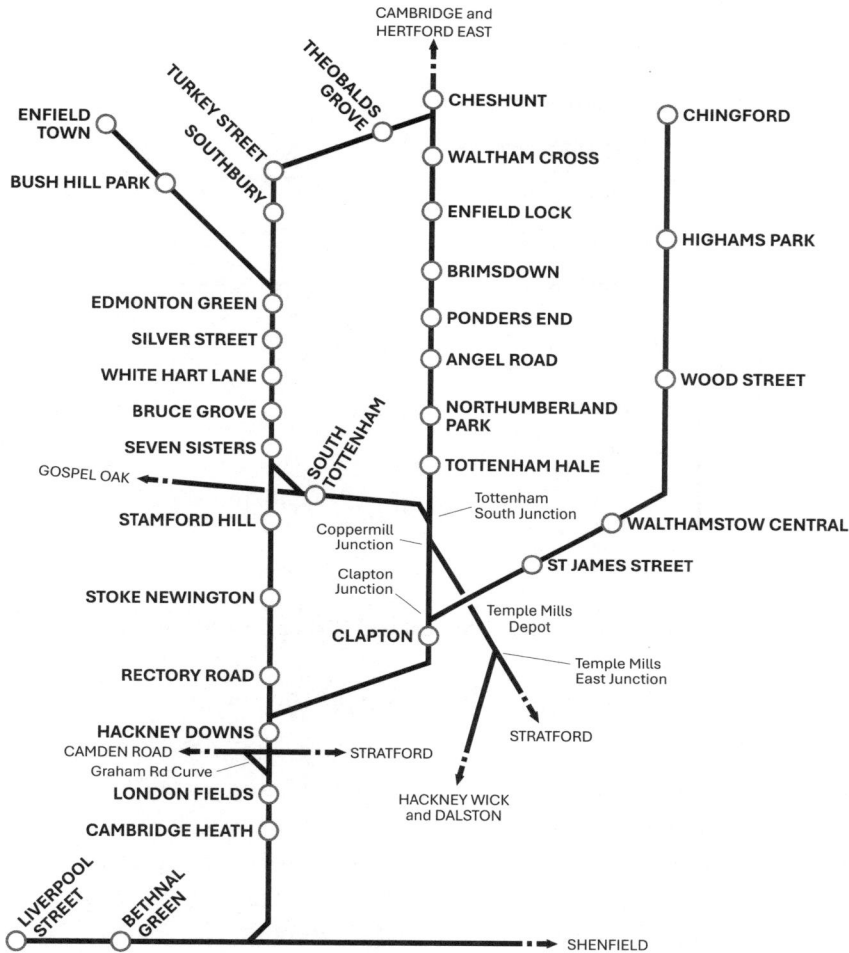

Bethnal Green Suburban Lines to Enfield, Chingford and Cheshunt

Lea Valley Main Line from Copper Mill Junction to Cheshunt was not included in the project. It is assumed that this was because the route had a lot of industry but very little residential travel. This section was, however, subsequently electrified at the end of the 1960s, utilising the BR Mk 2 OLE system.

The net result of the electrification was a significant increase in patronage. As part of this programme, although very much in second place, the London Tilbury and Southend (LTS) route was electrified from Gas Factory Junction, Stepney, through to Shoeburyness including the so-called Tilbury loop into Tilbury Riverside and then on to Pitsea. (As noted above, the line from Fenchurch Street out to Stepney had already been wired as part of the main Great Eastern scheme.) This work included the wiring of the East Ham

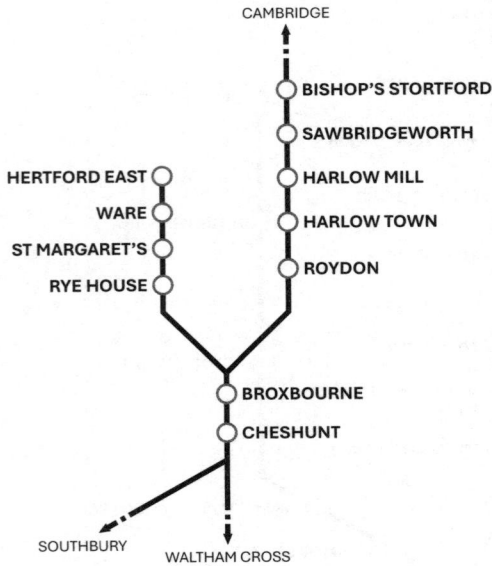

CAMBRIDGE

BISHOP'S STORTFORD

SAWBRIDGEWORTH

HERTFORD EAST   HARLOW MILL

WARE   HARLOW TOWN

ST MARGARET'S

RYE HOUSE   ROYDON

BROXBOURNE

CHESHUNT

SOUTHBURY   WALTHAM CROSS

Bethnal Green Suburban Lines to Hertford East and Bishop's Stortford

EMU depot and stabling sidings at Shoeburyness. The energisation dates for the various sections were:

Shoeburyness to Pitsea 14.5.61
Pitsea to Dagenham East 13.6.61
Upminster to Grays 18.12.61
Dagenham East to East Ham 18.7.61
East Ham to Stepney (just west of Limehouse) before 6.11.61

The total length of route wired was 109km (68 miles)

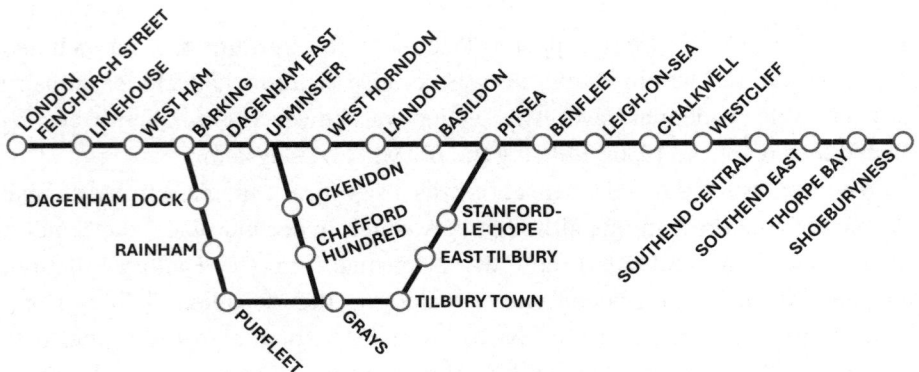

London, Tilbury and Southend Railway

The main route was available for the new EMUs working the lines from 6 November 1961 although steam trains also ran some services until 18 June 1962. The delay in converting to fully electric working was caused by the poor availability of EMUs. This was because the units that had been built for the line (AM2s, later Class 302s) were being substituted for (AM5 later Class 305) EMUs on the North East London route that had been withdrawn for repairs due to rectifier failures.

However, looking at the benefits of electrification, the positive results have been very marked indeed: between 1955 and 1968 the increase in patronage on the GE route was nearly 300 per cent and the LT&S route more than 100 per cent (see Appendix 5).

Kent Coast Electrification

The Southern Railway's plan to electrify many routes east of the Portsmouth Direct Route had been abandoned by BR upon nationalisation in 1948. However, the Southern Region's senior management continued to hold aspirations for further electrification schemes. The management's lobbying of the British Transport Commission eventually resulted in approval in principle being given in that organisation's 1952 annual report for an ongoing programme of electrification of the lines from Gillingham to Ramsgate, Sevenoaks to Dover, and Tonbridge to Hastings. Over the following four years, however, there was no progress.

As already noted, one of the five electrification schemes included in the 1955 Modernisation was for the 750V DC Kent Coast electrification, and final approval for this project was given a year after the publication of the plan in February 1956.

This was the first BR major works scheme to combine comprehensive infrastructure improvements with electrification, which became the sine qua non for most electrification schemes thereafter.

Track layouts were improved, and colour light signalling installed between Victoria and Ramsgate (via Chatham) and Hither Green and Dover (via Tonbridge). A number of stations were rebuilt or modernised, including some on the inner suburban area of the route that had been electrified before the war. A new depot for the new electric locomotives and EMUs that were to be introduced to run the services was constructed at Stewarts Lane, London.

The works were progressed in two stages, namely:

Phase 1 (288km/178 track miles)
Gillingham to Faversham & Ramsgate

Sittingbourne to Sheerness
Sittingbourne (Western Jn–Middle Jn) to Faversham & Dover Marine
Stewarts Lane area

Phase 2 (210km/132 track miles)
Sevenoaks to Ashford, Dover & Ramsgate
Maidstone East to Ashford, Canterbury West & Minster junctions
Paddock Wood to Maidstone West
Folkestone Harbour branch

The estimated cost for Phase 1 was £25M (£625M at 2025 prices) and for Phase 2, £20M (£500M at 2025) prices. The two phases were completed to programme in June 1959 and June 1962 respectively.

Phase 2 of the scheme included electrification of the Ashford to Ore line (now known as the Marshlink) and some preparatory work was carried out, but the plan was later dropped due to declining patronage.

The electrification scheme delivered a major increase in traffic on the lines covered. During the nine months following the introduction of Phase 1 services, traffic levels as measured by numbers of tickets sold increased by 32 per cent. Certain stations showed quite spectacular increases in ticket sales – at Canterbury West, for example, sales increased by 49.5 per cent comparing the first nine months of electrification with the previous nine months, while at Teynham Station, between Chatham and Faversham, the corresponding figure was 155 per cent.

Glasgow Suburban Services 10/61 and Onwards

In the 1920s the LNER had given some thought to electrifying its Glasgow suburban services. In the event no formal decision was taken and it was not until the early 1950s that a scheme was brought forward and developed. The genesis for this was the report by the Glasgow and District Transport Committee chaired by Sir Robert Inglis, a former LNER Scottish Division General Manager. Among other things, this recommended the phasing out of the Glasgow tram network and electrifying all of the suburban lines. Arising from this, the Scottish Region, recognising that it would be necessary to do one route at a time, selected the former LNER suburban line from Helensburgh to Airdrie as the first one to be converted. This decision reflected the fact that it had the highest revenue while ameliorating the appalling conditions in the Queen Street low level central tunnel section for both staff as well as passengers. Further routes would follow in due course.

The First Route

This comprised the Helensburgh to Airdrie line, including the branches to Balloch, Milngavie, Bridgeton and Springburn, a total of 82.5km (50 route miles), this being the first scheme to utilise the BR Mk 1 OLE system in Scotland. In order to serve the shipyards, the ex-LMS route was used between Dumbarton East and Bowling, with a new connection being built at Bowling joining back to the former LNER line at Dunglass Junction. Both routes between Dalmuir and Hyndland were included. Because of the restricted vertical clearances in the tunnels between Finneston and High Street, the central section was energised at 6.25kV, which complicated the traction equipment on every multiple unit. As built, they were fitted with mercury arc rectifiers, which proved to be very unreliable.

Work on the project started in 1957. In October 1959 Sir Brian Robinson, Chairman of the Railway Executive, visited the Scottish Region and in an address to the staff, said that the new trains being built by Pressed Steel at Linwood,

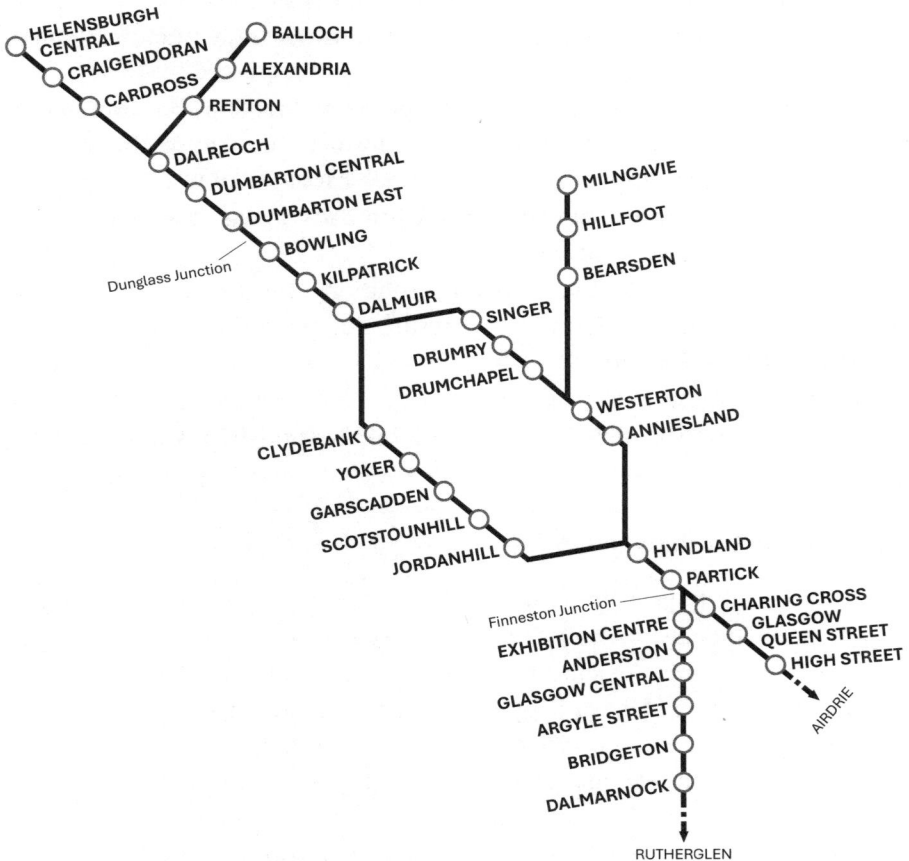

Helensburgh Central to Glasgow

Glasgow Queen Street to Airdrie Line, including the branch to Springburn

near Paisley, would be the best EMUs on the national system. The new electric service was introduced in September 1960, but there were quickly large number of failures in traffic, practically all caused by faults in the mercury arc rectifiers. These potentially caused danger to passengers and staff from explosive failures, and the decision was taken to reinstate the original steam service as from 30 December, albeit with some modifications due to the infrastructure changes that had already been required.

The faults were due to inadequate design of the traction transformers for the applied mechanical stresses on the transformer cores caused by frequent back firing of the mercury arc rectifiers. In some cases the failure of the voltage sensing and changeover equipment resulted in 25kV being applied directly to the 6.25kV transformer windings.

It took nine months to find a workable solution – the problem being solved by the redesign of the transformers. As a result, electric working was not reintroduced until 1 October 1961.

Following the tests at Crewe, mentioned later (see section on the West Coast Main Line), it proved possible to convert the 6.25kV section to 25kV thus removing this source of unreliability.

Subsequent Routes

Once the reliability problems with the EMUs had been resolved, the Scottish Region moved quickly to obtain the benefits of electrification south of the Clyde by converting a number of suburban services to electric working. Not only were the new trains much cheaper to operate but they gave improved journey times and a much cleaner environment for customers and staff alike, resulting in significant increases in revenue. They also helped to solve the staffing problems that had become endemic towards the end of the steam operation era, when the railway was perceived to be an unattractive place to work.

The first services to be converted, again using BR Mk 1 OLE, were those from Glasgow Central on to the Cathcart Circle and to Neilston and Motherwell (via Newton and Blantyre) on 27 May 1962 – a total of 44km (27¼ route miles). These were followed by the route to Paisley and on to Gourock and the branch to Wemyss Bay, which were completed for start of electric service on 5 June 1967. However, the full service was not introduced until 4 September 1967 due to problems with the train describer equipment in the Paisley PSB. The work here was carried out from October 1964 through to June 1967. These projects resulted in the wiring of Glasgow Central Station, Shields Depot and Corkerhill Carriage Sidings. This scheme included the reduction to two tracks of the 'Inverclyde' main line between Shields Junction and Paisley which, 50 years later, was partially reversed by the reinstatement of a third line in the period

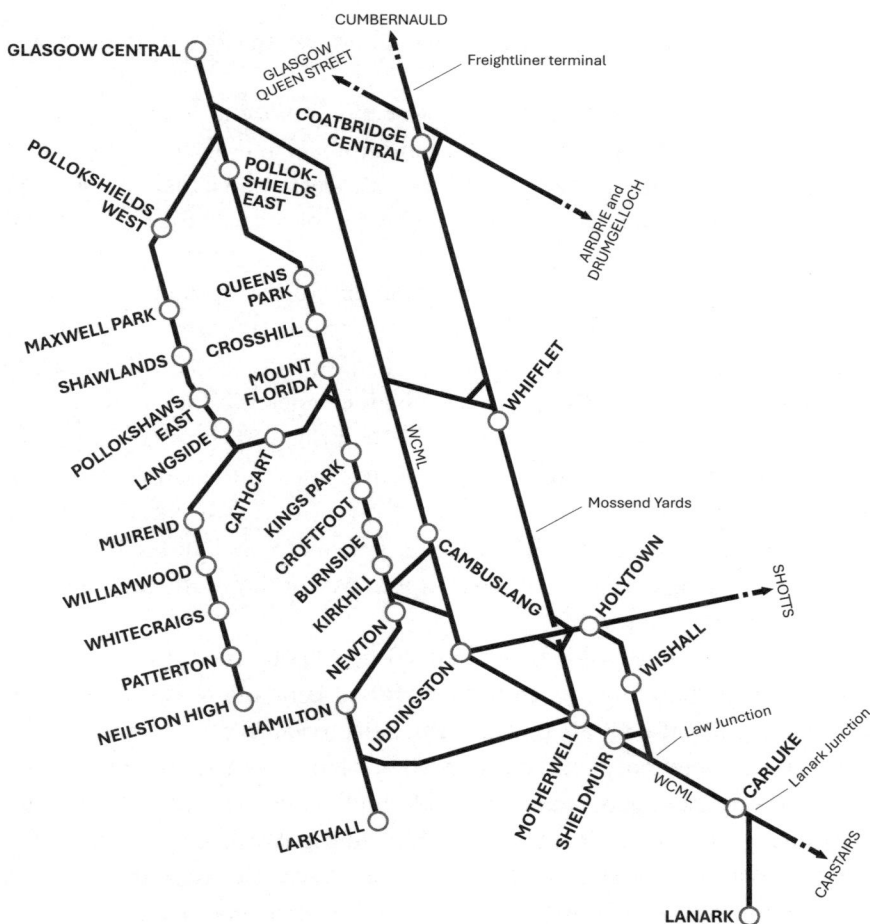

Glasgow Central on to the Cathcart Circle and Neilston and Motherwell

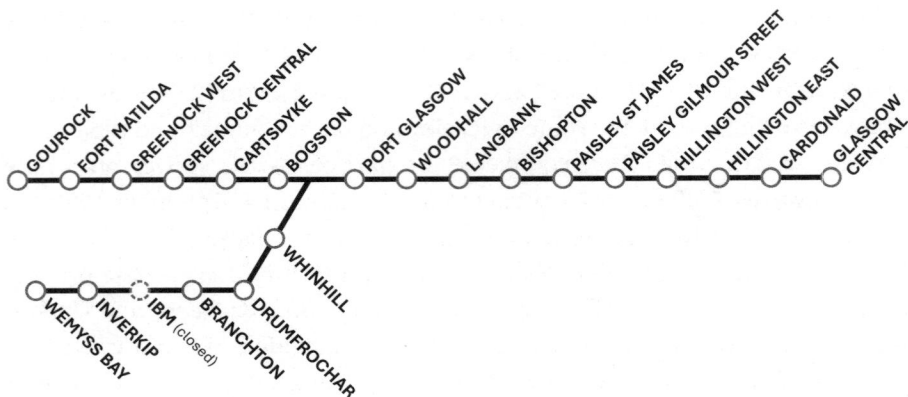

Glasgow Central to Gourock and Wemyss Bay

2015 to 2017 together with the resignalling of all three tracks for bi-directional working.

The works to Gourock and so on cost £6.9M (£140M at 2025 prices) and included £1.4M (£28M at 2025 prices) each for civils and electrification fixed equipment, £1.7M (£34M at 2025 prices) for S&T works and a total of 59km (37 route miles) were wired and 19 AM11 EMUs were purchased for £1.6M (£32M at 2025 prices).

In 1970 work had started on the electrification of the northern part of the WCML, which, within the greater Glasgow area, included the wiring of Polmadie Depot, Carstairs to Newton Junction, the Lanark branch, Law Junction to Uddingston Junction via Wishaw, the Wishaw Connecting Line and Motherwell to Gartsherrie South Junction for the Coatbridge Freightliner Terminal, including the Mossend Yard and the connecting lines to the Holytown route. This was all finished by 1974 and a further 57km (36 miles) of route was wired. All these sections utilised the BR Mk 3 OLE system. (Note that the route kilometres for the wiring of the rest of the northern part of the WCML are included in the main WCML section.)

The Hamilton Circle route was extended in 2005 by the reinstatement of the 5km (3 miles) Larkhall branch from Houghhead Junction West, which was electrified, utilising the SICAT OLE system, at its reopening.

With the completion of the electrification of the West Coast route, work started on the reopening of the former LMS line under the centre of Glasgow, again utilising the BR Mk3 OLE system. Named the Argyll Line, it linked the residential area to the west of the city on the North Clyde coast to the southern suburbs. Running for 7.5km (4.7 miles) from Partick via Glasgow Central Low Level to Rutherglen, it provides a cross Clyde link servicing the central

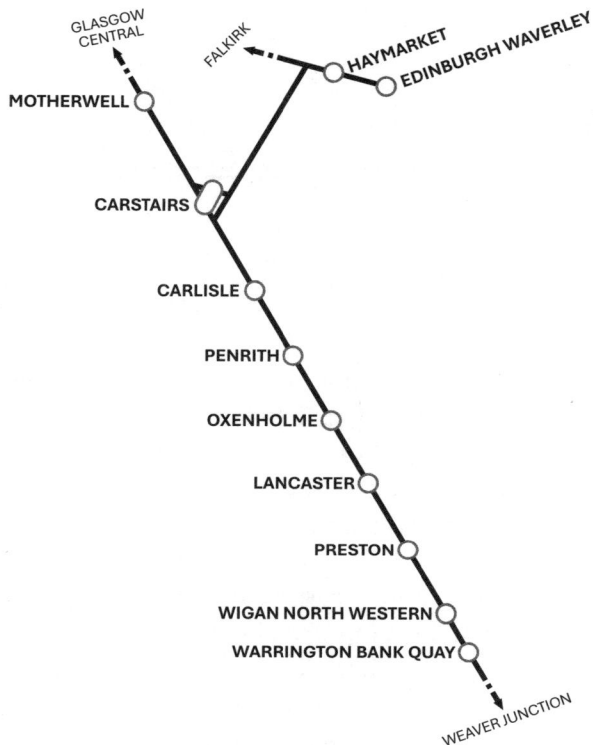

Glasgow Central and Edinburgh Waverley to Weaver Junction via Carstairs

city shopping area, and the new service started in November 1978. The tunnel section was flooded by the major storm which hit the country in December 1994, and it remained closed for nine months while the debris was removed. All the low-voltage equipment had to be replaced, and the tunnels and stations completely cleaned.

Isle of Wight – 3/67

Following a review of the Isle of Wight network in the early 1960s, it was decided to keep only one of the lines. The 13.5km (8.5 miles) of the route between Ryde Pier Head and Ventnor was retained and electrified at 630V DC, albeit with the line shortened back to Shanklin. It was planned to use cascaded electric rolling stock from the mainland, but because of the restricted structural clearances no existing BR electric stock would be suitable. It was therefore decided to convert stock of a design originally built for the London Electric Railway, later London Underground, between 1921 and 1931. Following conversion at Stewarts Lane,

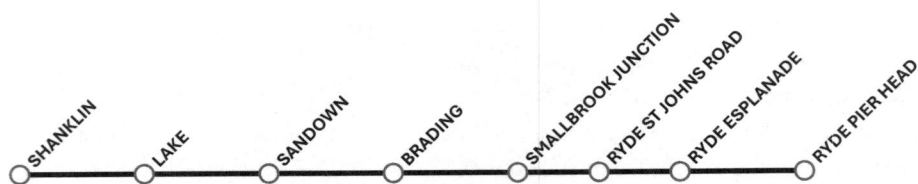

Island Line – Ryde Pier Head to Shanklin

the refurbished four-car and three-car units were designated 4Vec (later Class 485) and 3Tis (later Class 486).

After a quarter of century of service, these trains, some of which were more than 60 years old, were replaced between 1989 and 1992 by Class 483 units which were rebuilt from London Underground 1938 stock. These were, in turn, replaced, from 2021, by Class 484 units rebuilt by Vivarail from London Underground D78 rolling stock. The newly refurbished trains were a key element of a £26M project to improve Island Line services. This also included track and infrastructure enhancements, including a new passing loop at Brading and gauge clearance work to accept the larger profile of the replacement units. These works were undertaken with a planned line closure of two months commencing in January 2021 but that eventually was extended to ten months due to a succession of problems. Furthermore, the implementation of the new service from 1 November 2021 proved troublesome with the Class 484 units experiencing many teething problems, including software issues and severe wheel wear, some of which were still being resolved in 2025. As a result, plans to increase the service to two trains per hour had to be temporarily abandoned. On top of this, further infrastructure works were also found to be necessary. Accordingly, a further full closure of the line was implemented during September and October 2024 to upgrade the signalling at Ryde, refurbish a footbridge at Brading and repair a bridge at Sandown. On completion of these works, the section between Ryde Esplanade and Shanklin was brought back into service. However, the pier section was to remain closed for a further seven months for a £5M refurbishment of pier structure and track.

WCML Electrification from Euston to Manchester and Liverpool (5/66) and Birmingham (5/67)

The Route

The WCML was not conceived as a single trunk route as an entity; rather, it was formed over time by linking up lines constructed by a number of different railway companies. Often, the railway promoters side-stepped opposition to their railway projects by avoiding rural towns and landowners' estates and, in addition, tended to follow natural contours through hilly areas. As a result, the linespeed on many of today's main lines, and in particular the WCML, are constrained by the curvature of the route.

The WCML also features significant gradients, making the line expensive to operate and handicapped by extended journey times for passenger services prior to electrification.

A description of the WCML route is outlined in Appendix 2.

Background

The WCML was seen as an ideal candidate for electrification and the London, Euston to Birmingham, Liverpool and Manchester section was the first 25kV AC scheme to be authorised following the publication of the Modernisation Plan. Work began in 1957 utilising the newly designed BR Mk 1 OLE system.

The electrification of the West Coast Main Line was a major part of the Modernisation Plan published in 1955. In the event the work was carried out in two phases, the first was the electrification from London to Manchester and Liverpool, including the West Midlands network; and the second, which followed some five years later, was the northern section from Weaver Junction, north of Crewe, to Glasgow (Newton Junction). The first phase was carried out from 1957 through to completion in 1967. A new high-speed regular interval timetable of electrified services from London to Liverpool and Manchester commenced on 18 April 1966. These services slashed previous journey times (for example, London to Manchester 2 hours 30 minutes, saving nearly an hour) and successfully launched British Rail's highly regarded Inter-City brand. The West Midlands section was completed in early 1967, enabling the full electrified service to commence in May of that year.

The second phase ran from 1970 to 1974 and cut the fastest London to Glasgow journey time down to five hours.

Description of the Works and Traction

The first phase electrified a total of 657km (410.5 route miles) utilising the BR Mk 1 OLE system, and the second extended this by a further 395km (227.5 route miles) utilising the BR Mk 3 system (see Appendix 1).

In the first phase, the fixed works were carried out from north to south. The first section of route to be energised and handed over to traffic was that from Manchester, Oxford Road to Crewe (Basford Hall), including the Styal line and the North Staffs route from Cheadle Hulme through to Macclesfield

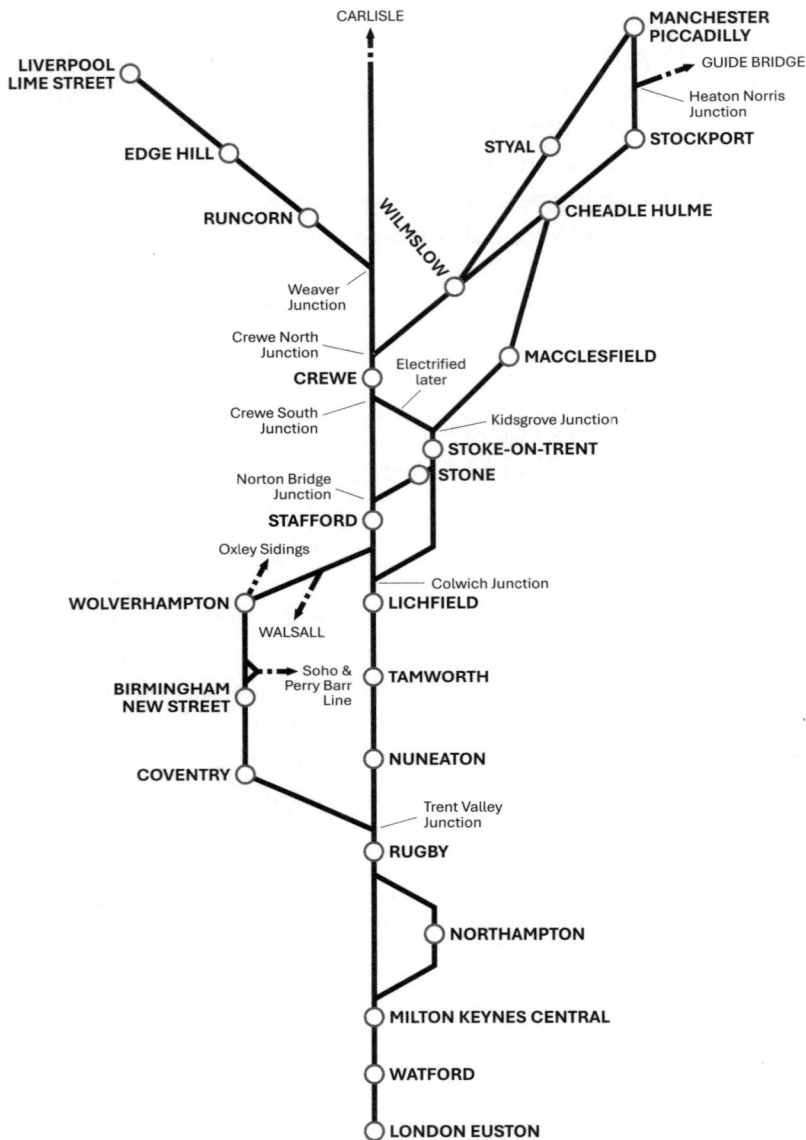

CARLISLE

MANCHESTER PICCADILLY

GUIDE BRIDGE

Heaton Norris Junction

LIVERPOOL LIME STREET

STYAL

STOCKPORT

EDGE HILL

RUNCORN

CHEADLE HULME

WILMSLOW

Weaver Junction

Crewe North Junction

Electrified later

MACCLESFIELD

CREWE

Crewe South Junction

Kidsgrove Junction

STOKE-ON-TRENT

STONE

Norton Bridge Junction

STAFFORD

Oxley Sidings

Colwich Junction

WOLVERHAMPTON

LICHFIELD

WALSALL

Soho & Perry Barr Line

BIRMINGHAM NEW STREET

TAMWORTH

COVENTRY

NUNEATON

Trent Valley Junction

RUGBY

NORTHAMPTON

MILTON KEYNES CENTRAL

WATFORD

LONDON EUSTON

London Euston to Manchester and Liverpool, including the West Midlands network and Weaver Junction, also later extension Wolverhampton to Oxley Carriage Sidings

(initiated September 1960). This enabled the Manchester suburban service to be converted to EMU working along with all of the freight traffic from the Manchester area to Crewe (loco changing was carried out at Heaton Norris and Crewe, Basford Hall). Subsequent sections to be energised were Liverpool Lime Street to Crewe (January 1962), Crewe to Stafford, Stafford to Rugby and, finally,

Rugby to London, including the Northampton loop in November 1965. A full list of energisation dates is shown in Appendix 2

In the following twelve months the West Midlands area was energised and commissioned into service. This comprised the route from Rugby through to Stafford via Birmingham New Street and Wolverhampton and the Stour Valley lines through Bescot with branches to Walsall and the Soho loops. It also included the lines from Colwich and Norton Bridge to Stone and on to Macclesfield via Stoke, thus completing the main route between Manchester and London.

In addition to the power supply system and the erection of the overhead wires, the fixed works associated with the electrification comprised the provision of clearances for the high-voltage overhead wires, the immunisation of the signalling equipment to cope with the earth return currents, in many cases the remodelling of layouts bringing them up to date before major signalling alterations or complete resignalling and the provision of maintenance depots for the new locomotives and the multiple units for the suburban services. It was also necessary to provide facilities for maintaining the overhead wiring system.

Initially, the clearances specified for the erection of the overhead lines were considered to be somewhat excessive and a series of special tests were carried out in the tunnels on the Crewe, avoiding lines in the late 1950s, which demonstrated that these could be significantly reduced, thereby making major reductions in the number of bridge reconstructions, awning alterations and so on.

As well as electrifying the route, the project also involved major permanent way remodelling and resignalling, also major engineering works. The associated alterations to the signalling were a mixture of brand-new power signal boxes and immunisation of the existing signalling. The latter was necessary because the industry just did not have the capacity to completely re-signal the whole route (within the required timescale) in the first phase, although it was able to provide new power signal boxes at Manchester (London Road), Wilmslow, Edge Hill, Weaver Junction, Stoke-on-Trent, Wolverhampton, Birmingham New Street, Coventry, Norton Bridge, Nuneaton, Rugby, Bletchley, Watford, Willesden and Euston.

Major engineering works were needed in the Kidsgrove area, on the Stoke-on-Trent to Manchester section, where there existed three separate Harecastle tunnels. Harecastle North was the shortest of the three railway tunnels at just 130 yards, and this was opened out in 1966. The structure gauge of the other two – Middle (180 yards) and South (1,766 yards) – was too tight to accommodate OLE so a diversionary route was constructed through Bathpool Park; this itself included a new 243-yard Kidsgrove Tunnel.

During the first phase of the electrification works in the early 1960s, it became apparent that the original project, as then specified and authorised, was going to become significantly overspent. The newly formed British Railways Board ordered a review of the situation and a re-estimation of the outstanding work. The result of this was a reduction in the scope of the then further work to completion, so as to bring the project back within the previously agreed budget. This resulted in a number of secondary lines being deleted from the project, which had a major impact on the economies that would have been gained had they been carried out. This was even more unfortunate in that towards the end of the scheme it became clear that, as a result of favourable tenders and so forth, there was going to be an underspend on the currently stated scope, which would, ironically, have enabled most, if not all, of the deleted items to have actually been carried out.

For WCML Electrification implementation dates, see Appendix 2.

New electric locomotives and coaches were procured for Phase 1. There were six types of locomotives ordered, with an initial build of twenty each of the first five types (AL1–AL5) and 100 of the final design (AL6). The first five classes were specified as prototypes and, based on the experience of running these different designs, a final design emerged in the form of the AL6, which, under the BR renumbering scheme, became Class 86. All these earlier designs had unsprung nose-mounted traction motors driving directly on to an axle and the resulting damage to the track was very significant indeed. The fleet was further extended as part of the second phase of the electrification through to Glasgow by the purchase of thirty-five Class 87 locomotives built to an enhanced speci-fication to eliminate these limitations. The new Mk 2 coaching stock included new air-conditioned Pullman cars.

Depots for the maintenance of the electric locos were provided at Longsight, Crewe and Willesden with maintenance depots for the electric multiple units at Longsight, Birmingham Soho, Bletchley and Willesden.

From the 1950s onwards, there were major changes in the freight traffic flows. These had not been foreseen during the detailed planning stages, and it resulted in a large number of sidings and goods lines being wired that, by the completion of the second part of the project in the mid-1970s, were no longer required.

Unfortunately, there then followed a hiatus of some four years while the gov-ernment dithered before authorising the completion of the project through to Scotland. This meant that every freight train going north to Preston and beyond had to change locomotives in Basford Hall marshalling yard and every passenger train had to re-loco in Crewe station. Not only were these major additional costs but they added significant time into the overall journeys, which, between London and Scotland, were a meaningful disincentive for people to travel by train.

The failure to continue electrification north of Crewe (Weaver Junction) caused the disbanding of the electrification teams across the whole of the supply industry as well as within BR itself. This resulted in a significant loss of expertise, which had to be rebuilt when the northern extension to Glasgow was finally authorised in 1970. It also had a significantly adverse impact on the people involved in railway electrification (industry as well as railway), resulting in a major loss of morale.

The Tortuous Path from Steam to Electric Traction on the WCML

A particular shortcoming of the introduction of electrification on the West Coast Main Line was due to the consequence of another major element of the Modernisation Plan. This was the 'dash for diesels' philosophy, pushed through by the Railway Executive to rapidly eliminate steam traction in the mistaken belief that this would solve the railways financial problems.

The powerful former LMS Princess and Princess Coronation steam loco-motives, which had operated the main line services since the 1930s, were very reliable and had always been maintained to a high standard. These locomotives were between seven and seventeen years old at the announcement of the elec-trification scheme and would have been more than capable of reliably operating the main line services during a progressive transition period.

The logical traction transition plan for the WCML route's core services should simply have been a phased transition from steam to modern electric power. This is what was originally envisaged, but it was not to be.

The plethora of diesels available on the LMR Western Division from 1959, delivered through the Modernisation Plan, had huge consequences for the West Coast Main Line. These locomotives were less powerful than the steam locomotives that they were to replace and were far from reliable when first introduced. It was, however, considered necessary for the new English Electric Type 4 locomotives to be found work on the WCML route for both political and staffing reasons. First, it was contended that the early introduction of diesel locomotives would demonstrate that modernisation was underway. Secondly, it was considered that the new traction would make it easier to retain and recruit personnel in the London area, where there were staff shortages, by affording better working conditions for those involved in their operation and maintenance. The validity of this latter claim is open to question as the Western Region faced similar issues, and overcame them by 'lodging' staff in and near the capital who had formerly been based in Wales and the West and so achieved an orderly transition to modern traction. In the event, the case for the interim WCML dieselisation was not proven as the labour issues didn't disappear. In the end, there were still many steam locomotives retained in service on the WCML alongside the diesels, almost to the close of the transition period of the London and Liverpool/Manchester electrification.

In connection with the introduction of these diesel locomotives, interim arrangements were considered in an attempt to utilise the new traction efficiently prior to the completion of the electrification. This included providing new operating and maintenance facilities, training staff in new skills and negotiating new terms and conditions of employment with the trades unions to replace well-established lodging and mileage bonus arrangements. In addition, a complete recasting of the timetable would be desirable.

It was soon realised, however, that to implement the envisaged changes would require a significant reorganisation, imposing a further huge strain on management and staff already coping with the degraded operational conditions resulting from the implementation of the electrification works.

Understandably, the LMR General Manager decided that it would be better to implement such a major change once, at the end of the electrification, rather than twice, in order to incorporate the proposed intermediate arrangement, that is, steam to diesel and diesel to electric.

Accordingly, during the transition period between 1959 and 1965/6, only a modicum of revised operating arrangements was 'patched-in' to accommodate the perceived need for diesel traction requirements, thus enabling diesels, as well as steam locomotives to operate the main line services.

Initially, the diesels worked alongside steam to the steam locomotive diagrams, so there was little or no immediate benefit from dieselisation. Latterly, as the electrification progressed, diesel haulage to 'change-over' points where electric traction took over became possible, allowing the diesels to be used somewhat more effectively. However, the return on the estimated £245M (2025 prices) invested in the English Electric Type 4 diesel locomotives was negligible. It is noteworthy that the total number of traction steam and diesel traction units had only reduced by 18 per cent five years after the introduction of the first English Electric Type 4 diesel, of which 104 had been placed in service on the WCML by 1964.

The lack of bespoke maintenance facilities meant that diesel locomotives often had to be maintained at steam depots. Inevitably, this resulted not only in the poor utilisation of the traction resources, but also a deterioration of both steam and diesel traction units, which had only reduced in number by 18, five years after the introduction of the English Electric Type 4 diesels. It has been estimated that the region's revenue fell by £250M (£5.5Bn at 2025 prices) during the period in question, and a large proportion of this would be due to the poorer WCML performance.

In addition, the late imposition of these 'interim' partial dieselisation arrangements was also far more than just a marginal distraction to the LMR managers whose main role was maintaining WCML train performance while implementing a complex electrification project. Thus, the circumstances that forced this stop-gap decision triggered a tortuous path to the eventual electrification of the route, resulting in widespread negative public relation consequences during the transition period.

Fortunately, the new high-speed, regular interval timetabled electrified services between London and Liverpool/Manchester, introduced from April 1966, quickly recovered the 'lost ground' of the preceding six years or so.

The key lesson learned from this episode was not to impose significant additional, un-costed change at short notice on a major plan that is in progress.

Sturt Lane (near Frimley) to Bournemouth – 6/67

Predominantly a passenger line serving major conurbations at Southampton and Bournemouth, the 231km (143 miles) South Western main line between London Waterloo and Weymouth was the last route to the capital operated by steam. The Surrey Section as far as Pirbright Junction, where the Alton line branches off to the west of Brookwood station, had been electrified before the Second World War. The Southern Region were keen to modernise the route both to improve the service and to eliminate steam traction.

In the light of the government's increased scrutiny of the railway's investment proposals, plans were advanced in the early 1960s to modernise the line to Bournemouth and Weymouth. Proposals for the scheme were exhaustively worked and re-worked comparing the economic advantages of diesel and electric traction. Eventually, in 1964, authority was granted for the electrification of the line to Bournemouth at 750V DC from the then ending of the electrified system at Sturt Lane, just to the west of Pirbright Junction.

The financial analysis failed to justify electrification to the end of the line to Weymouth. Through passenger services on the section of route between Bournemouth and Weymouth were to continue by diesel traction in coordination with the electrified section in, what then was, a novel way. Special forms of EMU capable of push–pull operation in conjunction with un-motored multiple units were developed and the latter units were also suitable for working off the electrified lines with diesel-electric locomotives equipped with push–pull control.

The method of operation consisted of a four-car powered EMU (4REP) pushing four or eight carriages (that is, one or two unpowered 4TC units) from Waterloo to Bournemouth where there was a waiting diesel-electric locomotive equipped for push–pull working. This loco then attached to the unpowered TC unit(s) that were de-coupled from the powered unit. The diesel-electric locomotive then hauled the TC units to Weymouth. The return service was operated with the loco propelling the TC unit(s) to Bournemouth where they were

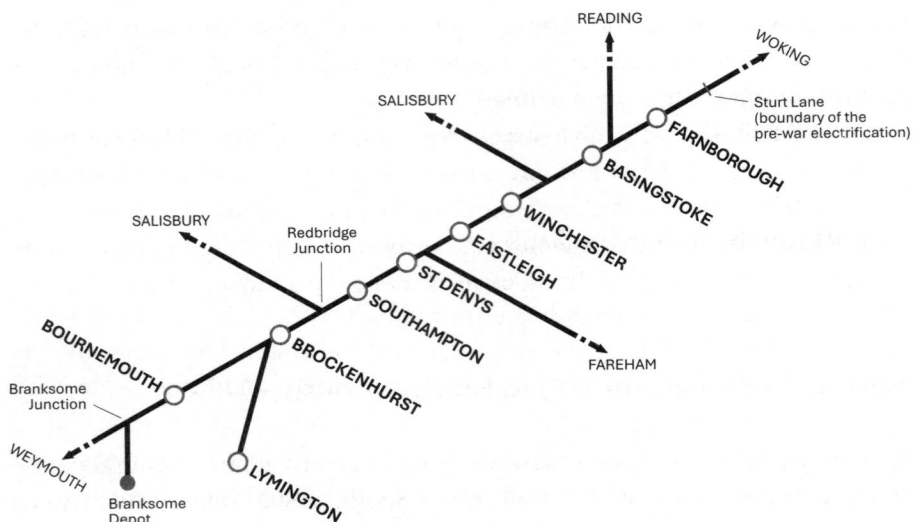

Sturt Lane (near Frimley) to Bournemouth and Branksome Depot

coupled to a 4REP powered unit. The loco detached and the powered 4REP unit hauled the TC carriages to Waterloo.

For stopping and semi-fast services between Waterloo and Bournemouth, 4VEP units were introduced, with more external doors per vehicle compared to the main line stock to reduce dwell times at busy stations. The VEP units were limited to a maximum speed of 160kmh (100mph). This limit was determined by practical tests, which demonstrated that at any higher speed the shoes that collected the current from the conductor rail would have an unacceptably short service life.

The electrification work proper was accompanied by a large programme of re-signalling, station rebuilding and track upgrading, all of which proceeded while maintaining a full train service. Major infrastructure enhancements to remove speed restrictions mainly at the London end of the route were, however, avoided both to save cost and to minimise disruption to commuter services during the works. The actual electrification was carried on to Branksome in order to access the new stock maintenance and berthing facility constructed on what had been the line to Bournemouth West Station, which closed in October 1965. The overall scheme included the branch to Lymington Pier and totalled 136km (85¼ route miles).

While various line speed restrictions precluded exceptional reductions in journey times, competitive times of 70 minutes for the 159km (79.25 miles) to Southampton and 100 minutes for the 177km (108 miles) to Bournemouth were achieved, gaining 10 minutes and 20 minutes respectively on the previous steam schedules. The new electrified service, the first high-speed passenger push–pull operation in the UK, commenced on schedule in June 1967.

Electrification was extended to Weymouth in 1988 in conjunction with the introduction of new BR Class 442 five-car EMUs (5-WES) rolling stock.

Lea Valley Main Line – 7/69

This was a very early infill scheme from Stratford to Cheshunt, utilising BR Mk 3 OLE (see Appendix 1). It provided an alternative route for the North East London suburban services to Chingford, Hertford East and Bishop's Stortford. It is interesting to note that as far as is known, this was the first electrification project to use 'computer networks' for the planning and the management of delivery activities. The 'network' was put on to the Research Department's mainframe computer at Derby and was updated at regular (weekly) intervals to give the project manager the latest information as to what still needed to be done to complete the project and the most likely end date.

An interesting aside is that this was also used to monitor very much more closely the activities of contractors, particularly those supplying small and medium-sized items of plant. Several contractors found, to their cost, that their duplicate claims for the same item of plant being used at different locations simultaneously were rejected as a result of the evidence provided by the historical element of the network computer programme printout.

Wolverhampton to Oxley – 5/72

This very simple scheme was the first extension to the original West Coast project. Carried out in 1971/2, its genesis was derived from the alteration to the original train service plan. In 1967 at the completion of the EMLB scheme, the basic service pattern consisted of a return working each hour from London to Manchester via Stoke, from London to Liverpool via Crewe and, hourly, London to either of Manchester or Liverpool via Birmingham, Wolverhampton and Crewe. Because the new service was so successful, the LM planned to enhance the part to the West Midlands in May 1972. The plan was to run a half hourly London to Birmingham service with every other train continuing on to Wolverhampton. The through services from the West Midlands to the North West would be provided by the Cross-Country services.

With trains turning round at Wolverhampton on a regular basis, it was operationally necessary to provide somewhere for them to lay over while they were serviced prior to their return working south. In addition, five of the coaching stock diagrams covering the London to Wolverhampton service were based at Oxley carriage sidings and this required diesel haulage each morning and evening to and from Wolverhampton station before they could become electrically worked. This created the situation where it was appropriate to extend the wires by approximately 2.5km (1.5 miles) from Wolverhampton, North Junction to Oxley and to wire most of the depot there so as to facilitate the aforementioned operation. At Birmingham, so as to facilitate the turning back every hour of the train from London, an additional facing crossover was provided in the down and up Coventry lines on the London side of New Street South Tunnel. This work was completed in time for the May 1972 timetable change.

West Coast Extension – Weaver Junction to Glasgow (Newton Junction) – 5/74

In April 1968, the British Railways Board published a report on the costs and returns to be expected from the electrification of the northern half of the West Coast Mainline from Weaver Junction (where the route to Liverpool diverges) to Glasgow. This project covered the northern part of the route and closed the gap between Weaver Junction and Newton Junction, just south of Glasgow. It was the first major scheme to utilise Mk 3 OLE. For passenger trains, it avoided the need to change locomotives in Crewe station, and freight trains did not have to be re-engined in Basford Hall sidings.

This report recommended that authority should be given for an outlay of £54.6M (£1.3Bn at 2025 prices) for implementing the scheme 'as soon as possible'. This would convert a total of 395 route km (227.5 miles) to electric operation.

The government approved the £25M (£560M at 2025 prices) electrification and the associated £29.6M (£670M at 2025 prices) resignalling works and so forth between Weaver Junction and Glasgow in March 1970. The announcement was made two days before the writ was issued for a by-election in South Ayrshire and caused a major political scandal. Accusations were made that the decision was politically rather than financially motivated.

In addition to the electrification of the line, the scheme involved a comprehensive upgrade of the track and signalling. This route modernisation involved improving bridge clearances at 154 locations and replacing the life-expired Victorian signalling equipment with new equipment controlled from four power-boxes at Warrington, Preston, Carlisle and Motherwell.

The cost of the electrification came in at £36M and the route upgrade at a further £38M, both figures at 1974 prices. While there had been some small changes in scope, the project was delivered during a period of very high inflation. BR estimated that, allowing for these factors, the project had been completed within 3 per cent over the estimates made in 1968.

The newly electrified service commenced on 6 May 1974. The number of through daily services was increased from five to eight and daytime journey times were substantially reduced compared to those prior to the start of the project by up to 100 minutes between London and Glasgow. The flagship 'Royal Scot' service achieved a five-hour journey time between Euston and Glasgow with one stop en route. In addition, other services between Manchester/ Liverpool and Glasgow saw a 60-minute improvement in journey time.

BR issued a promotional booklet to mark the completion of the entire West Coast project entitled 'Electric all the Way – London to Glasgow'. This publication extolled the benefits of the electrification itself, including a few words on the environmental benefits, and, not unfairly, made great play of the much-improved services offered.

Overall, the electrification of the WCML was a great success, resulting in a doubling of passenger traffic between 1962 and 1975. The Lanark Branch and the Hamilton Circle lines were electrified concurrently with the WCML scheme. As a result of this project, the length of electrified lines in Scotland increased to 368km (230 miles).

For WCML Electrification implementation dates, see Appendix 2.

Liverpool Loop and Link Schemes – 1974 and 1977

In the late 1960s the Labour government of the day decided to establish Passenger Transport Authorities for the four conurbations of the West Midlands, Merseyside, Manchester, and Tyneside. These authorities were composed of representatives from the local Councils of the areas that they covered, and they had specific remits in connection with responsibilities for providing public transport ensuring that, as far as was possible, road and rail were properly integrated. Both the Liverpool and Manchester authorities started work by carrying out land utilisation studies that were dubbed 'MALTS' and 'SALTS' where 'MALTS' stood for Mersey Area Local Transportation Study and 'SALTS' stood for the SELNEC Area Local Transportation Study. (SELNEC standing for South East Lancashire and North East Cheshire because the area covered by the PTA was much wider than the then Manchester local government area.)

One of the major results of the MALTS study was that the existing line under the Mersey into Liverpool Lime Street low level should be converted into a terminal loop so that trains served Moorfields (replacing Liverpool Exchange),

Liverpool Lime Street and Liverpool Central, thus providing a much greater distribution around the city centre and avoid the need to reverse underground. This work was carried out under the authority of a Parliamentary Bill promoted by British Railways on behalf of the PTA and was completed in 1974.

The second major recommendation of the study was that the lines from the north of Liverpool running into Liverpool Exchange should be extended to serve the south-east part of the city by means of a tunnel linking Liverpool Exchange to Liverpool Central, and this work was duly completed in 1977. Subsequently, the tunnel was extended to connect into the line from Liverpool Central (surface) station as far as Garston and was electrified on the third-rail DC system.

Pic-Vic – A Major Proposal that Did Not Happen

Following on from the results of the studies for the Mersey area, the SELNEC PTA were equally keen to action the results of the studies that were carried out for their area. This resulted in the Passenger Transport Executive developing a scheme for linking the railways on the north and south sides of Manchester together by means of a tunnel under the city centre with three underground stations serving the then key centres of the city. This scheme, which was officially called the Manchester City Area Railway Tunnel, became known as 'Pic Vic', and was pursued under a Parliamentary Act obtained in 1972.

The scheme consisted of a tunnel linking the electrified railway to the south of Manchester based on Manchester Piccadilly running via Whitwell Street, St Peter's/Albert Squares, Market Street to Manchester Victoria and then rising to the surface to link in with the line to Bury. The scheme depended entirely on central government funding and, following the financial crisis that developed in 1975, there was no money available to finance the works. Thus, it did not proceed.

Although it was not included in the original Pic-Vic scheme, there was an advantage of electrifying the beginning of the line from Stockport to Buxton out as far as Hazel Grove and this was done in the late 1970s. This included the provision of two stabling sidings at Hazel Grove and the rebuilding of Hazel Grove station with longer platforms.

Chapter 9

Schemes Promoted by British Railways between 1974 and 1994

Great Northern Suburban Electrification – 4/77

The idea to electrify the Great Northern suburban services was first put forward in 1903. It also featured in the first edition of the Modernisation Plan. However, beyond some very preliminary works, the scheme had not been pursued.

In the early 1970s, the BR passenger business established a commercial requirement to reduce journey times requiring the use of higher performance trains than the existing DMUs. The detailed financial appraisal for the two options considered (new DMU v EMU) showed a benefit to electrification of £5M over the service life of the project.

This appraisal and a number of factors coinciding resulted in the scheme being brought forward and authorised in August 1971 at a cost of £80M (£990M at 2025 prices). The budgeted electrification fixed equipment element was £12M (£144M at 2025 prices) for 364 single track kilometres (STK), equating to £33k (£440k at 2025 prices) per STK. According to the paper presented by Norman Howard to the IEE Proceedings in March 1982, the outturn cost achieved even bettered this at £23k per STK (£280k at 2025 prices).

The scheme involved, among other things, the need to completely renew the track in the King's Cross Station throat, and project savings were budgeted to be obtained by dequadrifying the North London Line between Dalston Western Junction and Broad Street. In addition, replacing the life-expired DMUs and the opportunity to provide a better link to the City by taking over the underground line from Finsbury Park to Moorgate also helped the financial case.

The project consisted of electrification from King's Cross to Royston, including the Hertford loop and the former underground line from Drayton Park to Moorgate (Northern City). The tunnels on this section had been built to surface rolling stock loading gauge but were not big enough to accommodate overhead electrification at 25kV. By using dual voltage stock equipped for 25kV AC and 750V DC power supplies, trains could run direct to the City with a power supply change at Drayton Park, which is a station in the open.

The scheme consisted of 365 STK of 25kV infrastructure and the severing of 750V power supplies from London Underground to the Drayton Park–Moorgate tunnel section. The LT fourth rail was retained to augment the DC return path.

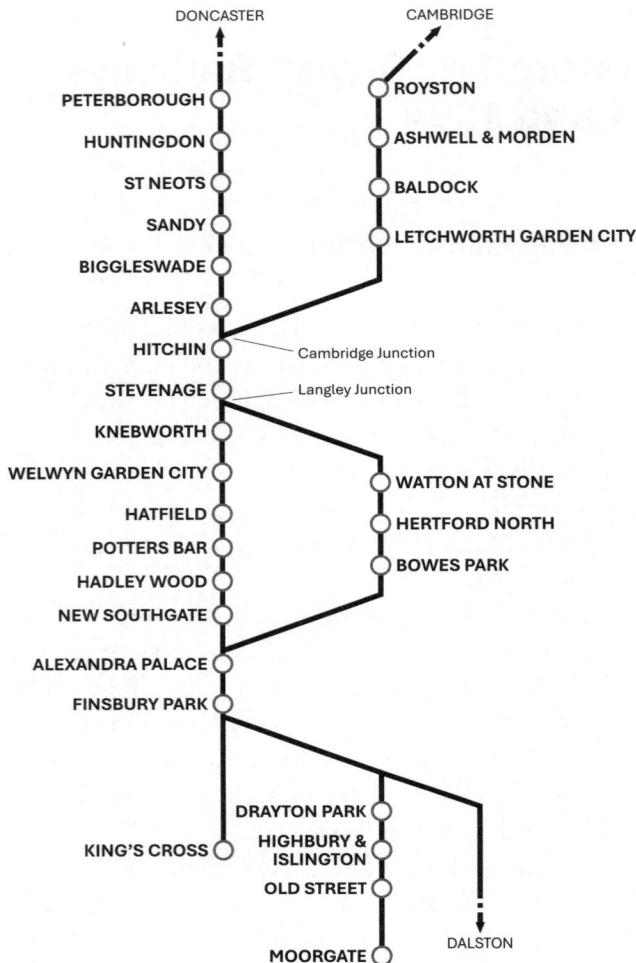

Great Northern Electrification (King's Cross to Royston, including the Hertford loop and the former underground line from Drayton Park to Moorgate)

New power supplies from the London Electricity Board were established at Poole Street. New transformers and rectifiers were provided at Moorgate constructed in an abandoned lift shaft. A novel feature was the design of the AC/DC interface at Drayton Park. Simply put, AC traction current is returned to the supply point in the running rails and the neutral connection is deliberately earthed. DC traction current also is returned to the supply point in the rails but because of the catalytic corrosion effect of DC, the negative connection is not deliberately earthed so that an engineering solution as effective as circumstances would permit was implemented. For further details of the electrification OLE and third-rail systems, see Appendices 1 and 4.

The remodelling of King's Cross included the closure of the eastern bores of Gas Works and Copenhagen tunnels and the conversion of the goods line between Copenhagen Junction and Holloway into passenger lines thus enabling up direction suburban services into King's Cross to fly over the down direction main line departures by virtue of the flyover at Holloway South, which was rebuilt to improve its alignment and hence raise the speed across it.

Associated with the electrification was the resignalling of the area right out to north of Sandy on the main line and beyond Royston on the Cambridge branch with a new power signal box at King's Cross and the closure of some sixty-seven mechanical signal boxes. The connection to the widened lines was abandoned as GN suburban passenger services were diverted to the former underground line to Moorgate and freight to South London had ceased in the late 1960s. The project included the reopening of Watton-at-Stone station, which had been closed at the outbreak of the Second World War together with the reinstatement of a through service from Hertford North to Stevenage. A total of sixty-four Class 313 dual-voltage three-car EMUs and twenty-six Class 312 four-car outer surburban EMUs were procured and a maintenance depot was built on the site of the former up marshalling yard at Ferme Park. The electrical control for the network was sited at Hornsey immediately adjacent to the new maintenance depot and was later extended to oversee the power supplies to the Midland Surburban electrification.

Initially, the project scope only specified the wiring of the slow lines from Woolmer Green to Hitchin. Clearly, this would be extremely restricting and prevent following trains overtaking preceding ones when that was appropriate either because it was in the timetable or because of out-of-course operation. The overhead system design (OSD) required the provision of dummy weights to the headspan design in place of the fast line equipment, in order that the headspans would hang correctly. The original scope also had the wires terminating at Hertford North until it was realised that wiring through to Stevenage would save an additional feeder station as well as providing an ECML electric diversionary route. Fortunately, therefore, before the work north of Welwyn Garden City was far advanced, the Investment Committee approved the wiring of the fast lines as well.

Midland Suburban Electrification (St Pancras to Bedford) – 7/83

In the early part of the 1970s the economy was 'roaring away' at the start of Chancellor Barber's 'boom and bust' period and, out of the blue, the Department of

Transport approached the Railways Board and asked if it had an electrification scheme on the drawing board that could be implemented quickly. There was not anything prepared in great detail, but there was a pressing need to replace the DMUs working the suburban services on the Midland Main Lines between London and Bedford. So, this scheme was put forward on the basis that the London Midland Region had the expertise to implement it, and resources were available because the northern extension of the WCML was approaching completion.

Civil Engineering Works

The LMR HQ Bridge design section was charged with leading the design process. There were only a handful of technical staff who were put under huge pressure to do the bridge designs to an impossible timescale. There was a lot of 'buzz' in the design office and exchange of ideas and views as to advance solutions. Standardisation was key wherever possible, so most existing road overbridges, which were four-arch structures, typically in a shallow cutting, had the middle two arches and pier blown out, the outer arches strengthened with brick supporting walls and BR SR hog back beams installed spanning all four tracks. There were a few non-standard bridge designs, such as at Flitwick Station and the very flat angle multi-span intersection bridge east of West Hampstead carrying the North London Line over the MML.

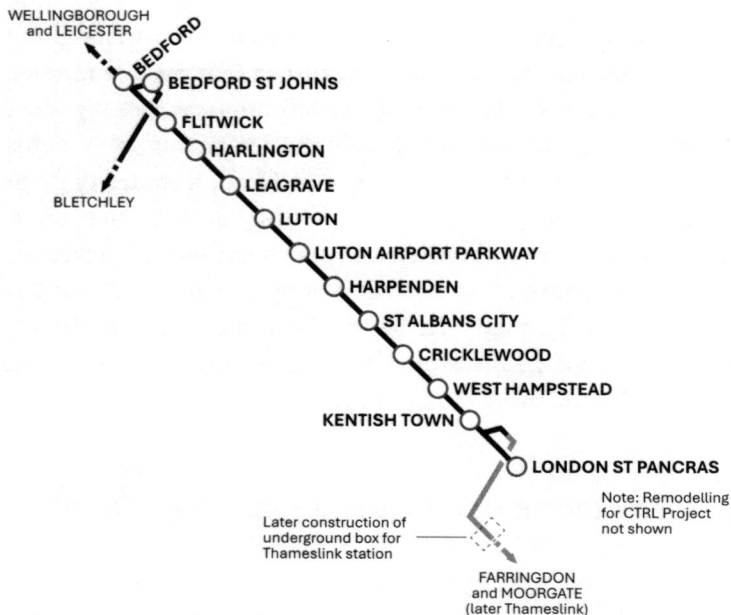

WELLINGBOROUGH
and LEICESTER

BEDFORD

BEDFORD ST JOHNS

FLITWICK

HARLINGTON

LEAGRAVE

BLETCHLEY

LUTON

LUTON AIRPORT PARKWAY

HARPENDEN

ST ALBANS CITY

CRICKLEWOOD

WEST HAMPSTEAD

KENTISH TOWN

LONDON ST PANCRAS

Note: Remodelling
for CTRL Project
not shown

Later construction of
underground box for
Thameslink station

FARRINGDON
and MOORGATE
(later Thameslink)

Midland Suburban Electrification (St Pancras to Bedford)

The station and platform designs were carried out by the 'LMR new works sections'. The Divisional office at Watford was responsible for minor alterations and security, including user worked crossings and fencing. In view of the very tight timescales the Watford Direct Labour Organisation (DLO) was employed to do much of the work, initially all of the platform works and increasingly the bridge reconstructions. This resulted in the role of a Resident Engineer being 'morphed' to that of a DLO site engineer.

The scheme included the conversion of the Metropolitan widened lines between King's Cross and Moorgate to slab track construction to accommodate the new Midland suburban electric services. It also resulted in the removal of the junction at Farringdon leading to Snow Hill and Blackfriars.

Signal Engineering

Until the mid-1970s, the route still had the signalling systems installed by the Midland Railway some fifty plus years previously. Before it could be electrified these had to be replaced in modern form based on a control centre (power box) at West Hampstead. This signalled the route from Moorgate right through to Sharnbrook (just north of Bedford) and from Cricklewood to Dudding Hill Junction (on the Acton Branch) and from Bedford Midland to Bedford St Johns.

Electrification

This was the first electrification project that was carried out by a central team created at BR HQ, responsibility for electrification projects having recently been transferred from the regions. All subsequent electrification projects up to the privatisation of BR were carried out by this team to a common set of design rules and processes. The teams were also responsible for maintenance instructions and procedures.

The standard Mk 3b design of overhead wiring was installed on the route. (A basic difference between the Mk 2 and 3 designs was that Mk 2 used imperial-sized components whereas from Mk 3a onwards, metric components were specified – see Appendix 1.) A construction depot was sited at Bedford and the erection of the overhead went ahead as per the programme.

A novel feature introduced on this scheme was the creation of what became known as 'special reduced' electrical clearance. Hitherto, BR electrical clearances were based on the UIC recommendation, which for 25kV were 270mm static and 200mm passing clearance from contact wire to kinematic load gauge. In 1962, following tests and service experience BR revised these clearances to 200mm and 150mm. The use of slab track produced a fixed-track position enabling the passing clearance to be further reduced to 125mm as there was no longer any need to allow for the usual track maintenance tolerance allowance. Accordingly, 2km of concrete double line slab track was constructed between the Midland curve under St Pancras station and the approach to Farringdon Station.

The AC/DC changeover was at Farringdon Station where tests showed that several hundreds of amps of DC current from south of the river wanted to flow in the AC traction return rails north of the river. The solution was to install IBJs and contactors to connect the track sections to the DC or AC return system as required.

On completion of the works, the train drivers' union, ASLEF, caused disruption by striking every week on Tuesdays and Thursdays in a protest about the introduction of the driver-only operation (DOO) of the new trains. Nevertheless, the scheme was completed on time and within budget. The new electric services were to be introduced in October 1982, but the industrial dispute lingered on into 1983. The first of the new electric four-car 317 trains were introduced in March but it was mid-1983 before the last old diesel Class 127 DMUs were taken out of service.

Rolling Stock

A total of 48 Class 317 four-car EMUs were ordered from British Rail Engineering. These units were state of the art at the time and consisted of two driving trailers, a pan/motor car with guard's brake van and an ordinary trailer vehicle. They were all second class and the 'units' could be coupled to another one to form an eight-car set. They were built at the Derby, Litchurch Lane works. Their loading gauge was such that they could traverse the Metropolitan Widened lines to Moorgate. The new Class 317 EMUs brought a significant reduction in journey times together with major improvement to the comfort and performance of the service.

The former Cricklewood up yard was used as the site for the EMU maintenance and servicing depot. Stabling sidings were provided adjacent to it as well as immediately to the south of Bedford station in the angle between the lines to Bedford St Johns and the return to the main line.

Sectorisation

At about the time of the completion of the Midland Main Line Scheme, it was decided that future electrification schemes would be sponsored by the new management units being put in place, termed 'Sectors'. As this was such a fundamental change, a brief history follows:

When Sir Robert Basil Reid (colloquially known as Bob Reid I) was appointed British Railway's Chief Executive in January 1980, he quickly realised that the geographical organisational structures that had been in place since nationalisation were less than ideal. Furthermore, the arrangement did not give him sufficient information to enable meaningful decisions to be made at the highest level. This led him to decide to reorganise the company on business grounds by line of route, and he split the whole of the operating railway into five businesses or Sectors as he termed them. These were InterCity, London and South East, Provincial, Freight, and Parcels. As an ex-LNER man, he was familiar with that company's 'Lines' management system and the Sectors were an enhanced version of that arrangement. In December 1981, he appointed directors to each Sector and these individuals had the authority to decide what services would be run and what infrastructure they would support to provide these services. At last, the railways would have a management system where income, expenditure, profit and loss were transparent on a line-by-line basis! This was a fundamental and much needed transformation.

The geographical based regions were initially left in place but, over a relatively short period of five years, the management of the railway was changed from the regional approach to one where everything was controlled by the Sector in charge of that particular section of the network. The Sectors had a primacy with InterCity at the top. Other Sectors, further down the pecking order, wishing to make use of assets owned by a prime Sector, had to negotiate arrangements to do so and they had to pay for the privilege. This ensured, for the first time since nationalisation, that assets were only retained if they were going to contribute to a particular business outcome and that changes had to be sponsored and paid for by the Sector requiring to make them. If only somebody had had the foresight to effect such a system at nationalisation, the railway industry currently would be in much better shape today!

The Sectors were responsible for the maintenance, renewal and enhancement costs by means of internal contracts with internal suppliers. This included investment electrification schemes. Therefore, if a sound business case could be made for wiring a route, the Sector could develop a proposal for authorisation.

Proposals to electrify the East Coast Main Line had been promulgated for many years. Eventually, a strong business case was put together and so InterCity secured the lion's share of electrification funding through the 1980s and early 1990s after this major scheme, together with extensions, was authorised.

In addition, Sectorisation, with its far more astute financial systems, also made it much easier for Network SouthEast, Provincial Services (later Regional Railways) and Railfreight to justify short to medium length in-fill electrification schemes. The first two Sectors in particular developed and implemented a significant number of such projects in this way. Furthermore, both Sectors were adept at initiating schemes that they could persuade third parties and stakeholders to finance.

Network SouthEast

The London and South East (LSE) Sector was set up with thirty staff at Waterloo as BR progressively moved from functional and regional-led organisations, as initiated at nationalisation in 1948, to business Sectors. The Sectors initially worked with the existing BR Regions and Functions, who initially continued to deliver the overall service and day-to-day operations and maintenance, including staffing and timetabling.

On 10 June 1986, the new Sector director, Chris Green, relaunched LSE as Network SouthEast. A bright new red, white and blue livery was introduced, and the rebranding was underpinned by the intention to invest in new trains, brighten up the stations and improve service standards.

While Network SouthEast did not at this first stage control or maintain infrastructure, it began to have a say in the scope and specification for all core service functions. As sectorisation progressed, Network SouthEast expanded its staffing structure to exercise more control over scheduling, marketing, infrastructure enhancements and rolling stock specifications.

In April 1990, Sir Robert Paul Reid (colloquially known as Bob Reid 2), the Chairman of British Rail, declared that the Regions would be disbanded within two years with the Sectors taking full control and becoming directly responsible for all operations other than a few strategic and standards functions.

Thus, in a relatively short space of time, Network SouthEast expanded from a BR business unit of about 300 staff to a major business operation with 38,000 staff, operating the inner and outer suburban and local passenger services within a hundred-mile radius or so of London.

Under its inimitable leader, Network SouthEast pioneered new 'Networker' trains and initiated and project managed route upgrades, including a significant programme of 'infill' electrification. Many of the 'branch' line services were marginal and, on their own, rarely justified the outlay for electrification. However, by replacing ageing DMUs with 'free' electric stock, it was possible to make sufficient savings in rolling stock maintenance and crewing costs to generate the requisite surplus to justify the outlay on the infrastructure works. The 'free' electric units were found by much more efficient diagramming of the various fleets that were then in operation within the NSE family of routes. By using electric trains the operating costs of running DMUs to servicing depots for fuelling and day-to-day maintenance were avoided as well as the frontline maintenance costs themselves.

The Sector thus became renowned for cascading rolling stock to meet changing demand and introducing innovative services such as the Thameslink service through the reopened Snow Hill Tunnel.

The Thameslink Project involved both 25kV AC and 750V DC electrification, including initiating an innovative voltage change-over arrangement to enable rolling stock to automatically change between AC and DC systems and vice versa. Further electrification of the local routes between Eastleigh and Fareham, the Oxted Line (East Grinstead branch) were progressed. In addition, main line schemes were also tackled with electrification of the Hastings line and the extension of electric services from Bournemouth to Weymouth.

The result of all these projects meant that virtually the whole of the old Southern Railway to the east of Basingstoke and Southampton and so forth had been converted to electric operation.

Many electrification schemes were successfully achieved therefore, and brief details of these projects follow:

Dalston to North Woolwich (DC) – 5/85

The section of the North London line between Dalston and North Woolwich was opened in stages as follows:

- The East and West India Docks and Birmingham Junction Railway (later renamed the North London Railway), opened between 1850 and 1853.
- The North London Railway section of route between Victoria Park and Hackney Wick Station to Stratford opened in 1854.

The steam worked services were replaced by diesel multiple units in the 1950s and 1960s.

In 1979, it was extended to Camden Road and branded as 'Cross-Town Link' services.

Following on from the closure of the Palace Gates to Seven Sisters line in 1964, the service to North Woolwich was cut back to a DMU shuttle between there and Stratford Low Level. The proposed closure of Broad Street presented the opportunity to link the then truncated eastern end of the North London Line service to the North Woolwich service at Dalston as well as reopening some long since closed stations at Hackney, Homerton and Hackney Wick. To do this, it needed the route from Dalston, Western Junction through to Stratford

Low Level and on to North Woolwich to be electrified on the third-rail system allowing through running from Richmond.

This electrification (from Dalston, Western Junction through a new Dalston station on to Stratford and North Woolwich) was carried out in 1983/5. It was financed by the redevelopment of the site at Broad Street station. The service from Richmond was diverted to North Woolwich at the May 1985 timetable change. The connection to the City was maintained by the interchange with the Great Northern Electric service at Highbury & Islington.

Going back to pre-grouping days, there had been a number of through trains to Broad Street from the Great Northern and the West Coast Main Line suburban areas in the peak hours. Notwithstanding what was arranged for the core North London route, it was considered essential to maintain such a facility. This was done by retaining a small number of these trains and diverting them into Liverpool Street by building a new connection from the North London Line to the Bethnal Green to Enfield Town line where the two routes intersected. This link was named the Graham Road Curve and the junctions at each end were called Navarino Road and Reading Lane respectively. This was electrified on the AC system.

There was considerable concern that this link would provide a path for the DC return currents from the trains running on the North London Line to end up in the Bethnal Green to Enfield Town line and hence reach the signalling

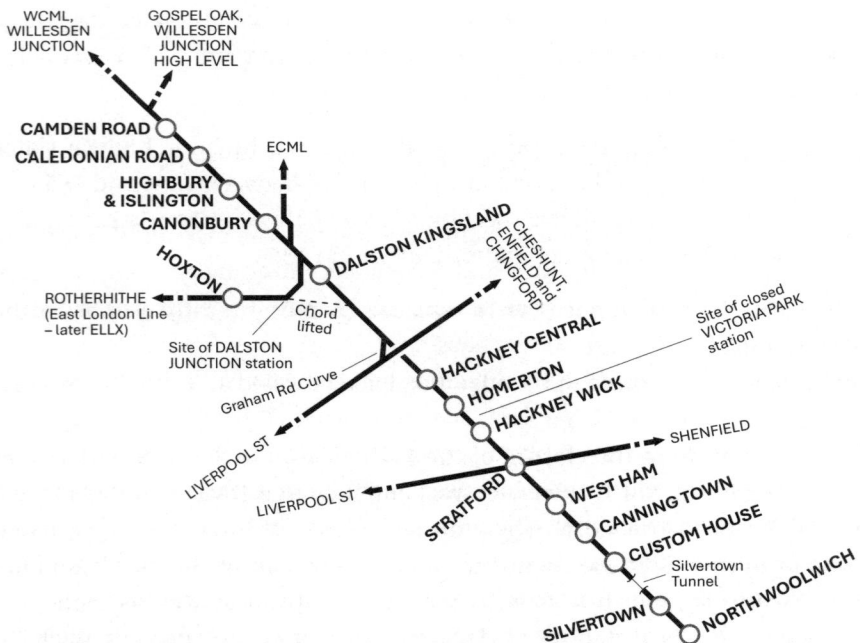

Dalston to North Woolwich

system in the Bethnal Green area, which was not immunised against such interference. Double-insulated block joints were installed in both rails at both ends of the new link to prevent this happening. In the event, this did not occur. Yes, the return currents did enter the rails of the North East London lines but measurements found that these currents ran in the opposite direction away from Bethnal Green towards Hackney Downs and through Clapton Tunnel to join the Lea Valley Line at Copper Mill Junction. They finally went to earth in the linings of the London underground tunnels, which joined the Victoria Line to its depot at Northumberland Park alongside the Lea Valley main line.

The diversion of these services on to this route extended the journey times to the City significantly with the result that the vast majority of people who used them pre-diversion found alternative routes to work, and they were withdrawn at the next timetable change.

Class 313 dual-voltage units were cascaded to the route in 1988 as these EMUs could take advantage of the 25kV AC OLE for specific sections of the route, by means of traction current changes.

The line from North Woolwich to Stratford closed on 9 December 2006 to permit its conversion to a Docklands Light Railway (DLR) line.

In 2004, it was announced that the former East London Underground line would be upgraded as part of a plan to create an orbital 'heavy rail' network round London. The upgrade was extended to Dalston Junction and, in addition, a connection would be made with the South London line at a later stage (see Chapter 14).

Tonbridge to Hastings – 5/86

The Hastings line is a secondary main line linking Hastings and Tunbridge Wells with London via Tonbridge and Sevenoaks. Running through difficult and hilly terrain in Kent and East Sussex, the railway was built by the South Eastern Railway (SER) in the early 1850s.

Due to lax supervision of the contractors during the construction, the linings of six tunnels, Bopeep, Grove Hill, Mountfield, Strawberry Hill, Wadhurst and Wells tunnels were not built to specification. In March 1855, part of the brickwork of Mountfield Tunnel collapsed and it became clear that this tunnel lining had been constructed with too few layers of brickwork. Subsequently, it was found that the other tunnels were in a similar state, the worst, Grove Hill Tunnel, having been constructed with just a single ring of brickwork!

As a result, it was necessary to strengthen the linings of these tunnels by adding additional brick rings or cast-iron segments. This restricted the loading

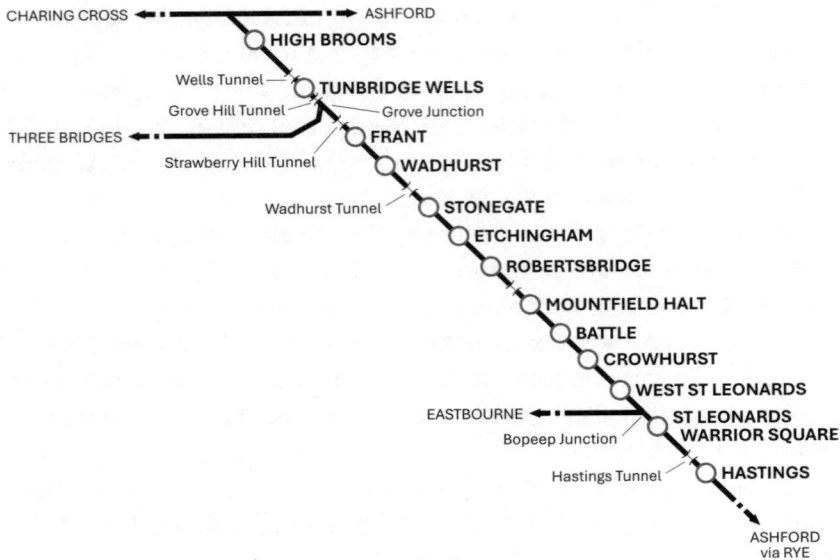

CHARING CROSS ◀━━━━━━━━━━━ ━▶ ASHFORD
◯ HIGH BROOMS
Wells Tunnel ◯ TUNBRIDGE WELLS
Grove Hill Tunnel ━ Grove Junction
THREE BRIDGES ◀━ ◯ FRANT
Strawberry Hill Tunnel ◯ WADHURST
Wadhurst Tunnel ◯ STONEGATE
◯ ETCHINGHAM
◯ ROBERTSBRIDGE
◯ MOUNTFIELD HALT
◯ BATTLE
◯ CROWHURST
◯ WEST ST LEONARDS
EASTBOURNE ◀━ ■ ST LEONARDS
Bopeep Junction ◯ WARRIOR SQUARE
Hastings Tunnel ◯ HASTINGS
ASHFORD
via RYE

Tonbridge to Hastings

gauge of the line, requiring the use of special rolling stock with a reduced profile, later known as 'Restriction 0'.

The SER first considered electrification of the line to Hastings in 1903. A shortage of finance meant that no progress had been made by the First World War. After the Grouping, the Southern Railway confirmed the plan to electrify the route, but the scheme was deferred, partly due to the fact that non-standard rolling stock would have been required.

The scheme thus failed to gain favour before the Second World War broke out, although two electric traction units built to the 'Hastings line' loading gauge were ordered by the Southern Railway in 1937.

As noted in Chapter 3, in 1946 the Southern Railway announced a programme to electrify all lines in Kent and East Sussex in three stages. The Hastings line between Tonbridge and Bopeep Junction (St Leonards) was to be part of the third stage of the plan. It was proposed that the permanent way be slewed within the affected tunnels to permit the use of stock with the standard profile. Due to the sub-standard clearance between the tracks, normally only one train would be permitted in the tunnel at any time. In an emergency, two trains could pass in the tunnel at the same time, but the speed would be restricted to 40 km/h (25 mph).

After nationalisation, the Southern Region cancelled all electrification plans and concentrated on catching up on deferred infrastructure maintenance and building steam locomotives. A further review of the utilisation of standard rolling stock on the Hastings line was again considered in 1952 involving singling the

Key

——	Unelectrified Route
——	Electrified by 2050
●	Charging Stations

1 Railway Industry Association (RIA) future electrification proposal endorsed by the co-authors. *Railway Industry Association (RIA)*

2 The first commercial electric railway – the Volks Electric Railway of 1883 at Brighton. *C J Marsden archive*

3 LBSCR senior staff and invited dignitaries sample a new electric train ride prior to the commencement of public service on 1 December 1909. *K Robertson archive*

4 Linesmen erecting the 6.6kV AC overhead line at Balham Junction just prior to the First World War. *K Robertson archive*

5 Midland Railway electric motor car and trailer traverses the River Lune Viaduct on the Lancaster–Morecambe–Heysham line in LMS days. Note bespoke portals design used on bridge. *C J Marsden archive*

6 North Eastern Railway EF1 locomotive No. 3 hauls a coal train, probably bound for Newport from Shildon. *C J Marsden archive*

7 Two three-car Metropolitan units bound for Altrincham run beneath a double portal on the 1500V DC Manchester South Junction and Altrincham Railway in the 1950s. *C J Marsden archive*

8 A train of three Great Eastern line three-car EMUs (later Class 306) runs beneath the Ilford flyover on a Liverpool Street to Shenfield service in April 1956. Note that the 1500V DC OLE is similar to that provided on the MSW route. *C J Marsden archive*

9 EM2 locomotive heads an express service on the Woodhead Line (Manchester, Sheffield and Wath route) in 1956. Note the 1500V DC MSW-style portal structure.
C J Marsden archive

10 D78 District Line London Underground train leaves Richmond on 14 October 1986 on the fourth-rail underground system. The third-rail BR SR system to the left of the picture is the Waterloo to Reading line. *C J Marsden archive*

11 BR Class 501 EMU bound for Euston (right) passes a train of London Underground 1938 stock on the shared fourth-rail BR/London Underground section of the North London line. *C J Marsden archive*

12 Networker Class 466 EMU on a Cannon Street to Orpington service on the third-rail 750V DC lines at Lewisham. Note the gaps in the conductor rail in the switches and crossings. *C J Marsden archive*

13 Class 501 EMU leaving Gunnersbury on a Broad Street to Richmond service on 19 September 1978. Note the fourth-rail electrification system. *C J Marsden archive*

14 Locomotive E3026 (later Class 83) heads a parcels train on the West Coast Main Line under BR Mk 1 compound 25kV AC OLE. *C J Marsden archive*

15 BR AM10 (later Class 310) EMU set 047 on a test WCML run beneath BR Mk 1 compound 25kV AC OLE. Note the catenary running over the boom which reduced steelwork costs but caused reliability issues. *C J Marsden archive*

16 Class 86 locomotive No. 86220 and train in BR InterCity livery at Colchester on a Liverpool Street to Norwich service on 5 April 1990 beneath BR Mk 1 simple OLE. *C J Marsden archive*

lines through the affected tunnels. However, the Operating Department objected to the singling of the tunnel sections, so the 1930s steam-hauled passenger stock was refurbished with the aim of extending its service for another ten years.

While the Southern Railway's electrification schemes were revived with the 1955 Modernisation Plan, this did not include the Hastings line. Instead, the Southern Region announced in 1956 that a fleet of 'Hastings Gauge' diesel-electric trains was to be built pending electrification sometime in the future.

By the 1980s the diesel-electric units were life expired and the London and South East Sector submitted a proposal to electrify the Hastings Line. Government approval for the scheme, involving the 50km (31 miles) of route between Tonbridge and Bopeep Junction, was given in October 1983. The scheme, managed by the London and South East Sector, later Network SouthEast, was costed at £23.93M (£80M at 2025 prices) and included the singling of the line through three tunnels to permit the passage of normal width rolling stock together with re-signalling of the route.

A full electric service was implemented from 11 May 1986, with the hourly fast services taking 84 minutes and the alternate semi-fast services 99 minutes for the 100km (62¼ miles).

No new rolling stock was procured, the services being operated by the re-diagramming of the South Eastern Division's indigenous 4-CEP, 4-CIG and 4 VEP units.

Wickford to Southminster – 5/86

This is a classic example of the LSE policy described earlier. By clever re-diagramming of the Great Eastern fleet and the avoidance of the start and end of service ECS movements to fuelling points, it was easy to justify the outlay on the electrification infrastructure and the start of the electric service was at the May 1986 timetable change. A total of 25km (16.3 miles) of route was electrified utilising Mk 3b OLE (see Appendix 1).

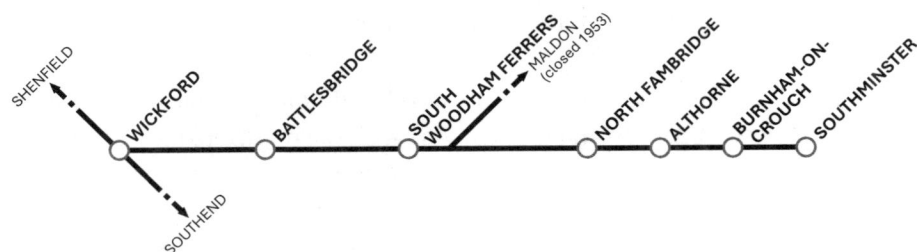

Wickford to Southminster

Romford to Upminster – 5/86

The Romford to Upminster line was constructed as a branch of the London, Tilbury and Southend Railway (LTSR). Services commenced on 7 June 1893 giving a link with the Great Eastern Railway at Romford.

Patronage declined during the 1950s and the line was proposed for closure in the 1960s but it nevertheless managed to survive. The same arguments as described for the Wickford to Southminster line applied equally to the Romford to Upminster line and so wiring of this short 5km (3.3 miles) line went ahead.

Electrification in 1985/6 put an end to speculation about the line's future. Electric services commenced on 17 April 1986. The Romford to Upminster electrification is of special note, because this is the route where the overhead mast driven tubular steel pile foundation was trialled and proved. Mk 3b OLE was suspended from these masts (see Appendix 1).

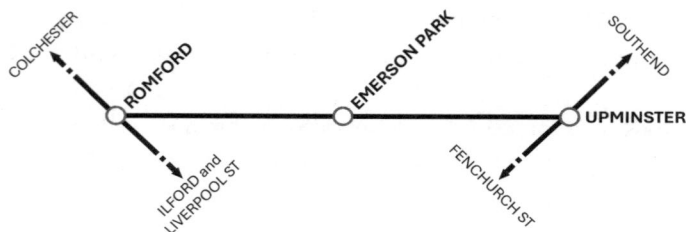

Romford to Upminster

Following privatisation in 1997, services were initially operated by First Great Eastern until 2004 when the 'one' franchise, run by National Express took over until 2012. This franchise was then taken over by the Dutch Abellio company in 2012 and renamed Abellio Greater Anglia. On 31 May 2015, operation of the Romford to Upminster line was transferred to London Overground (see Chapter 14).

Sanderstead to East Grinstead – 10/86

The former London, Brighton & South Coast and South Eastern & Chatham joint line between Croydon and Oxted, along with the branch to East Grinstead, was electrified within a similar timeframe to the Hastings line, partly using second-hand materials. A total of 28km (17¾ miles) was electrified.

The Southern Railway had electrified the Woodside and South Croydon Joint Railway in 1935, but passenger numbers never reached expectations. BR

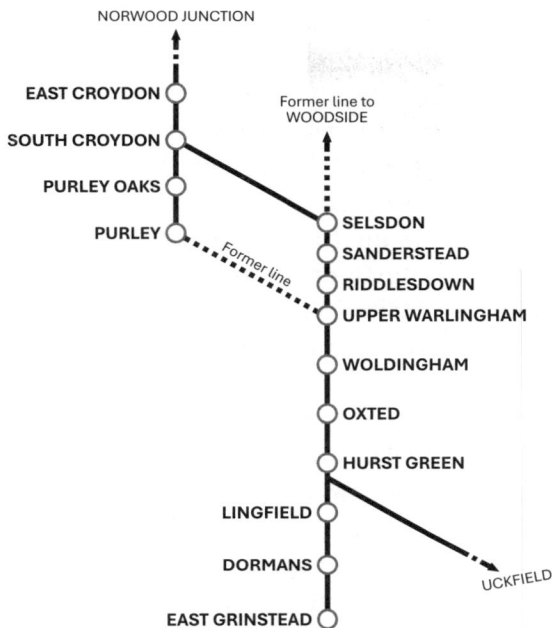

NORWOOD JUNCTION

EAST CROYDON

Former line to
WOODSIDE

SOUTH CROYDON

PURLEY OAKS

PURLEY

Former line

SELSDON
SANDERSTEAD
RIDDLESDOWN
UPPER WARLINGHAM

WOLDINGHAM

OXTED

HURST GREEN

LINGFIELD

DORMANS

UCKFIELD

EAST GRINSTEAD

East Grinstead Electrification

closed the joint line between Selsdon and Woodside on 16 May 1983, with the expectation that the route might become part of a tramway project in the future. A number of senior BR personnel were very supportive of such a change. Subsequently, in 1987, under Chris Green's stewardship, Network SouthEast published plans for a wide-ranging scheme to include a tram service from Croydon along the formation of the old joint line.

In 1983, BR set about electrifying the short section between South Croydon and Selsdon making use of some of the redundant materials from the Woodside line. A modest service of electric trains then operated from London to Sanderstead via East Croydon.

Subsequently, in 1987, the section beyond Sanderstead to East Grinstead was also electrified at 750V DC third rail. Again, rolling stock was provided by re-diagramming the Sector's existing fleet of EMUs.

Bishop's Stortford to Cambridge – 1/87

With the original North East London Lines electrification stopping at Bishop's Stortford, all London (Liverpool Street) to Cambridge and beyond trains were diesel-hauled and passengers wishing to travel from south of Bishop's Stortford to north thereof had to change there into the DMUs providing the local services to Cambridge.

Neither of these arrangements were very attractive commercially. The locomotives and DMUs were ageing and had very slow point-to-point running times, which, of course, also restricted the capacity of the route.

Therefore, a financial case was made for extending the wires from Bishop's Stortford to Cambridge and replacing all of the DMUs with EMUs from the North East London fleet to enable through services to be provided between the various local stations. It was also foreseen that the construction of a new railway to Stansted Airport would be required in the future and that this would have to be electrified in order to offer a service that would be attractive to air travellers.

In 1981, therefore, a proposal was submitted to the government to extend electrification of the Great Eastern Lea Valley Main Line from Bishop's Stortford to Cambridge (Anglia West). Approval was granted for the scheme costing £9.3M (£37M at 2025 prices) on 16 January 1984.

The wiring and clearance works were carried out in the mid-1980s utilising Mk 3b OLE, and that section of the route was energised in time for an electric service to be provided in January 1987. The length of route wired was 28km (17¾ miles).

Bournemouth to Weymouth – 5/87

Electrification was extended to Weymouth, a further 54km (34¾ miles) in 1987. To keep the costs down, the spacings of the substations was stretched to the limit, aided by the utilisation of a new, heavier conductor rail (150lb/yd, previously 106lb/yd). In addition, this more rigid conductor rail enabled the spacing of the supporting insulating pots to be extended from every four sleepers to every six, thus enabling further cost savings to be achieved.

However, the lengthened distance between substations limited the number of electric trains that could operate at any one time west of Poole and further

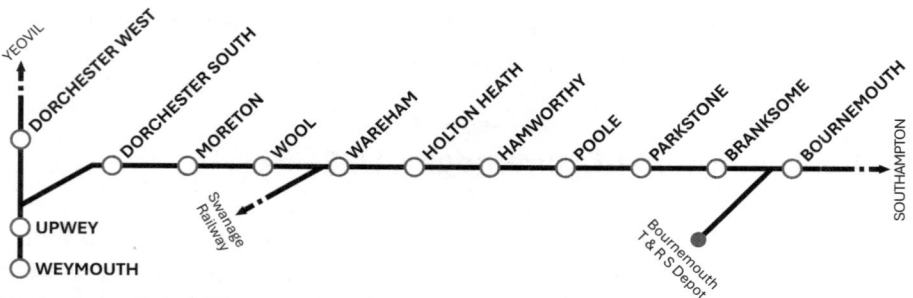

Bournemouth (Branksome) to Weymouth

restricted train length to no more than eight cars (2 × 4VEP/4CIG or similar or a new Class 442).

The electrification also saw the introduction of the new Wessex Electric Class 442 units. Twenty-four five-car units were built by British Rail Engineering Limited at Derby Litchurch Lane Works.

The elimination of the traction change at Bournemouth, together with the improved performance from electric operation resulted in a noticeable improvement in journey time west of Bournemouth. The new trains also had a greatly improved ambience for passengers and the net result was a significant impact on revenue. Dorset became commuter country!

Royston to Cambridge and the link to Stansted Airport – 5/87

The route from King's Cross to Cambridge in steam days had been much faster than that from Liverpool Street. With Cambridge station wired under the Lea Valley extension from Bishop's Stortford to Cambridge, the cost of extending the King's Cross to Royston Great Northern suburban system to Cambridge was minimal as it was an open route that had already been resignalled.

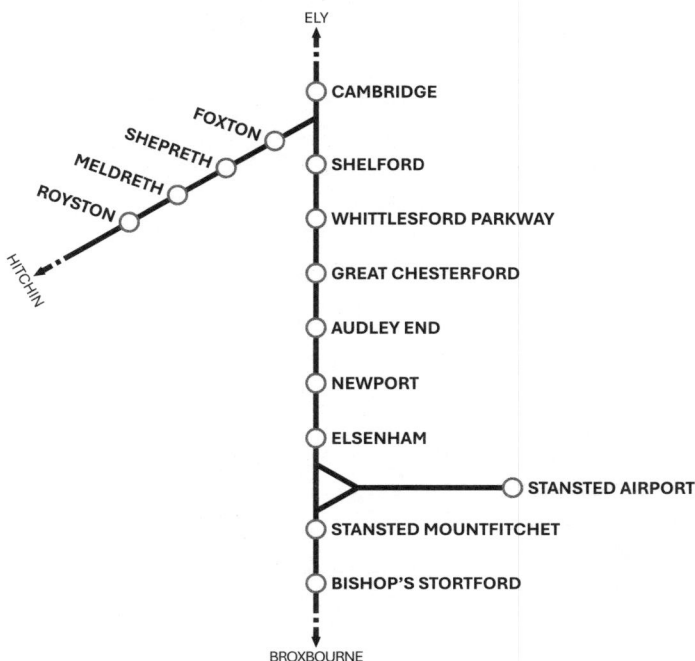

Bishop's Stortford to Cambridge and Royston to Cambridge (Stansted Airport)

The length of line to be wired was a mere 15km (9.5 miles) and, once authorised, the work was expeditiously undertaken utilising Mk 3b OLE (see Appendix 1). The new electric service was introduced at the May 1987 timetable change. This short length of electrification enabled the introduction of an hourly London to Cambridge non-stop service, which proved extremely popular. By the extension of the Great Northern Royston terminating services, it also facilitated the withdrawal of the Royston to Cambridge shuttle trains, which had provided the local service.

Great Eastern Extensions: Colchester to Ipswich, Harwich and Norwich – 5/87

Prior to the start of the Joint Review in 1978, a proposal for 25kV AC electrification, track rationalisation and resignalling between Colchester and Norwich (Anglia East) was submitted to the Department of Transport. The scheme was approved in December 1981.

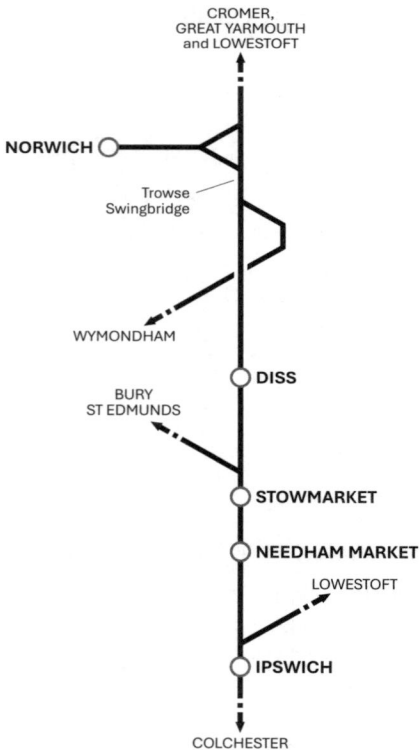

Ipswich to Norwich

The civil engineering work for Anglia East included the raising/reconstruction of 64 overbridges and the provision of slab track through Ipswich tunnel for an estimated cost of £85M (£340M at 2025 prices). Albeit under a separate authority, but completely associated with the main electrification project, was the reconstruction of Trowse swing bridge. This required a novel mode of electrification; the rest of the project utilised the Mk 3b system (see Appendix 1).

Testing, commissioning and crew training was underway in 1984, and the first electric-powered test train, consisting of two Class 305 EMUs, ran into Ipswich station on 9 April 1985. The first passenger carrying services, formed of Class 309 units, commenced on 17 April 1985.

From that date, through express trains to Norwich were hauled by Class 86 units to Ipswich, where Class 47 diesel-electric locomotives took over for the rest of the journey to Norwich. Work continued on the electrification of the line to Norwich for a further two years, with through electric services finally introduced in May 1987. Control of the newly electrified lines was exercised from Romford Electrical Control Room. The scheme also encompassed electrification of the branch from Witham to Braintree. The Anglia East project electrified 126km (78¾ route miles).

Thameslink – 4/88

In an article published in the London *Star* evening newspaper on 14 June 1941, George Dow, a well-known senior railwayman, proposed that new rail routes be built in tunnels under Central London. These included two north to south and an east to west route.

Following this, the post-war Central London review, referred to earlier, endorsed the north–south and east–west rail tunnels among a range of other transport infrastructure enhancements. The Underground Victoria Line constructed in the 1960s was the first line to fulfil George Dow's dream. More than forty years would elapse before Thameslink and Crossrail brought his other two proposals into focus.

The infrastructure for the Thameslink connection was already in place in the form of the original 'South London Line' from Farringdon through the Snow Hill Tunnel to Blackfriars. However, the line had been closed to passenger services since 1916, and freight services were withdrawn in 1970.

The Thameslink plan was to reopen the South London Line and divert most of the Midland suburban electrification services, then terminating at Moorgate, to run through the reinstated link and on to a number of destinations on the Southern Region using dual-voltage EMUs.

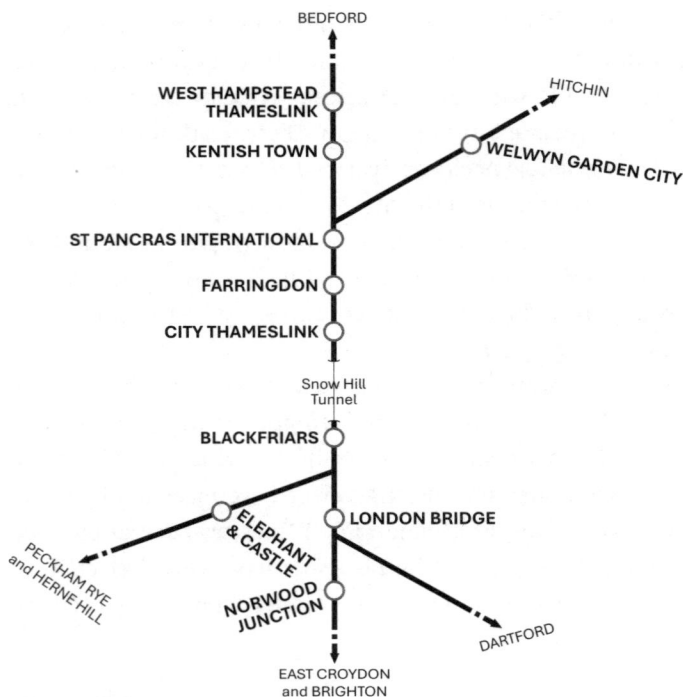

Thameslink

Accordingly, the Snow Hill Tunnel was recommissioned and the route reopened by Princess Anne on 25 April 1988. The first train was a school special from Bedford to Brighton.

The 750 yards long Snow Hill Tunnel cost £1.4M (£4.5M at 2025 prices) to reopen. Within a matter of a few months the route was carrying 20,000 passengers a day and it brought an initial 11 per cent growth to the Bedford–St Pancras route. A new station was provided approximately midway along this resurrected link, named City Thameslink, and this was opened at Easter, 1990. The station cost more than £50M (£160M at 2025 prices) to construct and was paid for by the developer of the property that was built over it. The platform works were completed during a sixteen-day possession.

Associated with this project was the closing of Holborn Viaduct high level station, replacing the old route to Farringdon and constructing a new low level route with a very significant gradient immediately to the north of Blackfriars.

The project also contained two stabling sidings to the north of City Thameslink station, equipped only for DC operation, which could hold four-car trains. The changeover from AC to DC traction power supply and vice versa was carried out at Farringdon station while the train is stationary, usually accomplished in less than two minutes.

A total of 86 dual-voltage Class 319 four-car EMUs were procured between 1987 and 1990 to provide this service. Initially, the route through to Moorgate remained operational with a few trains running over it in the peak hours. Subsequently, when the maximum permitted length of train was extended from eight to twelve cars, the route to Moorgate had to be abandoned so as to permit the extension of the platforms at Farringdon.

Watford to St Albans – 10/88

This is yet another example of the efficiencies of electric operation generating sufficient savings to pay for the outlay of the Mk 3b electrification infrastructure. A significant part of the saving came from removing the need to run an empty DMU from Bletchley maintenance depot to Watford and back each day – these moves being required for fuelling and servicing purpose whereas the EMU could be stabled overnight in a siding alongside Watford station.

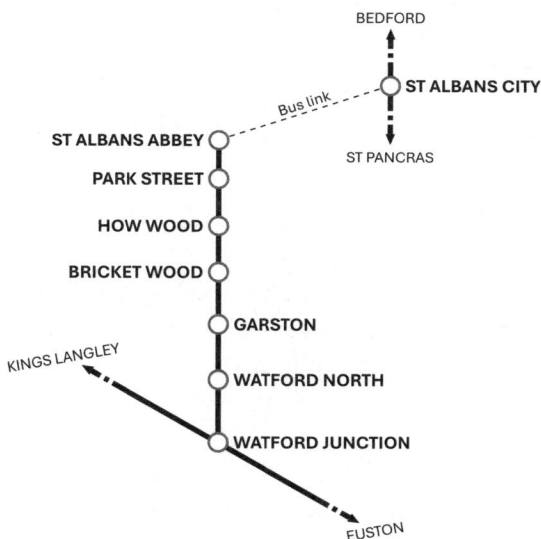

Watford Junction to St Albans

St Denys/Eastleigh to Portsmouth – 5/90

The Eastleigh to Fareham Line links Eastleigh on the South Western Main Line with Fareham on, what is now known as, the West Coastway line, joining Southampton to Portsmouth and Brighton. The lines were electrified under Network

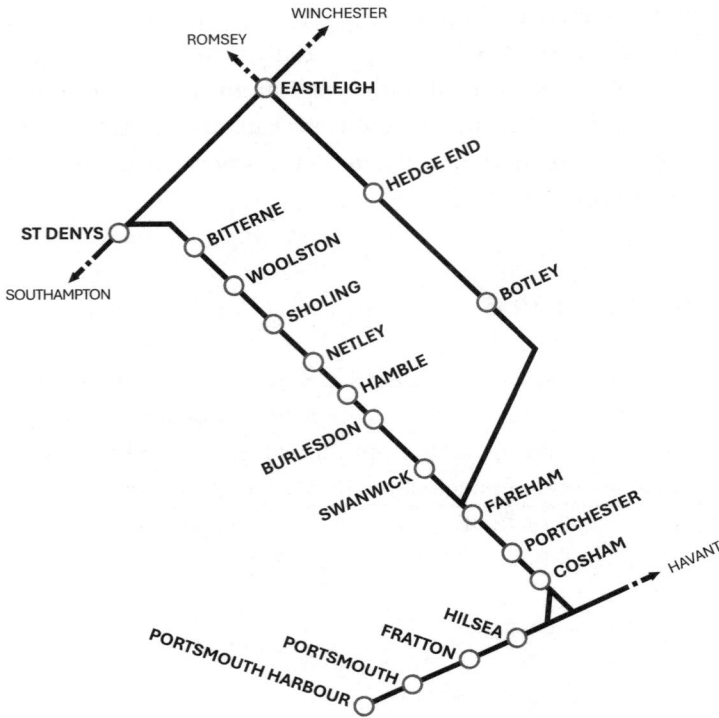

St Denys / Eastleigh to Portsmouth

SouthEast auspices with third rail at 750V DC in the late 1980s, and the electric service was introduced in the spring of 1990.

This scheme, termed SHES, infilled the Eastleigh/Portcreek Junction and the St Denys triangle, totalling circa 48km (30 miles). The original aspiration was to electrify to Salisbury via both Redbridge and Eastleigh, but the early 1990s financial recession and then the impending privatisation of BR scotched any hopes of achieving that goal.

A section of the scheme, between Eastleigh and Botley, was fitted with aluminium conductor rail with a strip steel contact surface. Prior to this, a short section of this conductor rail had been trialled on the Up Fast line at Woking. However, the section installed between Eastleigh and Botley had to be replaced as the steel contact surface peeled off in traffic. (Note, however, that this type of conductor rail has since been perfected and approved for use).

The route, usually served by an hourly service, is a useful diversionary route, when the lines around Southampton or the Portsmouth Direct lines are closed for engineering work or other out-of-course reasons.

Cambridge to King's Lynn – 8/92

With the electrification of both routes to Cambridge (from King's Cross and Liverpool Street), the only non-electric operation over either of these routes was the diesel loco hauled service to King's Lynn. This is 66.9km (41 miles 47 chains) north of Cambridge.

This was expensive to run, and the quality of the Mk II rolling stock was not of the level that NSE wished to provide for its customers. It also involved the running-round of locomotives at King's Lynn, which required extra staff.

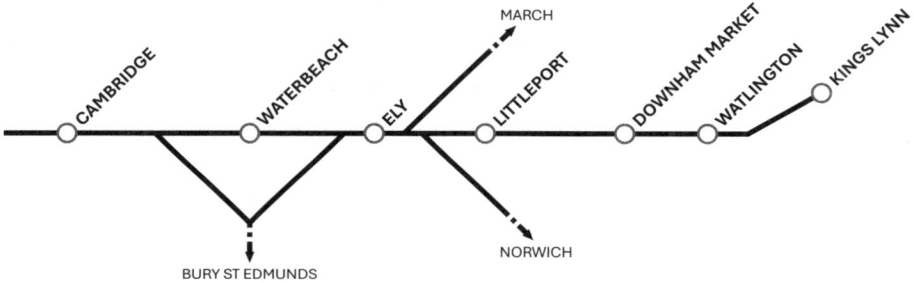

Cambridge to Kings Lynn

After an in-depth review of all of the costs associated with the operation, and taking the appropriate step to minimise these, it proved possible to make a suitable financial case for extending the overhead lines from Cambridge North through Ely to King's Lynn. Associated with this was the singling of two sections of the route north of Ely so as to reduce ongoing civil engineering maintenance costs and the outlay on the Mk 3b overhead line equipment. In addition, neutral sections were permitted at Milton Fen, 5km (3 miles) north of Cambridge, and just north of the Littleport bypass to save structure reconstruction costs. The supervision of the electrical power supply was provided from the existing Romford Electrical Control Room.

The new service utilising EMUs from the fleets already providing the King's Cross to Cambridge and Liverpool Street to Cambridge services opened in August 1992.

Stansted Airport Line – 9/91

After many years of campaigning for such a facility, an electrified rail link was provided to Stansted Airport opening in 1991. The new line utilising Mk 3b OLE comprised a triangular junction off the Lea Valley main line about 1 mile north of Stansted Station, which then went straight into a tunnel just under the main runway to a new station in the open air adjacent to the reconstructed terminal building. Construction work started in 1989 and was completed by the autumn of 1991. The tunnel was constructed to accommodate a second track should the traffic growth require it at a date in the future. The new line and station were constructed at the same time as the Airport Authority rebuilt its main terminal.

When the tunnel was completed, at the request of the Railway Inspectorate, a trial was held to check on the arrangements for escape of passengers and staff in the event it became necessary to evacuate a train in the tunnel. This was carried out with the aid of a large number of volunteers from NSE and the contractors involved in the construction of the new line and was completed to the satisfaction of the Inspecting Officer concerned.

The West London Line – 5/94

The West London Railway (WLR), linking Willesden to Kensington, was opened on 27 May 1844, but was closed to passenger services after only six months due to poor patronage, although it remained open for freight traffic.

Subsequently, powers were given jointly to the London, Brighton & South Coast Railway and the London & South Western Railway to extend the West London Railway to Clapham Junction. This section of line opened on 2 March 1863 with passenger services restored, together with a new passenger station located at Addison Road, which was later renamed Kensington Olympia. The section of route between Willesden Junction and Earls Court via Kensington Olympia was electrified in 1914/15 by the London and North Western Railway (see Chapter 2).

Patronage fell in the early part of the twentieth century due mainly to competition from the electric tramways and the Underground tube lines. In 1940, the line suffered significant bomb damage during the London Blitz so that regular steam and electric passenger services were withdrawn during October of that year. The route remained busy with wartime freight services, however.

The post-war public timetables showed a limited Clapham Junction–Kensington Olympia service, but a regular passenger service beyond there was not reintroduced until Network SouthEast did so in 1994 following third-rail electrification. At privatisation in 1997, the West London Lines, along with the Richmond to North Woolwich line, were taken over by National Express and marketed as 'Silverlink' (see also Chapter 14).

Heathrow Express – 6/98

This scheme was carried out by Network SouthEast for the British Airports Authority, which obtained the Parliamentary Powers to build a branch railway from Stockley Junction (just to the west of Hayes on the Great Western Main Line) to under the central terminal area within Heathrow Airport.

Heathrow Airport was a Second World War airfield that was converted into a civil airport at the end of hostilities. It quickly became the main airport for London and had a massive growth of traffic in the late 1950s/early 1960s. With an ever increasing range of destinations and numbers of flights it was not just the airport for London but became an international hub that resulted in very large numbers of passengers departing and arriving. Because of its location, road access was congested for much of the day for many passengers going to or from the airport. In addition, because of the ever increasing transit times from London, such journeys became very expensive. As a result of this, from the 1970s onwards, there was ever growing public pressure for a rail link, which was, in part, satisfied by London Underground extending its Piccadilly Line from Hounslow West into a terminal loop that served the then three airport terminals.

Notwithstanding the improvement that the underground link provided regarding access from Central London, the transit times of the tube and over-crowding in peak periods were unacceptable, and eventually the British Airports Authority was persuaded to act and to approach British Railways to provide a surface railway link. After consultation with BR, the BAA promoted a scheme for a new railway from a location designated as Stockley Junction (some 1.6km – 1 mile – west of Hayes) to the central terminal area (CTA), which was to be mostly in tunnel underneath the airport. The initial length of the tunnel was bored through a former municipal rubbish tip and there were concerns that the

Heathrow Express (Paddington to Stockley Junction)

decomposing rubbish would be emitting methane gas that would endanger the works and the tunnels.

The scheme included electrification at 25kV AC of the GW main line from Paddington out to Stockley Junction and of the new railway to the CTA. The wiring included all the running lines, utilising Mk 3b OLE, between Paddington and Stockley Junction but three of the platforms at Paddington were excluded due to concerns over electrical clearances. A maintenance depot for the fleet of special trains was built at Old Oak Common and fourteen Class 332 EMUs were purchased from Siemens. An interesting point is that because it was such a short length of running line, it was not worth establishing a separate Great Western overhead line maintenance unit and arrangements were made for the West Coast Main Line overhead line maintenance unit at Willesden to be responsible for this section. In addition, the power was obtained by the provision of underground cables from the main feeder station at Acton on the WCML.

Work on the electrification of the first 18km (11 miles) of the GWML together with the construction of new tunnels began in 1993. The new Austrian tunnelling method was adopted for the station works in the CTA and, unfortunately, despite all of the appropriate monitoring being in place, there was an incident when a section of the tunnelling collapsed. Fortunately, there were no injuries or loss of life but a large office building in the CTA collapsed into the resulting cavity together with a staff car park. As can be imagined, this caused a lot of disruption.

BAA were responsible for the five-mile underground section from Airport Junction to the Airport Terminals. In 1999, BAA's civil engineering contractor Balfour Beatty and their Austrian NATM tunnelling advisor, Geoconsult, were prosecuted.

The line was controlled from the Slough power signal box (PSB) and is now controlled from Didcot, Thames Valley signalling centre following the closure of the Slough PSB. The line is equipped with the Great Western ATP system.

There was one unique thing about this project on the main railway itself. In an attempt to reduce the environmental impact on adjacent residential housing to the west of Hayes Station caused by the noise from the trains, a 'Devon Wall' was built on the south side of the railway. (This is a noise barrier made from soil covered in turf.)

The initial service comprised four trains an hour non-stop between Paddington and the CTA, and the service was inaugurated on 23 June 1998 by the then Prime Minster, Tony Blair. Later, a local service stopping at all stations was provided principally for airport staff and this was called Heathrow Connect.

Network SouthEast Progress Cut Short by Privatisation

Network SouthEast were keen to continue with infill electrification schemes, but their efforts were thwarted by the government's privatisation programme. Thus, a number of isolated lines were left with diesel working. These were:

West of England Main Line Basingstoke, Worting Junction to Salisbury: 53km (33.25 miles)

Salisbury to Redbridge: 33km (21 miles)

Romsey to Eastleigh: 11km (7 miles)

The North Downs Line – this was already partially electrified. The gaps due to be converted were from Reigate to Guildford 28km (17.5 miles); and Aldershot South Junction to Wokingham 17.5km (11 miles), making a total of 45.5km (28.5 miles).

The Marshlink Line – Ashford to Ore: 40km (25 miles).

Oxted – Uckfield Branch: 41km (25.5 miles).

Railfreight and Eurostar
North London Line Electrification – 5/87

As described in Chapter 2, the North London Line consisted of three sections. The main or core part of the network was from Broad Street and North Woolwich to Dalston and on to Canonbury and Camden Town, joining the West Coast Main Line at Camden Junction. It had two significant branches from Homerton down to the Docks and from Camden Road through Kentish Town, under Hampstead Heath and on to Willesden High Level, terminating at Richmond with connections at Willesden down to the West Coast Main Line, and also on to the West London Line. There was also a branch from Canonbury to join the East Coast Main Line at Finsbury Park. The original DC third-rail scheme eventually ran from Broad Street through to Richmond with a connection to the LNWR main line at Camden and also at Willesden.

From 1983 to 1985 the DC system was extended from Dalston Junction to Stratford and on to North Woolwich with a major increase in the passenger service from two trains per hour to six trains per hour throughout the running day from 1985. (See also Chapter 2.)

Since the completion of the West Coast Main Line electrification in the mid-1960s, all the freight trains coming off the North London Line and going northwards from Willesden/Wembley had to change from diesel to electric locomotives for electric haulage going north and vice versa. This was obviously inefficient and wasteful of staff and resources.

With the completion of the electrification of the Great Eastern Main Line to Ipswich and the Harwich Branch, the opportunity presented itself for through electric working from the Haven Ports. All that was required was the wiring of the gap between Stratford and Willesden. This was the genesis for the North London Line AC Electrification Project.

The scheme also included the link to Finsbury Park and the North London incline at Camden, requiring a total of 16km (10 route miles) to be wired, all utilising Mk 3b OLE (see Appendix 1).

Subsequently, to facilitate the operation of Eurostar services from the Midlands and the North through to the Channel Tunnel, Railtrack were charged with providing an electrified route from the North Pole depot to the East Coast Main Line. To facilitate this, it was necessary for the section of the North London Line between Willesden and Camden Road via Hampstead Heath to be electrified and clearances to be provided for the Class 373/2 units so that they could move from the North Pole depot to the ECML. These works were carried out in two completely separate stages. The first one was done by British Railways in the period 1985 to 1989 and the second stage was done by Railtrack in the period October 1995 to September 1996.

For the first stage, based on saving of twenty-five diesel locomotives at the time of authorisation, together with avoiding certain train crew changes and shunting costs in the Willesden/Wembley area, Railfreight made a convincing case for the electrification of the route from Stratford through to Camden Road, including the line from Canonbury Junction to Finsbury Park, which followed twelve months later. The financial outlay was authorised in August 1984 at a cost of £12.1M (£47M at 2025 prices) with a further sum of £0.42M (£1.7M at 2025 prices) to cover the cost of the link to the East Coast Main Line in September 1985.

The scheme involved major civil engineering works, which were all carried out by contract except where track lowering was involved. In addition, there was the provision of the overhead line equipment, again by contract, whereas the immunisation of the existing signalling systems was carried out by internal labour. There was not a need for any power supplies, as there was adequate power available at both ends of the route. It included the purchase of four Class 87/2 electric locomotives (these were later reclassified as Class 90). It also included the wiring of the recently completed Graham Road curve together with the installation of a harmonic filter to minimise the flow of return traction currents between the North London Line and the Great Eastern North East London Suburban electrification systems.

For the second stage, undertaken between October 1995 and September 1996, the electrification works between Camden Road and Willesden were done to a minimum specification just sufficient to permit the 373/2s to operate but the opportunity was not taken to improve clearances to permit the running of most freight trains. The works required the closure of this section of route during most of this period so as to enable slab track to be installed in Hampstead Tunnel. At Willesden the route splits into four and the wiring was extended along three of them – to Mitre Bridge Junction on the West London Line, to Acton Central on the Richmond route and down the bank to the West Coast Main Line at Harlesden Junction. As a result, a further 21km (13 route miles) were wired.

It is interesting to note that with the progression of the AC works, piecemeal recoveries were made of the third-rail DC system, which required the dual-voltage Class 313 units to make a number of traction changes when operating on the lines between Richmond and Stratford. Inevitably sometimes drivers forgot to drop the pantograph so bridge strikes were not uncommon!

InterCity Sector

Electrification of the East Coast Main Line – 8/88 and 7/91

As one of the authors was the Project Director for this scheme, a somewhat more detailed account follows.

Background

Arising from the 1978–81 Review (see Chapter 7), the most favourable return on capital was the electrification of the ECML route from London to Edinburgh, including Doncaster to Leeds.

In early 1981 the then BR Chief Executive, Ian Campbell, decided that the board should act and, in April of that year, he asked the Major Projects Department to nominate a Project Manager to commence working up the investment business case. The Chief Projects Officer appointed Donald Heath who immediately set to work by obtaining a sponsor and developing the outline specification for the project.

It was clearly an InterCity scheme so the newly appointed Managing Director, Cyril Bleasdale, was asked to collaborate in the specification process. (Note: In pure project management terms, this was the wrong way round, but it was the way that the then Chief Executive wanted the process to be done.)

The summer of 1981 was spent defining the scope of the project and setting in hand the necessary studies to develop a timetable that was the fundamental

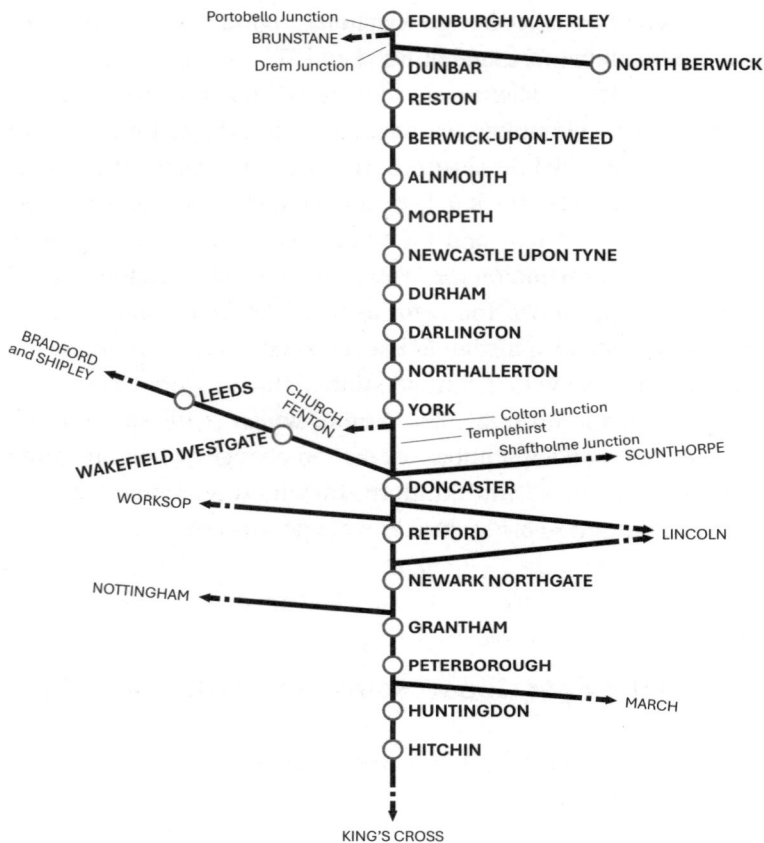

East Coast Main Line

building block for determining power supply capacity, rolling stock requirements and depot facilities. The timetable then also determined the lines to be wired and this in turn identified the structures and other items that had to be altered to provide electrification clearances.

The financial impact of all these various activities was worked out in detail by hand – it was done before the time of desktop computers! A total of more than one quarter of a million calculations were carried out by the Finance Officer in evaluating the overall financial effect of the proposed changes.

A comprehensive overall investment submission was prepared, and this was cleared through the Investment Committee in 1982 and then forwarded on to the Department of Transport (DoT). This produced the inevitable list of questions that the project team then responded to. In parallel, a second submission was prepared using the same financial information but based on only electrifying to Leeds and Newcastle. It was felt that this should be evaluated because it was a lower outlay, but it quickly showed that going right the way through to Edinburgh was a significantly more favourable option.

As is described in Chapter 16, the detailed arrangements for the authorisation of major projects were changed by the government early in 1984. The ECML electrification project was the first major project to go through under this procedure and the Secretary of State gave his consent just before the end of July that year.

Similarly, the electrification contractors' team that had been busily engaged on completing the St Pancras to Bedford project was about to be disbanded in late winter/early spring 1983/84, and it would obviously be an advantage to maintain the skills that this team had developed. To do this, an application was made to the Investment Committee to allow a small cohort of this group to be retained and to start installing foundations on the section from Hitchin northwards on to Huntingdon. Reluctantly, the Investment Committee agreed that half of this work could be carried out and the fact that this happened was subsequently of great benefit to the start of the main scheme. However, it completely failed in its purpose as the group ran out of work and had to be disbanded some three months before the main project authority was obtained.

Mobilisation

The main authority was received on 27 July 1984. The authority comprised the following:

Description	£M (Q4, 1983)	£M (Q4, 1983)
Infrastructure clearances and protection	35.400	
Immunisation	26.990	
Electrification fixed equipment	95.541	
Supply points	10.186	
Electrification construction depots	2.996	
Electrification maintenance depots	2.371	
T&RS Maintenance Depots	0.156	
Track and signalling alterations at Edinburgh	0.584	
Total Infrastructure		**174.224**
31 × Class 89 locomotives plus spares	32.719	
31 × Class 87/2 locomotives plus spares	29.155	
31 × Mk III DVTs plus spares	8.868	
31 × Mk III TRUB plus spares	8.238	
252 × Mk III Day coaches plus spares	49.057	
4 × 4-car 317 EMUs plus spares	4.038	
Total Traction and Rolling Stock		**132.075**
TOTAL AUTHORITY ISSUED		**306.299**
		(£1200M at 2025 prices)

The principal electrification contractor was awarded an initial contract under the allocated contracts procedure. In anticipation of authority for electrification works eventually being received, the Eastern Region Civil Engineer had carried out some site surveys, design work and preparation of contract documents for the demolition and reconstruction of overbridges to provide electrification clearances. This enabled contracts for a number of demolition projects to be awarded very quickly after authority was received with the first two demolitions taking place within a month of start of work. The Signal Engineer had collaborated fully in the preparation of estimates of the costs of carrying out the works but had not expected authority to be received so quickly. This meant that there was a lot of 'catching up' to do but, fortunately, the change from the proposed T121 track circuits to the 'single rail DC' type had reduced the department's workload.

Description of Works – Civil Engineering

The provision of electrification clearance at overbridges was by far and away the biggest workload. The Eastern Region Civil Engineer divided the route into southern and northern sections with his King's Cross design office responsible for the former and the York design office responsible for the works in the latter. The boundary was set at Shaftholme Junction. The King's Cross office had had a period of being under-utilised with the result that they had done a lot of preparation in the anticipation that the electrification project would eventually be authorised. This meant that a considerable amount of the pre-design survey work had already been completed so that detailed design and the preparation of contract documents could be put in hand immediately after the authority was received. The York office were not so far advanced; however, as the plan for the project works was to extend the energisation from London (Hitchin) northwards in stages, this gave York more than two years to obtain the necessary survey information to enable the detailed design work to be progressed.

In addition to the provision of electrification clearances at overbridges, there were also station awnings that had to be altered and, on the odd occasion, signal structures. This last item involved collaboration between the two departments. On top of this was the insertion of the insulated block joints where the aster track circuits were being replaced, which was another item requiring co-ordination. Finally, in the clearance works, was the cutting back of vegetation so as to provide a 2.85m (9ft 6") clear corridor from the centre line of the outermost tracks to be wired over the whole length of the route.

Generally, clearances for the overhead wires beneath overbridges were normally provided by raising the bridge. In one or two cases where this was not possible, such as an aqueduct, the solution would be to lower the track.

Another difficult location was overbridge number 4 on the approaches to Edinburgh. This structure carried a major road junction and was full of pipes and cables, so a track lower was the only practical solution. Fortunately, there was a short diversionary route available, which enabled the section of line under the bridge to be closed for several weeks while a track lowering was carried out and an acceptable drainage system installed.

Two other interesting solutions to the clearance problem were obtained at the aqueduct at Alconbury and at Bridge 325 immediately south of Doncaster Station. At Alconbury (bridge 158), it had originally been proposed to lower the track by 12 inches, which would have required a half mile plus temporary speed restriction and probably caused a long-term drainage problem. The track lowering was avoided by removing the girders that supported the actual trough carrying the water course. This seemingly impossible feat was achieved by making the trough the structural member itself. Bridge 325 was a seven-arch overbridge but only four of the arches were used. To avoid completely reconstructing the whole bridge, two of the unused arches were taken to build buttresses in which enabled the two centre spans to be demolished without affecting the stability of the remaining five spans. A further innovation was that part of the bridge was cut in half longitudinally and a half was demolished at a time to enable the bridge to be kept open to road traffic while the other half was being replaced.

Description of Works – Electrification

While the civil and S&T engineers were able to carry out all of their activities within their existing depots and design offices, the Electrification Engineer had no such readymade works infrastructure, and the first job was to establish construction depots. Four of these were eventually built located at Peterborough (Eastfield), Doncaster (Hexthorpe Road), Newcastle (Heaton) and Edinburgh (Millerhill). Each depot contained an office block divided into two – one part for the BR Electrification Engineer and the other part for the contractors' engineers. Outside there was a secure storage compound for high-value items such as cables and an erecting shed where the headspans could be laid out and built up before being loaded on to a construction train and taken to site. There were also bunkers for cement and aggregate and a water supply for servicing the foundation construction train.

The overhead system design (OSD) was carried out at the Electrification Engineer's Derby design office and utilised the Mk 3 OLE system. The completed designs were based on surveys of the track as it was at that time, with proposed alterations to track layout being encompassed within the overall survey data.

The completed designs were then sent to the relevant construction depot where the positions of the foundations were set out by hand along the route. Once this had been done for a section, then a foundation construction train would go out to dig holes into which a concrete block was cast containing a polystyrene former placed in the centre of the top half of the block. The polystyrene was then subsequently dissolved, providing the hole for the steel mast to be inserted; the mast was then grouted into position.

During the course of the first stage of the project a new type of foundation was developed for use in soft granular and cohesive soils. This involved driving a hollow circular pile into the ground with the mast being bolted straight to the top of the pile. Not only was this cheaper than the conventional excavated mass concrete type but it was much quicker to install – and typically it increased the daily output by a factor of three to four.

Once the mast was in place it was then dressed with the small-part steel work including cantilevers and fixings/insulators for the earth/return current wires. The final stage for the single cantilever mast was then the running out of the wires. For multi-track sections where headspans were going to be used, once the masts on both sides of the line had been erected, the headspans were then winched into position by the use of pulleys temporarily fixed to the masts. Once the headspans were in position it was then possible to run out the catenary and contact wires for each track in turn. The final stage was then the accurate registration of the contact wire in relation to the track beneath it and this was carried out from ladders. This was followed by the running of a test pantograph underneath it at slow speed to confirm that it was in the correct position.

In parallel, track sectioning cabins were built at the lineside along the route as required by the overhead system design, as were the foundations for the feeder stations that were going to take the current from the CEGB.

Description of Works – S&T Immunisation

To protect the telecoms circuits from the adverse effect of the 25kV return currents, the main cables were replaced with a fibre-optic cable. The railway required a total of four fibres to meet all its internal requirements and a decision was taken to install a ten-fibre cable leaving six fibres available to hire to outside parties.

The provision of an optical fibre cable enabled the use of a technology known as 'drop and insert (D&I)' to be used north of York. This had significant benefits in terms of reducing the amount of immunisation work that had to be carried out to the local circuits. The principle was to allow individual time slots and channels to be extracted or inserted without de-multiplexing the whole 2Mbit thirty channel system. This would occur at intermediate locations involving signal post telephones (SPTs), level crossing phones, SSI signalling circuits, NRN control lines and level crossing CCTV surveillance equipment.

The lineside equipment consisted of small REBs, complete with power supply, into which the D&I equipment was installed. These were spaced roughly 5km apart, thus enabling copper circuits to be connected from these points, having lengths sufficiently short so as not to be a problem with induced voltages from the 25kV overhead lines.

On the signalling side all of the equipment in the King's Cross (north of Hitchin), Peterborough, Doncaster and Leeds PSBs had to be protected from the 25kV return currents. The biggest task was the replacement of the existing aster track circuits and some of the lineside equipment that had been installed when the first three of these resignalling schemes were carried out on the route in the 1970s.

From Templehurst through to Colton Junction the signalling on the Selby diversion had been installed completely immunised so there was only minimal work to be done there. However, from Colton Junction through to the Scottish Border, the 1950s era signalling was replaced in modern form independent of the electrification.

Parliamentary Bill

During the detailed development of the project prior to authorisation, a number of opportunities were identified where cost savings could be made if certain bridges were closed. Also, there were items where it was going to be necessary to acquire small parcels of additional land to facilitate the works. To do these things, the acquisition of Parliamentary Powers was going to be very helpful, and appropriate items were inserted in the 1984 BR Parliamentary Bill. This received the Royal Assent in the summer of 1985.

Listed Building Consent

The 1948 Planning Act introduced legislation that resulted in buildings and structures of historical interest being listed. There were four classes of listing, namely, Grade I, Grade II*, Grade II and a building within a conservation area. The actual implementation of the act was carried out over many years, and it wasn't until the early 1980s that it was specifically applied to British Railways. The East Coast Electrification Project was the first major infrastructure project on BR to be carried out with this procedure in place.

For all listed building owners, the procedure was very akin to obtaining planning consent and, after consultation with the local authority and any interested parties such as historical societies, an application under the listed building procedure had to be submitted and the local authority had to give their approval before any works could be carried out. In the case of Grade I listed buildings owned by British Railways, an arrangement had been agreed whereby BR could bypass the local authority and go directly to the Royal Fine Art Commission (RFAC) for this consent. There were a large number of listed buildings on the East Coast Main Line and a number of these were Grade I listed. This resulted in the Project Director making a total of ten appearances before the members of the RFAC, seeking their agreements to the proposed alterations, which always involved the supporting of the overhead wires either on structures or through stations.

Listed Building Applications

The first application was in respect of the Royal Border Bridge at Berwick. For fixing the masts to the bridge, BR had selected the standard design, which the Electrification Engineer always used on such structures. This was a heavy underbraced steel portal fixed to the outside of the structure with the uprights of the portals over the centres of the piers supporting the structure. Although in prior consultation with the local authority Berwick District Council had been willing to accept the standard approach, the RFAC were having none of it and the proposed design was rejected.

One of the members of the Commission at that time was Ove Arup who was the proprietor of the eponymous engineering consulting firm. After the meeting he contacted BR and offered to provide a design that he was confident would obtain the Commission's approval. He duly did this and, when the proposal was resubmitted using the Arup design, it went straight through and was the first listed building for which consent was obtained for a Grade I structure.

Listed building consent was obtained for York station, Darlington station, Croxdale Viaduct, Durham Viaduct, Durham Station, Newcastle Central station, the High Level bridge, the Bothal bridge at Morpeth, the Royal Border Bridge, Dunglas Viaduct, Edinburgh Waverley Station and Princes Street Gardens. It should be noted that the proposals for the last three items were dealt with in Scotland by the Scottish equivalent body – Historic Scotland – in association with the local planning authority.

Dunglas Viaduct was interesting in that the boundary between Lothian and The Borders Councils ran down the middle of the river, which the viaduct carried the railway across. Ideally, there should have been a portal on each of the two piers but Lothian refused to agree to such a proposition, which resulted in there just being one portal on the centre of the pier in The Borders Council area. This is most unfortunate in that it looks decidedly unbalanced, but it was the only practical solution.

The erection of the electrification masts, on occasions, had an adverse impact on signal sighting and also on the viewing distances for unmanned level crossings. Once the problem had been identified in the first instance, great care was taken to ensure that this did not prove to be an issue, generally addressed by repositioning the affected signals.

About six miles south of Peterborough, the railway crosses Stilton Fen. The Fen consists of some 15m (45+ ft) of peat overlying the ground rock below. Being peat, it is very soft and highly elastic, and it required special measures to ensure that the foundations of the masts did not sink through it down to the bedrock below. A special design evolved that, in plan view, consisted of a very large block of concrete and the masts were fixed to these by means of feet, which were placed at the one-third and two-third points of the foundation block. This was a most satisfactory solution and caters very well with the fact that, as a train goes across the Fen, there is a bow wave in front of it with the track rising up and then going down to its original level again as the train passes.

Public Relations

Considerable effort was given to keeping the public along the line of route informed as to the progress of the project and, in particular, any adverse impact it might have on services where works were being carried out in a given area. A project newsletter was produced on a four to six monthly interval basis and lectures were given to local organisations along the line of route as requested. In addition, a programme of education for school children was instituted and every school within a five mile boundary of the route was visited prior to the overhead lines being energised. These visits were carried out by members of the British Transport Police and clearly had a very beneficial effect.

When the scheme was finished, a Royal opening was arranged, and a special train was run from London to Edinburgh with a stop at Newcastle. Before joining the train at King's Cross, a number of people who had worked on the project were presented to Her Majesty The Queen. Similarly, further presentations to her were carried out at Newcastle and at Edinburgh.

Some two months after the first electrically hauled train had run the length of the route, another special was run to ascertain what might be feasible in terms of reduced journey times. Travelling at a maximum speed of 140mph, albeit with a significant number of local speed restrictions to reflect the local track geometry, a time of 3 hours 29 minutes was achieved. This demonstrated that it would be possible, given the removal of certain other fixed speed limits, to run a regular service between London and Edinburgh, calling only at Newcastle, in 3 hours 40 minutes, which would be a significant improvement over the then current 4 hours plus 'best possible' journey time.

Traction and Rolling Stock

For InterCity services, the original proposal was to run eight-car trains of Mk III stock hauled by a Class 89 Co-Co locomotive. In the event, the sponsor decided that this was too pedestrian, and he came up with the concept of a Bo-Bo electric locomotive designated Class 91 hauling nine-car trains of Mk IV coaches, which were slightly longer than Mk III vehicles, capable of 140mph (225km/hr) and being adapted to tilt. (In the event, the maximum service speed was 125mph (200km/h) and the coaches were never fitted with tilting mechanisms.) At the other end of the train a driving van trailer (DVT) was provided, which, as well as a driving cab, contained a guard's office and space for luggage and bicycles. The Mk III rakes had a seating capacity of 468 persons, including 114 first class, whereas the Mk IV rakes have a seating capacity of 535 including 129 first class.

The Class 91 locomotive was based on a number of items that had proved to be very successful with the Advanced Passenger Train, including, in particular, in board traction motors driving through cardan shafts, which, in turn, engaged with both of the axles on the bogies. Subsequent analysis showed that eight coaches was not going to be enough to carry the peak flows and the standard train length was altered to nine cars with a driving van trailer at the opposite end to the locomotive such that the train could be driven in either direction as required. Once the decision had been taken to go to a Bo-Bo locomotive, tenders were invited from several European manufacturers. In the event Brown Boveri from Sweden made the most attractive offer, but it appears that the government of the day directed that the bid submitted by GEC should be accepted as it was a home manufacturer.

For freight and parcels, a total of 31 Class 87/2 Bo-Bo locomotives were chosen. (These were later re-designated Class 90.) To cater for the outer suburban services from Peterborough inwards, a total of four four-car Class 317 electric multiple units were ordered.

Using GEC electrical equipment, the Class 91 locomotives were built by British Railways Engineering Ltd at their Crewe Works. The locomotives arrived somewhat in advance of the Mk IV coaches, so special arrangements were made so that they could replace a power car on a standard HST set, and this enabled them to enter service much earlier than would otherwise have been the case.

The Mk IV coaches were built in the former Metro-Cammell works at Birmingham. All of the basic design work was carried out in Italy, and this was the first time that such a major design package had been done remotely using computer links. The coaches were designed to enable them to tilt, which resulted in a slightly narrower cross-section when compared with the Mk III vehicles.

The outer suburban service north of Hitchin through to Peterborough was the first section of the scheme to actually yield any revenue. The success of the improved service that the four units facilitated was such that, in the first twelve months, passenger carrying in this section increased by an overall total of 32 per cent with one station recording a 57 per cent increase. (This, again, is a classic example of the 'sparks effect'.) These increased flows required a doubling of the initial capacity, and it proved necessary to reallocate some of the NSE fleet to the Peterborough services just to cope with the phenomenal passenger growth that occurred in such a short time.

Some Observations

The overhead system design was undertaken at Derby based on site surveys that had been previously carried out. These were backed up with a 'walkout' by a team who inspected the position of every proposed mast to ensure that there were no problems that the survey had missed. In the event, this was extremely successful but, nevertheless, there was the odd occasion when the actual installation of a foundation discovered a problem that had been missed. In those circumstances, the engineer in charge of the foundation installation had the authority to relocate the foundation by a 'few feet' in either direction, reporting the changed position back to the OSD team so that the resulting wiring drawings could be amended to suit the new configuration. This was the benefit of having a joint railway–contractors' team where decisions could be taken on the spot rather than having to be referred back up contractual chains, which, as the subsequent Great Western electrification project showed, could result in many weeks of delay and much additional cost.

On the infrastructure side, the most significant changes were the extension of electrification in the Leeds area from Leeds station out to the Leeds Neville Hill depot and the associated wiring of the depot, together with the adoption of optical fibre cable for the main telecommunication transmission systems. While the reversion to single-rail DC track circuits in lieu of the TI 21 track circuits to replace the Aster track circuits simplified the signalling work, this had a significant impact on the civil engineers' workload. As a consequence, the Permanent Way teams had to install almost 1,400 additional insulated rail joints. As a post-authorisation change, this had a significant impact on the progress of the works. On the plus side, the use of steel tubular piles vibrated into position for mast foundations saved much time and expense.

Traditionally, the National Grid system, which supplies power directly to the railway. had to be protected from the adverse impact of the unbalanced electrical loads of the railway system. However, in the case of the section north of Morpeth, due to the presence of an electric arc furnace at Lynemouth, the railway had to be protected from the Grid! This required a special high-voltage (HV) electrical filter to be installed at Ulgham Grange. This was unique within the UK.

The submissions to obtain listed building consent were masterminded by a town planning expert. He helped the project in a number of other ways, including preparation for the input to the British Railways 1984 Parliamentary Bill, which sought powers to reduce the amount of clearance works required by closing foot-bridges, re-routing footpaths and acquiring land for feeder stations and track sectioning cabins. He also carried out a review of all of the occupation bridges on the route and was able to identify a number of cases where it was possible to avoid reconstructing them following demolition because the land on each side of the railway was now in separate ownership as opposed to when the railway was first built. (This process is known as 'closure as a result of severance'.)

The project required a large volume of training of signalmen, train crews and other critical staff for operating and maintaining the electrified railway, utilising the new traction, rolling stock and electrification equipment. This was a particularly difficult task in respect of the signalmen at Newcastle where the training equipment for the 'new-work' stations and associated telephone answering touch screens was not ready prior to commissioning of the actual installations. It is to the credit of all the operating staff concerned that they were able to acquire the necessary skill-sets to enable them to work the actual equipment on the day of the commissioning.

The scheme depended upon the use of engineering possessions to enable the various works to be carried out. During the course of the project the arrange-ments for taking possession were streamlined such that, towards the end of the implementation period, the time taken to take and give up possessions had been significantly reduced. This had been done without any added safety risk. In a similar vein, the amount of work carried out in a possession was increased by intelligent possession management, which enabled more than one activity to be safely delivered in a single possession.

Although not part of the main electrification authority, there were a number of other schemes that had to be carried out prior to the installation of the overhead line equipment and its energisation. The main ones were the six major resignal-ling schemes at York, Newcastle, Morpeth, Alnmouth, Tweedmouth and Miller-hill. On the civil engineering side, with the remodelling of the east end approach to Edinburgh Waverley station, the opportunity was taken to carry out major repairs to the Calton Hill tunnels, which included significant lengths of re-lining.

Experience Gained

The ECML project was a ten-year project. As with all other major investment projects, it went through many cycles of optimism and pessimism. It spanned a period when the national economy went from deep recession through to a peak of growth, which turned out to be unsustainable, with the result that tender prices fluctuated from the most attractive to the unacceptably expensive and the supply of resources, both internal and external, varied in a like manner.

After the completion of the project, a Lessons Learnt conference was held in Newcastle on 11 and 12 July 1991. It was attended by representatives of all departments that had been involved in the project but only a handful of these people had been with the project since the initial planning stage pre-authorisation. The churn of people had, over the ten years, been very significant and, in many respects, this reflected the statements that were made at the conference.

All these factors had a major influence on the development and implementation of the project, which resulted in some changes of direction to some of the business objectives. Nevertheless, the underlying plan and aim remained constant, and a notable feature was the equally single-minded determination of the lead sponsor to acquire a high-quality product on time and within the agreed budget. There were two lessons learned conferences and the second one was attended by 58 people, all of whom had been active in at least the previous twelve months and many of whom had been intimately associated with the project for several years beforehand. Notwithstanding this, it is inevitable that the events of the previous six to twelve months dominated the conference and its conclusions. In particular, the major problems that arose in connection with the completion of the Newcastle resignalling project dictated the views of the representatives present.

Prior to the problems arising from the Newcastle resignalling, there were five years of implementation and, before that, a further three years of development in which virtually every activity was successfully completed either on time or frequently early and, invariably, within the budget. It is inevitable that during a project of the scale and duration of the ECML electrification, there is a continuous learning process. The volume of material arising from the various working groups in the lessons learned conference is difficult to present in a manageable form. Overall, a comprehensive summary of the most significant features that arose during the project has been made and this follows below.

There was widespread agreement that the creation of a dedicated proactive project team with a guaranteed resource commitment would generate improvements in planning, safety management, integration of all dependent activities, including possession planning, financial control and reporting, all of which were exposed as weaknesses during the conference.

The conference had a clear perception that a major strengthening of project management and control was needed. Nevertheless, there was an acknowledgement of the quality of the teamwork that permeated the whole project and generated considerable pride and satisfaction in all of those involved. This was evident in all aspects of the work from the senior management through to the ground level resources.

Despite the many problems raised in the conference, the overall result of the project must be viewed within the perspective of the scale in which the works took place. The package of schemes at the time of their completion was the biggest single investment project undertaken by the InterCity business let alone the Railways Board within the previous twenty-five years. Overall, it was delivered to plan and to budget and the final completion was only eight weeks behind the original programme, which had been determined seven years previously. Bearing in mind the problems caused by the Clapham accident and the delays arising from the ensuing Hidden Enquiry, that must be a magnificent achievement by any yardstick. All the people who worked on this project earned the thanks and appreciation of InterCity and the other sponsors for their dedication, commitment, effort and achievement.

The main lessons to emerge were:

- A proactive approach to seeking listed building consent by meetings with the planning authority staff concerned at the earliest possible stage ensured that there were no delays awaiting these approvals.
- The use of fibre-optic cable for all of the main telecoms links avoided the usual interference problems at energisation.
- The ability to convert key lock level crossings to remote control by the use of the new Codex system was a major achievement and enabled significant staff savings as well as avoiding an increasing staffing problem.
- When the contractor on the Scottish section ran into access problems in the Princes Street Gardens area of Edinburgh, the bringing in of an experienced maintenance team from the Ilford EFE maintenance depot ensured that the wiring of this difficult area to wire was successfully completed 'just in time'.

The most troublesome problems were all on the traction and rolling stock side. Vibrations on the Class 91 locomotives caused significant problems for the drivers but were resolved by fitting of a resilient absorbing pad between the primary and secondary suspension. (They were affectionately referred to as the 'headache machines'!) On the coaching stock side, the biggest issues were the control of heating in the Mk IV coaches and the failure of the door shut and locked detectors to work properly. The heating problem was resolved by the contractor providing a completely new system. The door lock detection problem was finally identified as being due to an inadequate fixing of the detector switch and, once this had been resolved, there were no further problems. However, until then, trains were frequently being heavily delayed because the traction interlock system was being told that there was one or more doors still open.

The main achievements of the project were the arrival of the first electrically hauled train into Leeds one year early, the successful immunisation of all the signalling and telecoms circuits through the use of the latest technology and the adoption of the driven steel pile foundations for the EFE masts wherever the ground conditions permitted it. Additionally, the very early completion of the Eastern Region bridge raising programme must be commended due in part to the foresight to carry out many of the site surveys prior to the formal authorisation of the project in July 1984. Further notable successes were the novel solutions to the clearance problems at bridges 158 and 325 (Hexthorpe Road, Doncaster), and the closure of the Viaduct lines at Leeds, with the associated removal of Gelderd Road Junction.

Finance

The final project outturn was another success. The table below is a summary of the overall result. At constant prices it is seen that the main electrification project is £9.782M or 3.2 per cent overspent. When the associated projects are added in, the overspend is reduced to £7.173M or 2.3 per cent and, counting in the other projects, the overspend becomes £7.98M or 2.4 per cent. The excess is well within the permitted tolerance of authority plus 5 per cent. The overall actual outturn of £457.034M reflects the impact of inflation during the project life. Inflation over the period averaged 5.1 per cent per annum.

FINAL FINANCIAL RESULT							
	Authority £M Q4 1983	£M Q4 1983	Final Cost £M Q4 1983	£M Q4 1983	Actual Spend £M	£M
ECML electrification	306.299		315.917		421.042	
Associated projects	8.158	314.457	5.713	321.630	7.489	428.531
Other projects	7.490	331.947	18.297	339.927	29.503	457.034

From the above it is seen that the main electrification project is £9.782M or 3.2 per cent overspent.

Lessons Learned and Recorded at the Conference

Other 'lessons learned' were:

- The most disruption to train services during the period of the works was caused by route rationalisation and re-signalling rather than the electrification itself.
- Most cost increases or extensions of project timescales were the result of changes in business requirements, no matter how well justified.
- The closer the relationship the project manager has with business managers the more likely he or she is to succeed.
- Timescales and target dates must be underwritten by realistic and timely resource plans and schedules.
- Listed Building and Planning Consent requirements should be fully understood and allowed for.
- An effective change control process is essential.
- It is essential to draft and implement an effective safety plan from the very earliest stage of the project.
- Training should be treated as a major issue.
- The provision of an appropriate inventory of spares is essential.
- Adequate resources need to be allowed for to clear away redundant materials from the lineside to avoid creating a safety hazard and promote good 'housekeeping'.
- The processes to take and hand-back possessions should be streamlined to improve the time available for work, thus improving productivity.
- Interfaces between the main contractor and the Railway should be simplified, and railway labour should not provide the 'numerous services to the contractor' as happened on this project.
- The lessons learned on this project have a more general application and interest across the Railway.

Politics – post-ECML Project

BR had shown that it could manage major projects to time and budget! Despite this, the political mood was that the private sector could manage the railways better than BR management.

As a result, despite the project's very satisfactory outcome, the 'hoped-for follow-on' electrification schemes never materialised, due primarily to the government's pre-occupation with privatisation.

Electrification from Carstairs to Edinburgh Waverley – 7/91

The genesis for this 'add-on' scheme to the main ECML electrification project was the proposal that InterCity East Coast should continue to offer a through service to Glasgow. With the long-term vision for the service being full electric working, this would require a route between Edinburgh and Glasgow to be wired. Of the routes available, the one that was going to be the cheapest to electrify would be the Carstairs to Waverley link, as this was the shortest length of new wiring, the most open route and, most importantly, avoided the need to wire the Glasgow terminal (Queen Street). It also had the added benefit that it would provide a feeder station to the west end of Edinburgh (at Currie Hill) thus making Waverley independent of the feeder station at Portobello. Mk 3b OLE was used, and this installation was notable in that it was the first scheme to use the then newly designed structure mounted outdoor switchgear.

The Works

The major work on the project was providing electrification clearances through the Haymarket South Tunnel. The best way of doing this was to install a slab track thus minimising construction costs as well as ensuring that the stability of the tunnel walls was not adversely impacted upon. It required the tunnel to be taken out of service for six months and, because Haymarket station is immediately at the west end of the tunnel, it meant that access for the construction works had to be carried out from the east end on the south side of the line opposite Princes Street Gardens.

The other notable feature was Linhouse viaduct. The Scottish Region Engineers decided that access to the top of the piers for constructing the fixings for the masts would have to be carried out from ground level, which required the erection of scaffolding on seven of the piers. This, in turn, needed access at ground level. This gave the landowner an opportunity to exploit the situation financially for the right to go over his land. The cost of these foundations was further exacerbated by delays in carrying out the work once the scaffolding had been erected, which meant that it was in position for well over three months

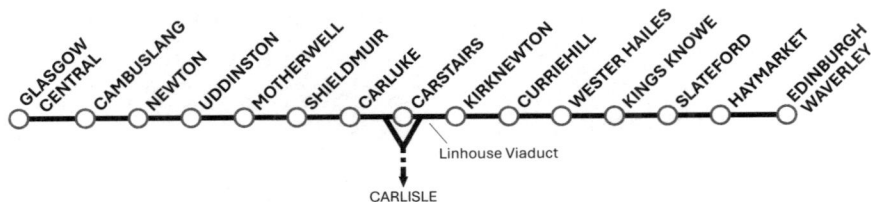

Glasgow Central to Edinburgh Waverley via Carstairs

rather than the two or three weeks that would have been necessary if everything had been done in a continuous sequence.

The most difficult part of the wiring was in Princes Street Gardens. This was because of the problems of acquiring access. Once the service trains had ceased running, this area was used for shunting trains between the various parts of Waverley Station: the shunting being necessary to facilitate overnight cleaning of the stock. The problem was further exacerbated by the need to send some diesel multiple units through to Haymarket depot for refuelling. In order to complete the project on time to meet the previously publicised energisation dates, it proved necessary to bring a team of overhead linesmen up from Ilford for two weeks to assist the main contractor in completing the work.

In addition to providing for the through electric working of East Coast trains to Glasgow, the scheme had the added benefit of being able to close the Carstairs motive power depot with the saving of some twenty-five drivers and a similar number of guards. It further permitted a change to the working of a small number of West Coast freight trains destined for the Edinburgh area. Instead of going to Mossend and being re-engined they could become electrically hauled throughout, via Waverley station to access Millerhill Yard.

This modest extension to the main East Coast scheme illustrated some interesting lessons for future schemes. It re-emphasised the need to have all the civil engineering structural works completed before the overhead wires were run out. It also highlighted the need for major projects to have total control – if not ownership – of the key major resources that are required for the successful completion of the project.

Provincial Services (Regional Railways) Schemes

Provincial Services, later re-named Regional Railways, was a BR Sector responsible for more than 50 per cent of the British Railways infrastructure, but being heavily dependent on subsidy, had only a very modest capital budget. With the 1950s built DMUs becoming life expired in the 1980s, most of these funds were being expended on the DMU replacement programme. However, good business cases were developed to enable the schemes described below to be implemented:

Ayrshire Electrification – 7/87

This was a scheme sponsored by the Strathclyde Transport Authority and was the first of only two electrification projects carried out by the Provincial Services

GLASGOW

LARGS ○

○ PAISLEY GILMOUR ST
———— Elderslie
○ JOHNSTONE

FAIRLIE ○
Hunterston ————
WEST KILBRIDE ○

○ MILLIKEN PARK

○ HOWWOOD

ARDROSSAN HARBOUR ○
ARDROSSAN SOUTH BEACH ○

○ LOCHWINNOCH

○ GLENGARNOCK

SALTCOATES ○

○ DALRY

STEVENSTON ○

○ KILWINNING

○ IRVINE

○ BARASSIE

○ TROON

○ PRESTWICK AIRPORT

○ PRESTWICK TOWN

○ NEWTON-ON-AYR

○ AYR

STRANRAER

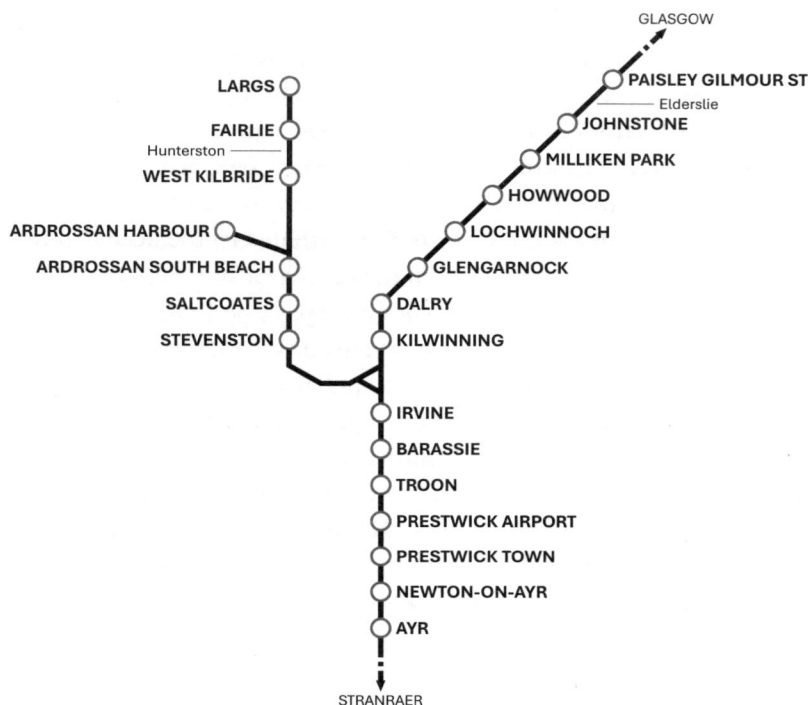

Ayrshire Electrification

Directorate (later Regional Railways) set up under the sectorisation of British Railways. It comprised the wiring of the main line, utilising Mk 3b OLE, from Paisley to Ayr and the branches to Largs and Ardrossan Harbour together with 21 new EMUs plus 12 refurbished ones. (The route from Glasgow as far as Paisley had been electrified some twenty-five plus years previously as part of the first tranche of electrification described in Chapter 8.) It also included complete resignalling based on a new power signal box at Paisley. A further 83km (52 miles) of route were wired and converted to electric operation.

The scheme was authorised in October 1982 subject to obtaining a European Development Grant at a total outlay of £80M. Work commenced in 1984, and the project was completed by September 1987 with the new electric services starting at that timetable change. Beyond Castle Hill Junction on the Largs branch, the down line was made bi-directional so that the up line from the iron ore/coal loading point at Hunterston did not have to be wired: freight trains continuing to be diesel hauled. The wiring was extended beyond Ayr station to Town Head carriage sidings.

With the closure of the branch to Kilmalcolm and the severance of the 'loop line' serving Paisley Canal at the Hawkshead oil sidings, the complex at

Elderslie Junction was greatly simplified. Uniquely, and so as to contain the project outlay, all the overbridge reconstructions required for the provision of electrification clearances were carried out by the local Highways Authorities. The overhead line works were carried out from a construction depot established at Barassie. At the completion of the construction works this was converted into an overhead line maintenance depot.

A total of 21 Class 318 EMUs were ordered that were fitted with end-connecting gangways thus enabling the Conductor to have access to the whole train when two units were coupled together in the peak periods. This was an important facility as, with many of the stations being unstaffed under the Strath-clyde Manning Agreements, the Conductor was responsible for ticket inspection, and, where required, sales. After railway privatisation, First Group – the initial franchise holder for Scotland – controversially removed these without any reference to the Strathclyde PTE who were, of course, providing revenue support for the service. The units were based at Shields depot for maintenance with overnight servicing at Town Head Sidings in Ayr. The fleet was augmented by the addition of 12 Class 311 refurbished EMUs.

Manchester Airport – 1993

A north-facing rail link was provided from the Styal loop to a new station alongside Manchester Airport in 1993. The OLE comprised the Mk 3b system. Services from the east side of the Pennines terminating at Manchester Picca-dilly were extended to the airport along with some of those from Liverpool, Southport, Preston and beyond that previously ran into Manchester Victoria.

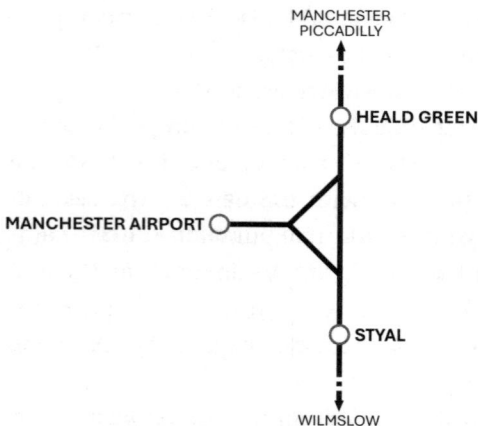

Manchester Airport

A south-facing connection was added in 1995 (Mk 3c OLE) thus facilitating direct running from the Potteries, Chester and North Wales.

In December 2008, Network Rail provided a third platform at a cost of £15M. While this increased operational flexibility, it did not increase capacity much, as operators took advantage of the extra line to layover trains. In addition, there was also more congestion at the throat. Accordingly, a further new platform, costing £23M was opened in May 2015 to alleviate the situation.

Birmingham Cross City Electrification – 6/93

The Cross City line in Birmingham runs from Lichfield to Redditch and on 7 February 1990, the Secretary of State approved the electrification of the route. This enabled the most intensive suburban diesel service in Britain to be converted to electric operation.

The scheme provided for the 25kV electrification of some 57km (33 route miles) utilising Mk3b OLE, between Lichfield Trent Valley high level and Aston Junction (exclusive) and New Street (exclusive) to Redditch via Selly Oak. It also included the wiring of holding sidings at Lichfield City and Longbridge but, on the four-track section between Kings Norton and Longbridge, only the slow lines were wired. At the north end two new feeder stations were provided from the National Grid at Winston Green and Galton Junction.

A construction depot was set up at Kings Norton and the main electrification contractor was Pirelli. The Mk 3b form of catenary was used throughout. Construction work started in May 1990 and was completed in time for the new electric service to be inaugurated on 6 June 1993.

Associated with the actual electrification were signalling alterations. These comprised resignalling and rationalisation between Aston and Lichfield and immunisation between New Street and Redditch.

A total of 17 × three-car EMUs were supplied to cover the peak demand. The trains, known as Class 323, were built by Hunslet TPL and were assembled at their Leeds Works. The vehicles were 23 metres long and had twin leaf sliding plug doors located one-third and two-thirds along the body side. The traction equipment comprised eight motors per unit with three phase drive and regenerative braking capacity. These new units resulted in journey time savings end to end of 16 minutes.

The introduction of the new service was far from smooth and there were significant problems with the units. It took a further twelve months for these all to be resolved. The new service was restricted by the single line between Barnt Green and Redditch and it was not until 2014 that the problem was solved by

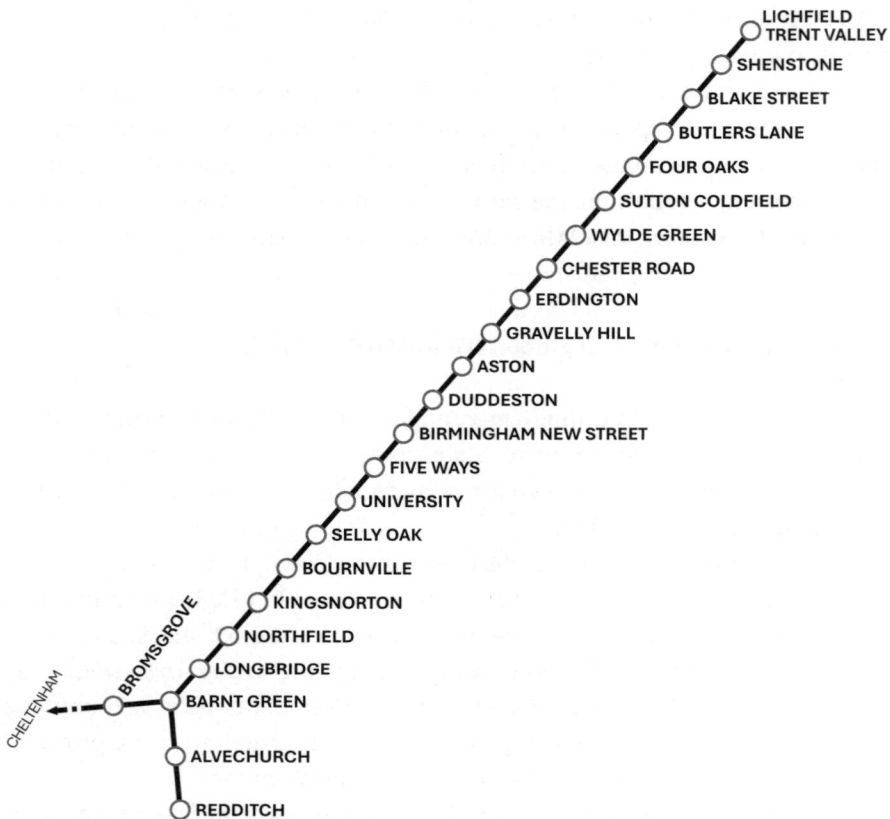

Birmingham Cross City Electrification

the provision of a new passing loop at Alvechurch – the work being carried out between July and September that year.

An extension of the electrification of the Cross-City Line from Barnt Green down the Lickey Bank to Bromsgrove, including the resiting of Bromsgrove Station, was first mooted in the early 2000s. Work commenced on the construction of the new station in March 2014. However, the discovery of an unknown culvert and issues with soil contamination delayed the opening of the £24M station until 12 July 2016. The station development included two centre platforms to turn-back Cross-City Services. It was located some 500 yards south of the original and was designed to facilitate cross-platform interchange between the terminating local services and the through trains to and from Worcester and so forth. In 2018, the wiring was extended from Barnt Green down the Lickey Bank to the new station at Bromsgrove. Electric services from Bromsgrove to Birmingham New Street and on to Lichfield Trent Valley of three trains per hour commenced on 29 July 2018.

West London Line – 1993

To permit Eurostar trains to run from their depot at North Pole to the then terminal at Waterloo, in 1993/4, 10km at the southern end of the West London Line from Shepherd's Bush to Clapham was electrified utilising the third-rail 750V DC system. In addition, the 25kV AC system from Willesden was extended southwards by a kilometre to Mitre Bridge Junction, enabling, after many years, suburban services to be reinstated between Willesden Junction and Clapham Junction. An innovative feature was the introduction of a system that enabled traction power changes to be undertaken while the trains were on the move. It was also proposed for Regional Eurostar trains from the both the West and East Coast main lines to run via the North London and West London lines en route to Paris and Brussels and vice versa.

Just to the north of Shepherd's Bush station, the railway runs through a very tight tunnel beneath the Westway built in the 1960s precluding any increase in vertical clearance. Due to the existence of a major sewer at a very shallow depth under the railway at this location, track lowering to provide for overhead electrification was not possible, preventing the extension of the 25kV AC system southwards. The solution adopted was to install the 25kV AC system from Willesden, West London Junction, southwards, with the 750V DC third-rail system running from Shepherd's Bush to Clapham. The third-rail DC system was thus extended by 10km (6.25 miles) and the overhead 25kV AC wires by 1km (0.625 miles).

The work facilitated the reinstatement of a local passenger service in 1994 and the reopening of two local stations that had been closed since October 1940 due to the aforementioned Second World War bomb damage. Further new stations were reopened at Shepherd's Bush, in 2008, and Imperial Wharf (formerly Chelsea Basin) in 2009 (see Chapter 14).

Leeds North West – 9/94

This was one of the last two electrification schemes carried out by the Regional Railways sector of British Railways prior to privatisation. The project was sponsored by the West Yorkshire PTE.

Following the electrification of the Doncaster to Leeds line as part of the East Coast scheme, the West Yorkshire Passenger Transport Executive (PTE) altered its specification for the local service on this route. It substituted electric multiple unit working in place of diesel multiple units, which were based at Neville Hill. This made a significant improvement in the quality of the offer to the travelling public in terms of reduced journey times as well as up-to-date rolling stock. As a result of this the PTE decided to examine the case for electrification of its local services based on Leeds station. This gave rise to a proposal to electrify the Aire Valley services running from Leeds to Bradford (Forster Square), Ilkley and Skipton. The electrification fixed works were carried out in the period 1992–4, utilising Mk 3b OLE, and the full electric service was introduced in the September 1994 timetable change.

The civil engineering work comprised ten overbridge reconstructions, eight footbridge reconstructions, three overbridge abandonments and seventeen track lowering schemes through bridges and tunnels. In addition to 57 bridges requiring parapet protection works, there was the remodelling of three junctions, the reinstatement of 1.5km of double track on the Ilkey branch and new and reconstructed platforms in association with the track lowering schemes. There was also a major track slew in the Kirkstall area where the tracks in use were moved over from the former Bradford Lines alignment to the Midland Main Line alignment. The work in the Kirkstall area resulted in a small speed improvement but the genesis for it was the better condition of the two redundant river bridges on the MML alignment. These two bridges were also refurbished.

Except for one occupation bridge (O/B 83 at Cononley) that became caught up in a legal battle over a land slip, October 1993 saw the substantial completion of all civil engineering works as programmed. The scheme comprised 70km (44 route miles) in total.

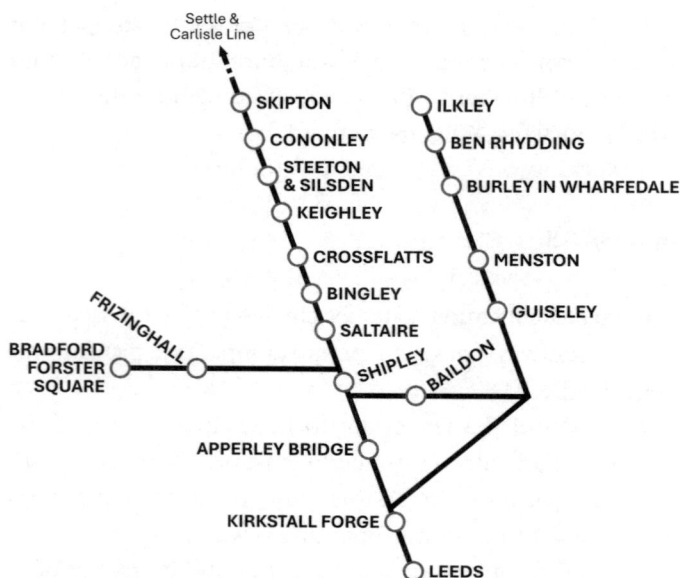

Leeds North West Electrification

The resignalling of the line was based on an extension of the York Control Centre, which had been provided in association with the electrification of the East Coast Main line, and included the transfer of control from the Leeds area PSB that had been immunised under that scheme.

This scheme was a major success with the work being completed on time and on budget within three years from project commencement. The financial breakdown was:

Costs (Q3, 91 prices):
Electrification – £30M (£70M at 2025 prices)
General civil and building £10.23M (including Skipton depot – in total £23.6M at 2025 prices)
Bridgeworks and track lowering £5.25M (£14M at 2025 prices)
Resignalling £23M (including Building Engineering services – in total £53.5M at 2025 prices)

Although the project included the purchase of 16 new EMUs, these were not available when the fixed works were completed. So that the PTE could exploit the electrification at the earliest possible moment, they agreed to use some old EMUs, which had been released from service on the lines out of Liverpool Street.

These Class 308 units were slam door rolling stock, and they also caused considerable higher levels of interference with the signalling and telecoms circuits along the route. The slam doors created delays because the passengers – particularly in the evening going home – refused to close them properly and, where platform staff were not available, the train guard had to run the length of the train doing this. The high levels of interference did cause a number of the solid-state interlocking trackside units to burn out, which led to significant delays while temporary working was introduced and the units were replaced. This meant that the new services had a very bad start in customer perception terms, and it wasn't until the new Class 333 trains came into use in 2001 that the major improvements to the service, which the PTE envisaged, were actually achieved. The EMUs were based on Neville Hill Depot.

Chapter 10
Channel Tunnel – 5/94 (Not a BR Scheme)

Proposals to construct a tunnel under the English Channel were first mooted in 1802. In total five schemes were proposed during the nineteenth century, but no actual physical progress was made until 1881. Following the establishment of an Anglo-French protocol in 1875, the British railway entrepreneur Sir Edward Watkin and French Suez Canal contractor Alexandre Lavalley proposed that a rail tunnel be constructed using Beaumont–English boring machines. Operating on compressed air, these 30ft long, innovative machines could bore up to 600ft per week in chalk. Two such machines, one on the French side at Sangatte and the other on the English at Shakespeare Cliff, each bored 7ft diameter pilot tunnels more than a mile long before the project was abandoned due to fears that the tunnel would compromise Britain's security.

Further ideas for a tunnel were put forward in the first half of the twentieth century but it was not until the late 1950s, when national defence arguments had become less relevant, that a serious proposal was again considered. Detailed geological surveys were carried out in 1964/5 and the British and French governments agreed in principle to build a Channel Tunnel in 1973. Work commenced in 1974 with the installation of tunnel boring machines on both the French and English sides but after a short trial boring, the project was again abandoned when the British Labour government unilaterally cancelled the project due to the doubling of cost estimates and the general UK economic crisis of the time.

Following this, in 1974 Sir Alistair Frame, who was the MD of Rio Tinto Zinc (RTZ) at the time, was extremely keen that the proposal for a Channel Tunnel should not be lost for another century. He spent a lot of time lobbying among leading industrialists and persuaded British Railways to collaborate with him, which resulted in a modest proposal being developed in the late 1970s. This was subsequently put forward in 1982 when a 'single-track' scheme was unveiled.

In 1981, Prime Minister Margaret Thatcher and the French President, François Mitterand, agreed to set up a working group to consider how a Channel Tunnel Project might be privately funded. This study group favoured a twin-tunnel scheme capable of accommodating conventional trains together with a vehicle shuttle service. In April 1985 promoters were invited to submit scheme proposals. Four submissions were shortlisted and eventually the proposal by the Channel Tunnel Group/France – Manche, formed of two separate French and British consortia, was selected to take the project forward. The two organisations were linked under a bi-national project organisation named TransManche Link (TML).

The Channel Tunnel Bill authorising the construction of the 50.5 km (31 miles) tunnel was passed by the British House of Commons in February 1987. Royal Assent for the Channel Tunnel Act followed in July 1987. Tunnelling commenced in 1988, with tunnel boring machines working from both the English and French sides of the Channel. 'Break through' was achieved without ceremony on 30 October 1990.

Completing the tunnelling and fitting out the tunnels took another three years. Vehicle shuttle terminals were constructed at Cheriton, connected to the M20 motorway, and at Coquelles, linked to the French A16 autoroute. There were serious problems during the latter part of the construction programme, which were caused by the fact that the contractors, who had formed a joint venture to build it, were also acting as the project's clients. The finance ran out and the project was only able to be completed by virtue of the banker, Sir Alistair Morton, coming in and raising several billions of pounds to finance the then outstanding work.

The tunnel was officially opened on 6 May 1994, a year later than originally planned, by Queen Elizabeth II and the French President, François Mitterrand. Initially, the tunnel only carried freight traffic; the first train, carrying Rover and Mini cars, ran on 1 June 1994. The first Eurostar passenger train ran on 14 November 1994.

The Channel Tunnel's OLE, electrified at 25kV 50Hz, has a high overhead clearance of 6.3m (20ft 8ins) to permit the passage of the tall shuttle vehicles. The OLE is fed from two 400kV substations, one at each of the two terminals. These substations also provide power for the tunnels' ventilation, lighting and plant.

On the French side, the 300km/h LGV Nord line was built in parallel with the construction of the Channel Tunnel from the outskirts of Paris to the tunnel terminal. In Belgium, another high-speed line was constructed, which opened in 1997, providing a direct link from the LGV Nord from the east of Lille to Brussels.

On the British side, although a high-speed link to the Channel Tunnel had been considered from the 1970s, only work to upgrade existing lines to London Waterloo Station had been authorised and implemented at the opening of the tunnel. These works included the upgrading of Ashford Station, which was renamed Ashford International, and a new flyover on the approach to London Waterloo linked to a western extension of that station. As a result, Eurostar trains were restricted to run at conventional line speeds on the existing railway's 750V DC infrastructure between the Tunnel and the London terminus.

The Class 373 Eurostar trains were based on SNCF TGV sets, equipped to operate on the three electrification systems used in France, Belgium and the UK (25kV 50Hz, 3kV DC and 750V DC respectively). To operate on conventional UK infrastructure, the Eurostars were built to the smaller British Loading Gauge. The Class 373 Eurostar trains are capable of operating at up to 300 kilometres per hour (186mph) on high-speed lines.

With Channel Tunnel safety in mind, the Class 373 Eurostar trains were designed to consist of two independent 'half-sets', each having its own power car, so that in the event of a serious incident in the tunnel affecting one half-set, passengers could be transferred into the other undamaged half and driven safely out of the tunnel. The downside of this is, of course, the fact that passengers and staff could not transfer between one half of the train or the other half while the train was in motion and there was a significant cost in providing the duplicate facilities.

In December 2010 Eurostar International Ltd ordered ten interoperable sixteen-car high-speed E320 train sets from Siemens, based on that company's Velaro design. This was followed by an additional seven ordered in January 2015. The new trains, designated as Class 374, have seating for 900 passengers providing in the region of twenty more passenger capacity than the Class 373 trains.

These 374 units started taking over services from the Class 373 sets in November 2015, although eight Class 373 train sets were refurbished and retained for operation alongside the Class 374 units.

Chapter 11

Schemes Started by BR and Finished Post-Privatisation

CrossRail (renamed Elizabeth Line) – 04/22

(Note: The first version of this project was called CrossRail and its subsequent version, that is, the one that was implemented, was designated Crossrail. In 2016 the Mayor of London announced that the railway would be called the 'Elizabeth Line'.)

Proposals for a cross-London railway in large diameter tunnels go back all the way to the 1940s (see also Chapter 9). The first east–west proposal was put forward by the well-known senior railwayman, George Dow, during the war years. East–west and north–south cross-London links were included in the 1944 Abercrombie Greater London Plan but due to a lack of funds, were not progressed in the early post-war years. The situation was reviewed by the London Plan Working Party in 1949 but, whilst endorsing the previous proposals, yet again no action followed.

In Paris, however, a similar concept utilising overhead line electrification was first put forward just before the war. Developed in the 1950s, the construction of an east–west cross-city link commenced in the 1960s. The project, which became the first of the Réseau Express Régional (RER) lines, served two purposes, operating as a 'suburbs-to-city-centre' suburban commuter line as well as a rapid metro service, albeit with fewer stops. It became a rapid-transit model that was emulated in other major cities around the world.

In 1989, the Central London Rail Study proposed a similar RER-type system, linking up to the existing rail network as the 'East–West CrossRail', 'City Crossrail' and 'North–South Crossrail' schemes. Following this, London Underground and British Rail deposited a Parliament Bill in 1991 for a scheme to build a new East–West 'CrossRail' underground line from Paddington to Liverpool Street. This had the support of the Conservative government, but was subsequently rejected by the Private Bill Committee. However, the government issued a directive to ensure that the route was protected for a future 'CrossRail' scheme.

While this CrossRail project was being promoted, two more ambitious alternative schemes, termed 'Super CrossRail' and 'Superlink', were put forward, but these were rejected due to concerns about cost and the operational capacity of

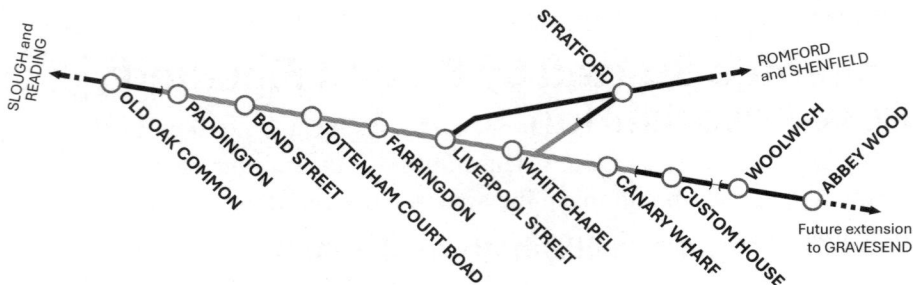

Crossrail (Elizabeth Line)

the existing network at either end. Following the Parliamentary rejection, British Rail and London Underground further continued to develop the CrossRail scheme utilising the safeguarded route but further refining the necessary enhancements to the east and west of the proposed tunnelled sections.

The new proposal was submitted to Parliament, and the Crossrail Act 2008 was given Royal Assent in July 2008. The act empowered Crossrail Limited, a wholly owned subsidiary of TfL, to deliver the project, Crossrail being jointly sponsored by the Department for Transport and TfL with Network Rail also taking the responsibility to upgrade their existing network at either end of the tunnel. The 117km route (73 miles) linked Heathrow Central and Maidenhead west of London to Shenfield and Abbey Wood to the east via 13km (8 miles) long tunnels beneath Central London. The line was to be electrified at 25kV AC 50Hz and seamless connections made with the existing lines that would become part of the route. (Note: The route followed was that in the 1992 Bill but amended by a diversion at the eastern end via Whitechapel and it included a branch to Abbey Wood via Canary Wharf.)

Construction on Crossrail commenced in 2009 involving not only the tunnelling under London, but also major enhancements to the Great Eastern Lines to Shenfield and the Great Western route, initially as far as Maidenhead in the latter case. In the tunnelled section, new stations were constructed at Whitechapel, Liverpool Street, Farringdon, Tottenham Court Road, Bond Street and Paddington – each with interchanges with London Underground and/or National Rail lines.

New bulk power supply substations located at either end of the Central London tunnels supply the main power to the track and services under Central London. The trains draw power in the tunnels from a rigid overhead aluminium, hollow profiled alloy 'conductor-bar' catenary system, which incorporates a non-tensioned copper contact wire. The technology had been previously used in the UK for the Thameslink platform lines at St Pancras and in the Severn Tunnel.

The system was adopted for Crossrail primarily due to space restrictions, and the need to avoid the risk of de-wirements in the tunnels. The system used is novel in that to allow for thermal expansion, the aluminium profile conductor bar sections are configured to slide, rather than allowing the supports to swing slightly as in earlier installations. This minimises the number of moving parts, which, it is hoped, should make the system more robust and reduce the need for maintenance.

The 2× 25kV overhead traction power system with associated autotransformer feeder stations provide sufficient power to run the proposed intensive service (25tph). Supplying sufficient power to the main tunnel for the lighting and ventilation systems together with the track electrification involved the installation of more than 1500km (932 miles) of cable. These two feeder stations also supply the Network Rail tracks at either end of the tunnels to boost supplies to the OLE.

Major civil engineering works were undertaken on Network Rail infrastructure on both the eastern and western ends of the main tunnels. On the Great Eastern side, a new spur off the main Crossrail route at Whitechapel was constructed running via Canary Wharf and under the Thames to Abbey Wood where it provides an interchange with the North Kent Line.

To the west of the main tunnels, the signalling of 19km (12 miles) of the Great Western Main Line between Paddington and West Drayton was upgraded, including the branch to Heathrow Airport. At Acton, a new 'dive-under' was constructed, removing the confliction between main line services and freight trains accessing Acton Yard. In addition, a new flyover incorporating an impressive 120m long steel span was built just west of Hayes & Harlington Station at Stockley Junction to enable Crossrail services to/from Heathrow Airport to have grade separated access to the Up Relief line. (Crossrail trains having replaced the previous Heathrow 'Connect' services).

The Crossrail project also encompassed the electrification of the Great Western route between Stockley Junction Flyover and Maidenhead utilising Series 1 OLE. In 2014 it was decided that Crossrail services would be extended to Reading. Beyond this point, with the exception of the Reading Station area, the Network Rail Great Western Main Line project team picked up the electrification work. At Reading, as part of the station rebuilding programme, a new grade-separated junction was provided to the west of the junction.

On 23 February 2016, Her Majesty Queen Elizabeth II visited the Crossrail construction site at Bond Street Station, where the Mayor of London, Boris Johnson MP, announced that the new railway would be known as the Elizabeth Line in her honour.

A series of technical issues and the Covid pandemic delayed the opening of the project and inflated costs. In May 2022, the National Audit Office estimated that the final cost for the Crossrail project to build the railway would outturn at £18.9bn (£21Bn at 2025 prices).

Crossrail was officially opened by Queen Elizabeth II on 17 May 2022 during her Platinum Jubilee year and launched as 'The Elizabeth Line'. With the exception of Bond Street station, the central section of the route between Paddington and Abbey Wood was opened to the public on 24 May 2022, with a service of twelve trains per hour. In May 2023, the line became fully operational with new nine-car Class 345 trains operating up to twenty-four services through the central section. Services split between Reading and Heathrow on the western side and Abbey Wood and Shenfield at the eastern end.

Passenger demand is above the post-pandemic expectations with 210 million journeys made in 2023/4 and rising. The railway's success has driven forward regeneration along the length of the route, particularly at Abbey Wood, one of the largest regeneration areas in the city, and also in the West London area. In view of the increasing ridership, a further ten trainsets have been ordered, which hopefully will help secure the future of Alstom's Derby Works (formerly owned by Bombardier).

Channel Tunnel Rail Link (CTRL Subsequently Renamed HS1) – 11/07

From the commencement of passenger services through the Channel Tunnel, the operation of Eurostar trains on the busy conventional railway between Folkestone and London Waterloo limited the number of services that could be run. In addition, service reliability was regularly compromised, and journey times were far longer than would have been the case had a dedicated high-speed line been provided in the first place.

Planning for the Channel Tunnel Rail Link (CTRL) had, however, actually commenced under the auspices of Union Railways, a subsidiary company of British Rail, in the mid-1980s. Following confirmation by the UK and France that a Channel Tunnel would be built, work to develop a route to London intensified.

Initially a proposal to construct a route approaching London from the south-east in tunnel to an underground terminus beneath the King's Cross/St Pancras stations was the preferred option. However, a late alternative proposal, put forward by the Consultants Arup and backed by the Conservative Deputy Prime Minister Michael Heseltine, was adopted instead. This proposed a route approaching London from the east and terminating at St Pancras.

Moreover, this alignment would enable significant 'brownfield' areas of North Kent and East London to be regenerated along the route. The Thames

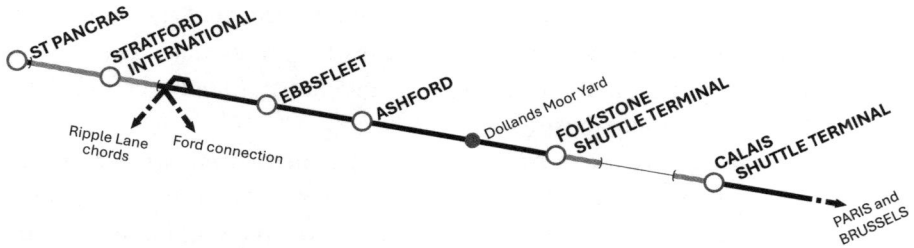

Channel Tunnel Rail Link HS1

and M25 motorway would be crossed by a new viaduct near Dartford. (In the detailed design phase, it was subsequently decided to cross the Thames by a tunnel.) The regeneration benefits made the cost of constructing the link to the Channel Tunnel by this route far more attractive to politicians and so this scheme was developed for approval.

The idea of an underground terminus was dropped by the government in 1993, and it was confirmed in 1994 that the existing St Pancras Station would become the terminus. Accordingly, a Parliamentary Bill seeking authority for the project was deposited in November 1994 with the full authorisation process taking two years, culminating in the passing of the Channel Tunnel Rail Link Act 1996.

The government also promoted a competition to find a private sector developer. In early 1996 it was announced that London and Continental Railways (LCR) was the winner. LCR's founder-shareholders were Virgin, National Express, Bechtel, Arup, Halcrow, Systra, London Electricity and SG Warburg. In June 1996, LCR took over Union Railways, the design project team, and Eurostar UK. The consortium was also vested with the development land around St Pancras and Stratford in East London.

LCR planned to float the company to raise funds for the construction of the new railway but by late 1997 it was clear that the Eurostar business was trading well below forecast, so the planned flotation had to be abandoned. The company therefore sought additional public sector funding to build the link. This was refused and the project was left teetering on the verge of collapse. John Prescott, the new Labour Deputy Prime Minister, however stepped in and brokered a deal between LCR and Railtrack.

Under the terms of the deal, construction of CTRL was to be staged in two sections. LCR would build the southern section between Folkestone and Fawkham Junction near Ebbsfleet, under the control of the reformulated Union Railways (South). Railtrack took control of this latter organisation, which became the 'client', and contracted to purchase this first section of the route at cost. Rail Link Engineering (RLE), a consortium of four of the LCR shareholders

(Arup, Bechtel, Halcrow and Systra) became the project management team, responsible for the design, procurement and construction management.

Railtrack was also granted an option to acquire Section 2 between Ebbsfleet and St Pancras on a similar basis.

Following this tumultuous period, Section 1 construction commenced in October 1998 only six months later than LCR's original programme. The civil engineering contracts were divided into geographical packages and awarded after competitive tender to joint ventures drawn from the major European and UK contractors.

Section 1 involved a 4km (2.5 miles) length of 'cut and cover' tunnel and a viaduct through the centre of Ashford with links to Ashford International Station, a 3.2km (2 miles) 12m (39ft 4ins) diameter twin-track tunnel deep beneath the North Downs and a 1.25km (0.78 miles) viaduct across the River Medway.

Most of the route was, however, a less demanding surface alignment, following the route of the M2 and M20 motorways for a significant proportion of its length.

Electrification of the CTRL was based on the proven technology as used on French TGV routes using the 2×25kV autotransformer (AT) system, which is described as follows:

The conventional 25kV system uses a single-phase transformer to step down the supply voltage at a grid supply point (usually 132/275 or 400kV) to 25kV. The primary winding is connected to two phases of the high-voltage system and the secondary winding is connected at one end to the OLE and the other end to the track and earth.

The AT system uses a single-phase transformer to step the supply voltage down to 50kV. The centre point of the secondary 50kV winding is connected to the track and is earthed; one end of the winding is connected to the OLE and the other is connected to a feeder wire run on insulators suspended from the OHL structures. The OLE is therefore at +25kV with respect to the track and is −25kV with respect to the feeder wire.

Auto transformers are provided at approximately 15km intervals along the route. They are a single phase 1:1 transformer with both the primary and secondary windings at 50kV and with their centre points connected to the track and earth. Insulated overlaps in the OLE separate each 15km electrical section and a train is always only able to 'see' the 25kV on the OLE. Inside the 15km section the traction current is drawn from the OLE and returned to the track while the action of the autotransformer ensures that an equal but opposite current flows in the feeder wire that also has the effect of reducing interference to other circuits. Outside the 15km section the power is being delivered at 50kV, which halves the current drawn by the train and effectively halves the transmission losses.

AT system supply points can be two to two and a half times further apart than the conventional 25kV system for equal performance levels. This system was subsequently adopted for Crossrail, much of the West Coast Main Line upgrade and for many of the Network Rail electrification schemes undertaken from 2007. Its use was considered for ECML electrification but on balance the initial electrical loadings did not justify it.

CTRL has a feeder station at each end of the tunnel, both supplied from 400kV grid networks.

While construction work was proceeding on Section 1, the decision to proceed with Section 2 was confirmed in 2001, although Railtrack, beset with many financial issues at that time, did not exercise its option to underwrite and acquire this section. Funding for Section 2 came from a mixture of government bonds, Railtrack's purchase of Section 1 and a £2.2bn grant.

Work on the construction of Section 2 began in July 2002. During the year, LCR bought out Railtrack's interests in Section 1 following the placing of the latter organisation into railway administration. Thus, the entire project was therefore brought back into to LCR's control.

This second phase of the project included extending and renovating the Grade 1 St Pancras Station while maintaining the conventional domestic train services operated by the Midland Mainline Franchise, and providing terminal facilities for new Kent high speed services. Three new bridges were also placed over existing railways on the approaches to St Pancras.

A significant part of Section 2 involved tunnelling, including the 19km (12 miles) of tunnels bored beneath East London, either side of a 1km sunken 'box' in which a new station, Stratford International, was built. A new train maintenance depot was also constructed nearby at Temple Mills. From the London Tunnels East Portal at Dagenham to Thurrock, a distance of 7km (4 miles), some 6,000 piles were driven to support the track bed across the inner Thames Marshes.

A further significant civil engineering challenge was presented in constructing the 1km Thurrock viaduct, which crossed both the conventional London, Tilbury and Southend railway line and the Dartford Road Tunnel exit lanes, before diving under the northern wing of Dartford Road Bridge and then dipping almost immediately into the Thames Tunnel.

At Ebbsfleet, part of the embankment carrying the existing North Kent line had to be removed to permit a new bridge to be slid in place to carry this line over the CTRL lines.

Track, signalling, power supply and OLE, termed 'Systemwide' contracts, were let on a whole route basis for both sections. The thinking behind the

'Systemwide' scope and specification was that the new CTRL railway infrastructure should be considered as an extension of the continental high-speed routes and that their 'tried and tested' high-speed railway technology should be adopted. This it was hoped would reduce programme risk and cost overruns in addition to delivering a reliable railway. Unsurprisingly, therefore, the design and construction of track, signalling and OLE mirrored French practice, as did the installation methodology.

Work on the 74km (46 miles) long Section 1 was completed in 2003, and on 16 September 2003 the Prime Minister, Tony Blair, officially opened the first section to Fawkham Junction. Regular passenger services commenced on this first phase on 28 September 2003, with Eurostar trains continuing to use the existing conventional suburban lines to London Waterloo International to complete their journey.

The construction of 39.4km (24.5 miles) Section 2 from the new Ebbsfleet International Station to London was completed early in 2007 with the first test run utilising a Eurostar train arriving at St Pancras Station on 6 March 2007. Testing and commissioning continued in stages during 2007 culminating in the hand-over of the new railway to Union Railways in July. The total cost for the railway outturned at £5.8Bn (£9.7Bn at 2025 prices).

On 3 September 2007, a special Eurostar train undertaking a test run arrived at St Pancras 2h 3m 39s after leaving Gare du Nord in Paris, setting a new UK rail speed record of 334.7km/h (208mph) on the way. On 6 November 2007, Queen Elizabeth II formally opened Section 2 at St Pancras International Station.

Normal passenger services, operated by Class 373 Eurostar sets, commenced on 14 November 2007, replacing the original slower link to London Waterloo. Today the route carries 300km/hr (186 mph) international passenger trains from London St Pancras to Paris Gare du Nord in 2 hours 15 minutes and Brussels in 1 hour 51 minutes, a saving of 20 minutes or more over the London Waterloo services.

Domestic high-speed commuter services, serving the intermediate stations and beyond in Kent, began on 13 December 2009 operated by Class 395 Hitachi units running at speeds of up to 225 kilometres per hour (140mph). Freight traffic is also run at night, hauled by specially adapted Class 92 locomotives permitting continental-size freight containers to reach London.

Of note is the fact that while official legislation, documentation and line-side signage have continued to refer to 'CTRL', in 2006 LCR had decided to adopt the brand name 'High Speed 1' (HS1) for the completed railway.

CTRL reverted to state ownership in 2009. In June 2010, the operation of the route (as HS1 Ltd) was put up for sale with a thirty-year concession awarded to a consortium of Borealis Infrastructure and the Ontario Teachers' Pension

Plan in November 2010. In July 2017 HS1 Ltd was acquired by a consortium of funds advised and managed by InfraRed Capital Partners Limited and Equitix Investment Management Limited. As for the Eurostar train operations, SNCF remains the majority shareholder.

Chapter 12
Post-Privatisation Schemes in England since 2003

Scheme Associated with West Coast Main Line Upgrade

The privatised railway, with its short-term passenger franchises and Railtrack, the infrastructure provider with a similar short-term outlook, provided little incentive to commit investment in electrification projects. In addition, there was a belief in government that other forms of energy, particularly hydrogen technology, would provide a sustainable way forward, obviating the need for electrification.

Consequently, hardly any new electrification schemes were progressed in the first decade following privatisation. The Virgin Group did succeed in securing a longer fifteen-year franchise in order to work with Railtrack to upgrade the West Coast Main Line. While this project included some four-tracking works, the only 'new' electrified route was the 13.5km (8.5 miles) between Crewe and Kidsgrove and this was only initiated to provide a diversionary route for the WCML to minimise disruption during the upgrading works.

The only other Railtrack scheme was in Scotland. This involved reopening and electrifying the 4.7km (3 miles) long Haughhead Junction to Larkhall branch in 2005 (see Chapter 13).

Crewe to Kidsgrove – 9/03

As part of the refurbishment of the West Coast Main Line electrification, undertaken from 1999 to 2009, it was realised that when the section between Stafford and Crewe was being dealt with, it would be extremely helpful if the line between Crewe and Kidsgrove was electrified. The OLE system utilised was the same as used on the elements of the WCML upgrade, namely, UK1 (see Appendix 1).

This was a scheme that had originally been proposed back in the early 1970s but, for various reasons, had not been progressed through to authorisation. It provided a route for WCML trains to bypass worksites located between Colwich and Crewe South Junction. Accordingly, the electrification of the Crewe to Kidsgrove route was included in the Strategic Rail Authority's 'West Coast Strategy' consultation document as an 'enhancement' element.

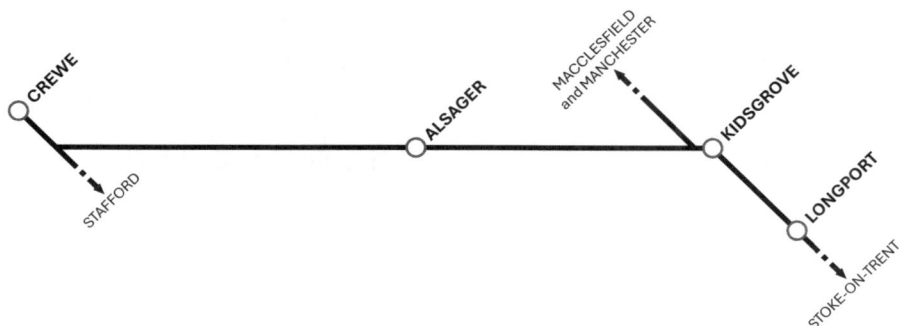

West Coast Main Line Upgrade, Crewe to Kidsgrove

The scheme involved the singling of the first three miles of the route between Crewe South Junction and Alsager and, due to very poor ground conditions, it proved necessary to install slab track over a 400m section just outside of Crewe. This requirement caused the implementation of the scheme to take far longer than had originally been envisaged.

The works were undertaken for a cost of £14M (£25M at 2025 prices).

2007 – Electrification Still off the Agenda

On 27 June 2007, Ruth Kelly became the first female transport secretary of state for nearly forty years under the then new Prime Minister, Gordon Brown. A month later she announced the Department for Transport's (DfT's) thirty-year plan for the future of the railways. The White Paper 'Delivering a Sustainable Railway' included funding for permanent way, signalling and station works at Reading and improvements at 150 stations across the network, including Birmingham New Street. Longer term investments were promised for Crossrail and the upgrading of Thameslink.

Focusing mainly on the Network Rail Control Period 4 (2009–14), the paper generally favoured making better use of existing infrastructure, rather than major new projects such as new high-speed lines and electrification. The White Paper assumed that alternative forms of traction energy, particularly hydrogen fuel cells, would be developed commercially and become available in the 2020s. The report 'glossed over' the fact that the properties of hydrogen, dictated by the laws of science, would limit its application as a fuel for rail use.

Rail electrification projects would only be progressed if strict 'pay-back' criteria could be met over a mere ten to fifteen-year period. Such schemes would only be considered on a 'case by case' basis – there was to be no network approach. Scant consideration was given to the community, journey time or environmental benefits that electrification would deliver. Incongruously, it was stated that electrification would be more cost effective following the migration to in-cab signalling.

The strategy was heavily criticised at the time for lacking in both ambition and vision.

The Changing Mood

The mood in government was, however, changing, as exemplified by the presentation given at the Railway Forum in October 2007 by Derek Chapman of the Department for Transport. The paper was entitled 'Choosing Sustainable Power'. In his presentation, he referred to the joint initiative that had recently got underway between the Association of Train Operating Companies and Network Rail. He hypothesised that a virtuous cycle of reduced capital costs might enable a rolling programme of electrification to go forward with a coordinated rolling stock plan. Such an approach would permit electrification resources and momentum to build in a managed way, and he considered that the case for electrification would continue to strengthen.

In 2007, the Rail Safety and Standards Board (RSSB) commissioned a report from W.S. Atkins for further electrification of Britain's railways (ref. T633: Study on further electrification of Britain's railway network). The report gave a methodology for preparing cost estimates for electrification and it contains worked examples of the estimated costs for several routes, including costs for the GWML and MML routes (see Appendix 4). The report was based on cost estimates that ranged from £500k to £650k per STK.

For the Association of Train Operating Companies, Chris Stokes, a former senior British Rail and SRA director, highlighted the benefits of electrification as follows:

- fuel savings, including regeneration
- reduced rolling stock maintenance costs
- reduced rolling stock lease costs
- revenue benefits, reflecting modest journey time improvements
- lower variable track access charges due to lighter electric train axle-loads.

The ball was at last starting to roll again in favour of rail electrification.

Crossrail Authorisation (see also Chapter 11)

The first scheme authorised involving electrification was the revamped Crossrail Project in July 2008, when the Crossrail Act received Royal Assent. This act gave permission for the new railway to be built, operated, maintained and integrated within the National Railway system. The 117km route (73 miles) was to be electrified at 25kV AC 50Hz and was to make seamless connections with the existing lines that would become part of the route. This included electrifying the Great Western Main Line from Airport Junction to Maidenhead as part of the scheme. Physical work on the project commenced in 2009. Subsequently, the project was extended to Reading and, as noted earlier, renamed the Elizabeth Line.

HS2 Phases 1 and 2

History

High Speed 2 (HS2) was originally proposed by the Labour government. On 5 June 2009, Lord Adonis was elevated to the position of Secretary of State for Transport, becoming a member of both the Cabinet and the Privy Council. Adonis was keen to encourage more expenditure on infrastructure and, under his stewardship, proposals were also first brought forward to build High Speed 2 (HS2). The new project was aimed at reducing journey times to Birmingham and the North-West and provide increased capacity to supplement the West Coast Main Line. Passenger traffic on that route had increased by 50 per cent and freight by 40 per cent over the previous decade.

As originally conceived, High Speed Two (HS2) Ltd was set up with the intention of building a high-speed line from London with a fork to Birmingham and a link to the West Coast Main Line. Sir David Rowlands was the first chairperson who was charged with identifying a route between the West Midlands and London, and options for connectivity with London Heathrow Airport, Crossrail, HS1 and the national rail network. The HS2 report, including options and supporting information, was published in March 2010.

On taking office in May 2010, the Conservative and Liberal government undertook a review of the plans for HS2 that it had inherited. Based on proposals put forward by Lord Adonis under the previous Labour government, a Y-shaped plan including branches to Leeds and Manchester, plus a connection to HS1, was published in December 2010. It was also decided that the priority for HS2 was to serve Northern England directly, and any consideration of a route via Heathrow Airport was dropped in preference for a branch to the airport.

Consultation documents were released on 11 February 2011, and the go-ahead was announced by Secretary of State for Transport Justine Greening in January 2012. The line was to have the capacity to carry 26,000 people per hour at speeds of up to 400 kilometres per hour (250mph).

The Chancellor, George Osborne, announced in November 2015 that the initial construction phase of the HS2 line would be extended to Crewe, which would become a 'hub station'. This section was to be built as Phase 2a, to follow on from Phase 1 between London and Birmingham.

Following the passing of the High-Speed Rail (Preparation) Act on 21 November 2013, the hybrid bill for Phase 1 was enacted as the 'High-Speed

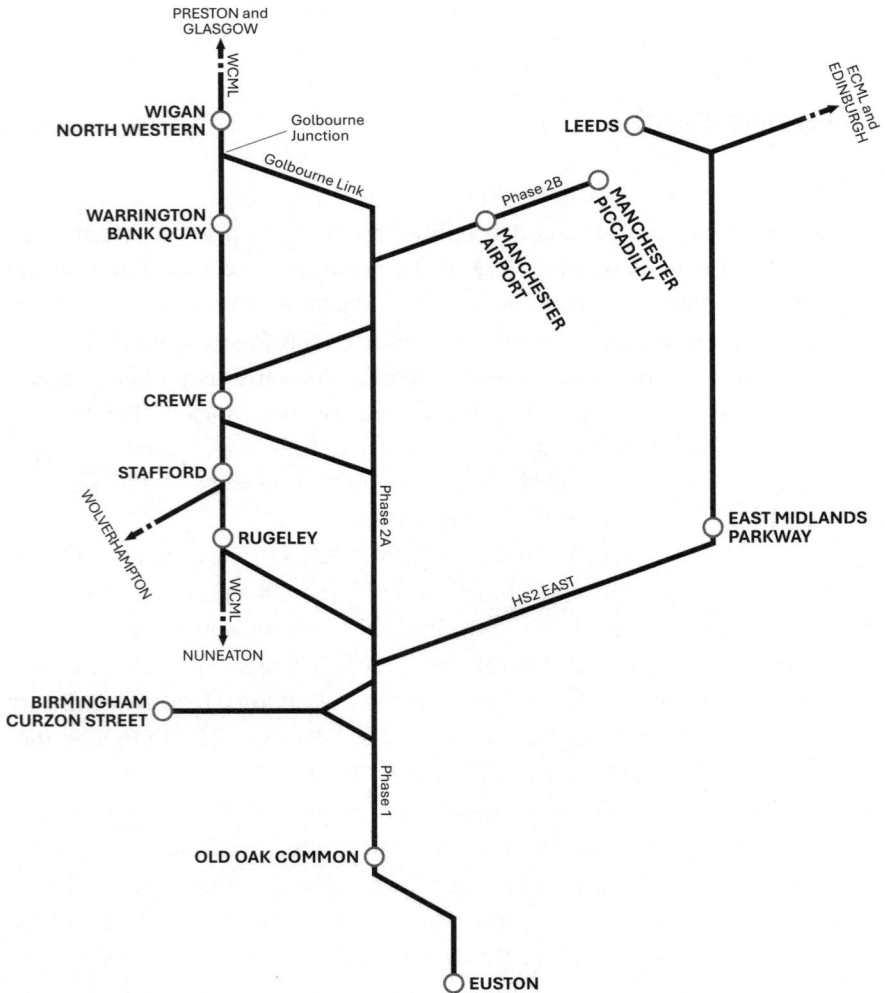

HS2

Rail (London – West Midland) Act 2017' on 23 February 2017. At the second reading of the latter bill, the two-kilometre-long link between HS2 and HS1 was dropped as it was considered that the £700M cost could not be justified for the limited intermittent service that would use it.

Advance Preparations

Preparation works then commenced, including demolishing the former carriage shed at Euston, together with numerous archaeological studies of affected ancient sites and the exhumations at St James's Gardens burial grounds. In addition, the project announced the intention to create extensive 'green corridors' for wildlife habitats.

Contracts were awarded to develop detailed plans for the four key stations for Stage 1 in February 2018 as follows:

- London Euston – Ove Arup & Partners International Limited (working with Grimshaw Architects LLP)
- Old Oak Common – WSP UK Limited (working with Wilkinson Eyre Architects Limited)
- Birmingham Curzon Street – WSP UK Limited (working with Grimshaw Architects LLP)
- Birmingham Interchange – Ove Arup & Partners International Limited (working with Arup Associates and Wilkinson Eyre Architects Limited).

The envisaged revamped Euston Station, it was announced, would comprise an additional eleven platforms for HS2 services, which would double the capacity of the station. A new concourse would be provided linking to an expanded Underground station with a link to the nearby Euston Square tube station. The Mace Dragados Joint Venture were appointed as construction partners, to build the new station, in February 2019.

However, by 2018 the HS2 project was running into difficulties. It had become apparent that the projected costs for HS2 were significantly greater than the original costings. This caused much negative media comment and the groundswell of criticism by groups opposed to the railway heightened the political difficulties that the government found itself in.

In addition, concerns about the limited interconnectivity between HS2 and the aspirations for Northern Powerhouse Rail were growing. More widely, the Conservative government had long considered that better rail services could play a key part in its so called 'Levelling Up' initiative, not only to improve links between the north and south of the country but also to bring communities across northern England and the Midlands closer together.

A review was undertaken by Douglas Oakervee, a former President of the Institution of Civil Engineers, in 2019 and the National Infrastructure Commission (NIC) carried out a 'Rail Needs Assessment for the Midlands and the North' in 2020.

Following these reviews, the government launched the Integrated Rail Plan for the North and Midlands (IRP), which substituted electrifying the section of the Midland Main Line route between Market Harborough and Sheffield as a partial replacement for the HS2 Phase 2b (East). Phase 2a (West) was to be retained. However, further cutting back was quietly announced by the government on 6 June 2022 when the 21km (13 miles) Golborne Link was dropped, primarily for local political reasons. This important link would have increased capacity and cut journey times to Glasgow.

Further cutbacks were announced by Prime Minister Rishi Sunak on 4 October 2023. He stated that the government now proposed to deliver a broad range of transport initiatives in place of investing in Phase 2 of HS2. It was also indicated that the route section between Old Oak Common and Euston would not proceed unless a significant level of private sector finance became available to fund the Euston Station enhancements. Furthermore, Parliamentary permission was to be sought to adapt the High Speed Rail (Crewe – Manchester) Bill to remove any elements required for HS2 purposes.

With the two new tunnel boring machines about to be delivered to site in October 2024, the recently elected Labour government announced that the section between Old Oak Common and Euston would go ahead as originally planned, even though the scope for Euston Station had still to be finalised.

Rolling Stock

In December 2021 HS2 Ltd confirmed that a £2Bn contract to design, build and maintain the new rolling stock was to be awarded to the Hitachi/Alstom Joint Venture. The new trains are to be constructed at factories in Derby and County Durham and are to be capable of speeds of up to 360kmh (225mph). These fully electric trains will also run on the existing network to places such as Glasgow, Liverpool, Manchester and the North-West. The rolling stock is to incorporate the latest technology from the Japanese Shinkansen system and European high-speed network. It is considered that the new trains will be the quietest and most energy-efficient high-speed rolling stock operating anywhere in the world.

On conventional track, as the new trains will not be capable of tilting, their maximum speed may be restricted compared to existing Pendolino sets, but they would be capable of operating at up to 225kmh (140mph), where track alignments permit when future signalling systems are implemented.

Permanent Way

Concrete slab track is proposed for most of the route, which it is hoped should reduce maintenance costs. A systems-wide contract is to be let with the contractor also responsible for the design, installation and testing of switches and crossings.

Power Supply, Overhead Catenary and Systemwide System (OCS)

The power system is to be the 0-25kV-0 AC, similar to that provided for HS1. Four feeder stations, connected to the National Grid, are to be provided along the Phase 1 route. Autotransformer stations are also required at approximately 10km intervals.

The high voltage (HV) power supply system contract, estimated to be worth £523M, includes the design, manufacture, supply, installation, testing, commissioning and maintenance of the HV power supply systems, both for traction and ancillaries.

The overhead catenary system (OCS) is to be the first system in Europe to be certified for speeds of up to 360km/h (225mph). HS2 have worked with SNCF to upgrade the 350km/h (218mph) SNCF Réseau system. The design team have found ways to reduce the steel and concrete that will be needed by increasing the distance between supporting catenary structures, and at the same time, succeeded in balancing the tensile load in the catenary. This should reduce the overall cost of the system as well as limiting its environmental impact.

The new design achieved TSI certification. In addition, the licence agreement negotiated opened opportunities for smaller suppliers in the UK by permitting HS2's contractor to tender the OCS components to secure the most competitive supplier.

Systems-wide contracts cover signalling, the overhead catenary that will provide power to the trains, mechanical and electrical systems, high-voltage power supplies, communications systems and the control centre at Washwood Heath, Birmingham.

Industrywide Issues

Skills Shortages

Prior to the commencement of Crossrail in 2008, approximately one-third of the national rail network was electrified. However, apart from the 'French-inspired' new build Channel Tunnel Rail Link (HS1), only the short sections of route between Crewe and Kidsgrove and the Larkhill branch had been electrified in the previous ten years. In the five years prior to this, there had only been the relatively modest Heathrow Express (1998) and Leeds North West (1994) overhead line electrification schemes. This stop–start nature of working meant that there had been a huge loss of electrification experience and expertise since the demise of British Rail's programme. Furthermore, the recent project management history of the West Coast (equipment renewal) Project did not inspire confidence in Network Rail's ability to effectively manage rail upgrade works.

The rail industry was therefore not in a great shape from either a technical or project management point of view to take on the challenge of a number of large electrification programmes. However, on the positive side, both the government and industry appeared to be in favour of a rolling programme of electrification that clearly should improve efficiency and drive down capital costs. The figures bandied around at the time for the cost of a single track kilometre at £550k to £650k were based on British Rail's experience, including an allowance for inflation (sixteen years later, these figures seem incredibly optimistic!) and were generally accepted across the industry, even though it was recognised that there were supply-side skills shortages.

The Railway Industry Association, for example, had reported in an annual business survey that in virtually all sectors of the industry skilled labour availability was generally the leading constraint on output. Network Rail's predecessor, Railtrack, had greatly weakened the organisation's skills base in its misguided policy of replacing competent and experienced engineers, who had been able to exercise their professional judgement, with project managers who were generally more junior in status and usually on lower salaries. Many of these new project managers were recruited externally having little, if any, first-hand rail experience. They were appointed to manage projects with little more than 'tick-the-box' processes assessed against a plethora of new, rigid standards and often unrealistic Gant charts. The initial salary savings made by easing out senior people and replacing them with more junior staff were soon overwhelmed by the spiralling costs of projects running out of control.

Cultural, Corporate and Labour Issues

At this time, the changing culture of the Railtrack organisation often made for an unpleasant environment for experienced railway engineers and operators. Some senior people in Railtrack viewed experienced railway staff in a negative light as 'Old Railway'. As a result, many experienced staff either retired early or left Railtrack under redundancy terms soon after privatisation to find new employments with other companies or to set themselves up as consultants.

World-wide, railways were booming and there was a huge international market for such scarce skills, especially for designers and project managers. Most had no trouble finding new employment with a significant proportion embarking on new careers overseas. When it subsequently became apparent that Railtrack needed experienced engineers and operators to put its house in order in the early 2000s, many of the people who the organisation had shed a few years earlier were re-employed as consultants with remuneration rates several times their previous salary.

Labour costs too were rising during this period, especially for artisan staff. British Rail had recruited its last apprentices in the early 1990s and virtually no training has been undertaken for key railway trades for more than a decade. Accordingly, skilled artisans with railway experience became a scarce commodity in the country's booming economy either demanding enhanced remuneration or leaving the industry for higher wages, thus exacerbating the rail industry's predicament.

In addition, the whole structure of the industry had changed massively since British Rail days. As an example, the nationalised industry often undertook all design work. Usually the specialised proportion of any project work was done in-house, and the organisation was self-insured against incidents, accidents, construction and operations risks. Post-privatisation, consultants, contractors and train operators generally took on all elements of their requisite tasks, including design, construction and operations as appropriate. As a result, each organisation had to insure itself to the tune of many millions of pounds at the outset. The insurance industry had little or no experience of rail risks and premiums were high. Following the serious rail accidents at Ladbroke Grove, Potters Bar and Hatfield, the insurers became more uneasy still about taking on rail business, and premiums increased further. These additional costs, along with a plethora of other inflated legal, financial, project management and enhanced safety requirement costs drove up project costs substantially.

Government-backed Electrification Again

By the close of 2008, the government appeared to be taking a more positive approach to electrification, at least as far as 'the heavily used parts of the rail network' were concerned. Ruth Kelly stood down as the Secretary of State for Transport in October 2008 to be replaced by Geoff Hoon who, on his appointment, said that he wanted to see more railway electrification. He set up a new National Networks Strategy Group, chaired by the Minister of State for Transport Lord Adonis, comprising senior officials from the Highways Agency,

Network Rail, the Treasury and other government departments to examine the nation's future transport options.

The Secretary of State for Transport announced in January 2009 that consideration was to be given to reviewing the case for electrifying the Great Western and Midland Mainline routes and that a new company, High Speed 2, would be set up to examine the feasibility of new rail routes to the West Midlands and Scotland. The Association of Train Operating Companies welcomed the government's announcement as 'an endorsement of the work carried out by train operators and industry partners and as a vote of confidence in rail'.

In June 2009, the new Secretary of State for Transport, Lord Adonis, made it clear that he was keen to encourage more expenditure on infrastructure including railway electrification. At about this time, Network Rail published its draft Route Utilisation Study: Electrification (RUS) Report.

The RUS costings were based on the earlier 2007 W.S. Atkins report (that is, £500k to £650K per STK) inflated to 2009 prices, although no actual figures were quoted by Network Rail.

The RUS indicated that there were opportunities for cost reductions using new delivery techniques for the electrification of the GWML. It went on to state that the project would have a benefit-cost ratio (BCR) between 'high value for money' and a 'positive financial case' over the appraisal period. Furthermore, the report added that the Midland Main Line scheme would offer a 'positive financial case' over the appraisal period (see Appendix 4).

Following publication of the draft RUS, Lord Adonis announced proposals for the 25kV electrification of the Great Western Main Line between London, Bristol, Cardiff and Swansea in July 2009. In addition, electrification of the Liverpool to Manchester Northern Route was also approved.

A number of infill schemes in North West England, including the lines between Manchester and Preston, were also subsequently put forward. Prospects in the North West were enhanced in December 2009 when Lord Adonis added electrification of the line between Huyton and St Helens Junction, effectively electrifying the entire Liverpool to Wigan route.

These schemes in the North West took cognisance of significant proposals to improve services in and around Manchester recommended by a steering group comprising Transport for Greater Manchester, Merseytravel, Network Rail, Northern Rail, First TransPennine Express, East Midlands Trains, Cross-Country, Freightliner, Deutsche Bahn and the DfT. These ideas developed into a full-blown project, termed the Manchester Hub, which included the construction of the Ordsall Chord, creating a new direct link between Manchester Piccadilly and Manchester Victoria stations. Network Rail did further work on the project and released detailed proposals in February 2010. Costs were originally estimated at £530M, later revised to £550M/£560M (£800M at 2025 prices).

Preliminary work then commenced on both the Great Western Main Line and Liverpool to Manchester electrification schemes with planned completion dates of 2018 and 2013 respectively. However, following a change of government in 2010, both projects were put on hold for further review, which delayed progress for a while.

Following the Labour government's defeat at the May 2010 General Election, Lord Adonis withdrew from frontline politics and had no further direct involvement with rail electrification until October 2015 when he took over the chairmanship of the National Infrastructure Commission (NIC). This body was tasked with carrying out a wide-ranging national infrastructure assessment.

Coalition Government (2010–2015)

The new coalition government, headed up by Prime Minister David Cameron, pledged increased investment in the northern provinces, particularly in transport infrastructure, identifying Liverpool, Manchester, Sheffield, Leeds and Hull as the core cities. The plan to improve rail journey times between these cities was developed from the 2009 Manchester Hub scheme. Later renamed the 'Northern Hub', the project was, in addition, to encompass TransPennine services operating from Liverpool to Leeds. These services were to be diverted from the Liverpool to Manchester Southern Route to the faster Northern Route running via Newton-le-Willows and Manchester Victoria. The latter station would become a key interchange point and the new Ordsall Chord line, between Manchester Victoria and Manchester Oxford Road, would permit rail services from the North East to operate to Manchester Airport via Manchester Piccadilly without reversing

Initially, however, in view of the UK's then poor economic situation, the Conservative–Liberal Democrat coalition government froze all major capital expenditure on public schemes, including the Liverpool to Manchester Northern Route and Great Western electrifications pending a full financial review.

As the results of the ongoing review became clearer, some elements of the schemes that had commenced prior to the election were given the green light to restart.

Significant support for the Northern Hub scheme had grown among local politicians and business leaders as well as the then Chancellor, George Osborne. Accordingly, £85M of funding for the Ordsall Chord was included in the Chancellor of the Exchequer's budget of 23 March 2011. George Osborne also stated that other aspects of the Northern Hub scheme were to be reviewed to ensure value for money. In his March budget the following year he approved a further £130M with government approval for the full £560M (£800M at 2025 prices) scheme following on 16 July 2012.

The Secretary of State for Transport authorised the electrification of the Maidenhead (that is, at the limit of the Crossrail works) to Newbury and also to Oxford via Didcot lines in November 2010 and the extension from Didcot to Bristol Temple Meads and Cardiff via Bristol Parkway in March 2011. A major step forward for the latter scheme occurred in March 2012 with the award of a £700M contract to Amey for the electrification of the Great Western Main Line.

Initial funding was also released during this early period of the Coalition's tenure, for the electrification of the TransPennine Manchester to Leeds line.

Following the financial review, the coalition government announced £4.2Bn (£5.9Bn at 2025 prices) worth of schemes for new 25kV electrification schemes in July 2012. This was to be the start of a £38Bn (£53Bn at 2025 prices) rolling programme of electrification and enhancements, heralded as the biggest investment in railways since the Victorian era. The drive by the administration to see more railway electrification was, in the main, to deliver the 'greener future' promised in the Conservative Party's election manifesto. These rail investments, it was stated, were vital and would also bring faster, cleaner and more reliable services.

The package of works largely reconfirmed the projects previously announced by Lord Adonis, including the Liverpool to Manchester Northern Route, other Northern infill schemes, which had expanded into the Northern Hub Project and the electrification of the associated TransPennine route.

Also re-confirmed was the Great Western electrification, which would now be extended to Swansea and include the Windsor, Marlow and Henley branch lines. The scheme would also encompass further enhancements to become the Great Western Main Line Route Modernisation Project. New schemes also announced included the Gospel Oak to Barking line and the West Midlands Suburban Lines Scheme. The Midland Main Line electrification to Nottingham and Sheffield was also reconfirmed and would, in addition, form a key element of a visionary concept, dubbed the 'Electric Spine'. This would form a 25kV AC electrified north–south route for freight between the Port of Southampton and major cities in northern and central England, including an inland port in the Midlands.

In August 2013, the Department for Transport announced that further routes in the North West of England would be electrified, including the North trans-Pennine route between Manchester Victoria, and York via Huddersfield and Leeds. Also, the Manchester to Preston Line and the Blackpool North and Windermere branches were to be wired.

There follows a summary of the early progress of these schemes.

Midland Main Line Extension and the Electric Spine

In 2006, the Strategic Rail Authority published a Route Utilisation Strategy (see Appendix 4). For the Midland Main Line, the plan took cognisance of the fact that traffic levels on the route were growing faster than the national average and that past rationalisations were constraining capacity. Network Rail subsequently published a plan to enhance capacity and electrify the route.

The Department of Transport announced plans on 16 July 2012 to increase the capacity of the Midland Main Line and electrify the route at an anticipated cost of £800M (£1.2Bn at 2025 prices). This included upgrading the existing electrified 'Bedpan' lines between London St Pancras and Bedford to increase capacity. Network Rail initially planned to undertake the work utilising UK Master Series OLE in stages as follows:

- Bedford to Corby by 2017
- Kettering to Leicester/Nottingham/Derby by 2019 and
- Derby to Sheffield by 2020.

On completion of the electrification, it was planned for ten-car Class 801 electric units to operate the long-distance services from London to Leicester, Nottingham, Derby and Sheffield. The north Northamptonshire towns of Wellingborough, Kettering and Corby were experiencing significant population growth and so a new fast and frequent Outer Suburban service to London St Pancras was also envisaged.

Further enhancements to deliver additional freight and passenger train paths were also included. To accommodate longer and heavier freight trains, two 775 metre freight loops were to be provided, one to the south of Bedford and the other between Kettering and Leicester. In addition, running lines that had been removed under 1970s rationalisations would be reinstated to further increase route capacity.

The Cameron government were keen to see an efficient electric north–south rail freight route that would take many lorry movements off the over-crowded road network. Accordingly, the Department for Transport encouraged Network Rail to promote a north–south rail axis, which, as already noted, was termed the 'Electric Spine' project. In part, this involved routing many of the proposed freight services over the electrified Midland Main Line. The two projects therefore effectively complemented each other.

The Electric Spine would provide a 25kV AC electrified route, with gauge clearance for large shipping containers, from the Port of Southampton to major cities in central and northern England. There would also be connections to adjacent electrified routes, depots and freight facilities including an inland container port.

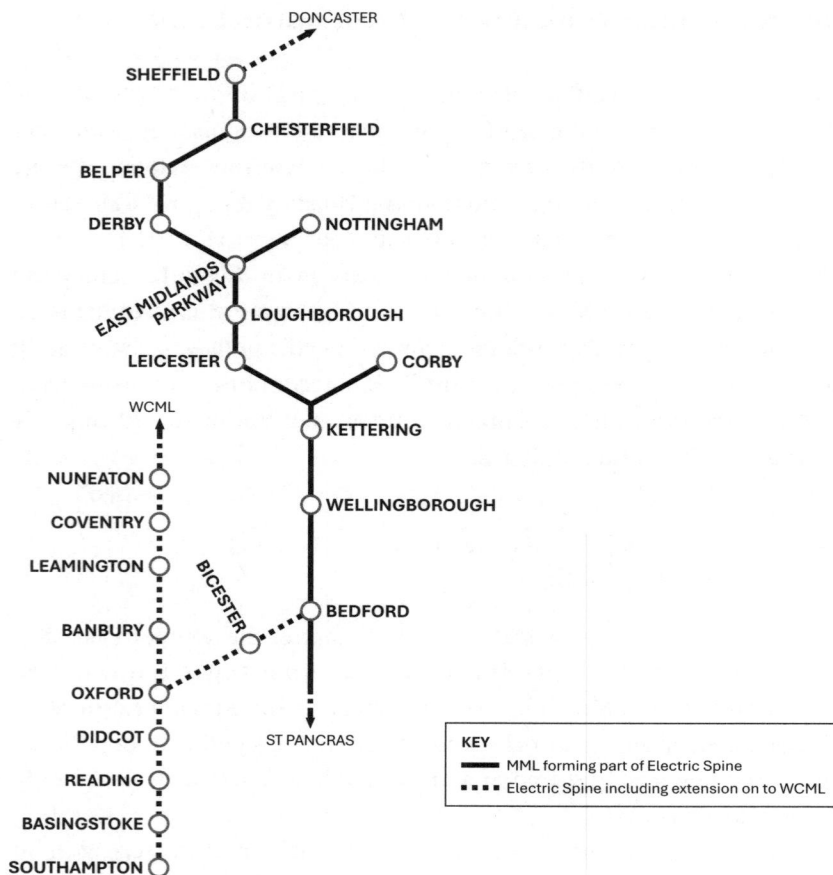

```
                        DONCASTER
              SHEFFIELD ○ ·····►
                         ○ CHESTERFIELD
          BELPER ○
          DERBY ○       ○ NOTTINGHAM
    EAST MIDLANDS ○
      PARKWAY      ○ LOUGHBOROUGH
        LEICESTER ○       ○ CORBY
  WCML
                   ○ KETTERING
      NUNEATON ○
      COVENTRY ○   ○ WELLINGBOROUGH
    LEAMINGTON ○
                BICESTER
       BANBURY ○      ○ BEDFORD
                  ○·····
        OXFORD ○···
        DIDCOT ○    ST PANCRAS
       READING ○
    BASINGSTOKE ○
   SOUTHAMPTON ○
```

KEY
— MML forming part of Electric Spine
▪▪▪▪ Electric Spine including extension on to WCML

Midland Main Line Extension and the Electric Spine

The project involved converting the third-rail 750V DC line from Southampton to Basingstoke to 25kV AC and electrifying the following route sections to form a continuous 25kV AC electrified route to the north linking:

- Basingstoke to Reading
- Reading to Aynho Junction and Banbury via Oxford
- Oxford to Bedford
- Banbury to Leamington Spa
- Leamington Spa to Coventry
- Coventry to Nuneaton, and
- Sheffield to Doncaster.

The project included the reinstatement and electrification of the 74km long East–West Rail Link from Oxford via Bicester and Bletchley to Bedford by 2019 as an integral part of the Electric Spine. Capacity upgrades would be implemented at Reading, Derby and Leicester, and the line between Leamington to Coventry was to be redoubled.

Excluding the reinstatement of the East–West Rail Link and the Midland Main Line, as noted earlier, the initial costing of the Electric Spine Project was put at £800M (£1.2Bn at 2025 prices). Subsequently estimates increased to close to £3Bn.

The electrification between Reading and Basingstoke was to be completed at an early stage from 2014 but, as technical and operational difficulties were foreseen in attempting re-electrifying the Southampton to Basingstoke line at 25kV AC, this section was deferred indefinitely. Not all rail freight companies were on-board with the Electric Spine proposals, and many considered that electrification funding should be directed towards the Felixstowe and Harwich routes. These factors, taken along with the ballooning costings, resulted in the Electric Spine Project being put quietly on the back burner.

The Midland Main Line work involved upgrading the original OLE installed between London and Bedford, as it was only fit for 100mph operation, in addition to extending the electrification north of Bedford. The works also included reinstating the fourth track between Sharnbrook and Kettering. The new electrification works would utilise the new Series 2 design of overhead line equipment, which is designed for 125mph running.

However, the work was soon slipping behind schedule and costs on the project were also overrunning. On 15 June 2015, just after the General Election, Transport Secretary Patrick McLoughlin paused the project for a review. Three months later the DfT revised the completion dates for the stages to:

- Bedford to Corby by 2019
- Kettering to Leicester/Nottingham/Derby by 2019 and
- Derby to Sheffield by 2023.

However, in November 2016 the Department for Transport announced that electrification of the Midland Main Line north of Kettering and Corby would be deferred. No timescales were given for completing the scheme.

A further announcement was made on 20 July 2017 by Transport Secretary Chris Grayling that the Kettering–Nottingham–Sheffield electrification project was to be cancelled and that bi-mode (electro-diesel) trains would be used on the route in the future. Cancellation of the Electric Spine, including the de-funding of the electrification of the East–West Rail Link was also formally announced at about this time. The decision, particularly regarding the East–West Rail Link electrification, was heavily criticised at the time.

On 23 January 2021, Transport Secretary Grant Shapps announced a £760M investment package for the delivery of the Bicester to Bletchley section of the East–West Rail Project, but, unfortunately, the decision not to electrify the route was left unchanged. In connection with the renovation of the Bletchley Flyover, a 'scrap' of electrification work was required to be undertaken by the East–West project team. This involved renewing 300 metres of the four-track West Coast Main Line OLE. The work was required because the track, formation and base slabs of the flyover need to be lifted to permit the refurbishment of the sixty-year-old structure to take place. The electrification of the WCML in the mid-1960s used fixed overhead wires on the flyover, but these have been replaced using new OLE masts supporting 'catenary' wires.

However, with the ongoing decarbonisation debate heating up (no pun intended), the pressure to restore the electrification element to the East–West Rail Link is likely to increase, particularly if the opportunity to operate electric freight trains over the route transpires.

In this regard, by 2021 the government did seem to be having second thoughts about the matter. When answering a parliamentary question on 18 February 2021, the Parliamentary Under-Secretary (Department of Transport), Baroness Vere of Norbiton, said, 'The case for the electrification of the East–West Rail Link is being considered, which includes consideration of full electrification along the whole route, as well as options for partial electrification using battery-electric hybrid rolling stock, or hydrogen traction.'

Reverting to the Midland Main Line, Network Rail announced on 6 November 2017 that Carillion Powerlines had been awarded a £260M (£340M at 2025 prices) contract for the electrification of the line between Bedford, Kettering and Corby. The same company was further awarded a £62M (£82M at 2025 prices) contract to upgrade the track on the route. The OLE system utilised was the UKMS25 system (see Appendix 1).

However, for practical reasons, on 26 February 2019 Parliamentary Under Secretary of State for Transport Peter Jones announced that a 15km extension of electrification would go ahead from Kettering to Market Harborough. This was primarily because it was realised that additional power supplies would be required, and the only suitable point for a substation was at a place called Braybrooke, which is some 3km (2 miles) south of Market Harborough. It was therefore decided that it would be sensible to take the overhead line equipment through to Market Harborough station and terminate there. After a pause, work began on extending electrification to Market Harborough in that year with firm plans to extend further to South Wigston, work on which was underway during 2023. The OLE system utilised was again the UKMS25 system (see Appendix 1).

Unsurprisingly, there has been much disquiet within local political and business circles concerning the cutting back of the Midland Main Line electrification. Many stakeholders called for the wires to serve Nottingham, Derby and Sheffield to be installed as originally planned.

A chink of light was detected on 17 September 2020, when during a House of Commons debate on transport, HS2 Minister Andrew Stephenson, responding to a question from Alex Norris (Labour/Co-op, Nottingham North), said, 'We are currently delivering the Midland Main Line upgrade, which includes electrification from London to Kettering, with additional electrification to Market Harborough being developed. Further electrification of the MML is currently at an early stage, but it is being examined by Network Rail.' Stephenson went on to say that the Department for Transport would continue to work closely with Network Rail to develop proposals to advance the delivery of electrification across the route.

In May 2021, electric services began operating to Corby and Kettering utilising Class 360 EMUs, the works having been further delayed due to the Covid pandemic from the spring of 2020. On 21 December 2021 the DfT officially announced that work would start on 24 December 2021 to electrify the 13km (8 miles) section of line between Kettering and Market Harborough.

In November 2021, as noted earlier, the government launched the Integrated Rail Plan for the North and Midlands, which included electrifying the section of the route between Market Harborough and Sheffield as a partial replacement for the HS2 Phase 2 East (see Chapter 16).

Subsequently, with the Prime Minister cancelling HST Phase 2 on 4 October 2023, a statement following the announcement intimated that some of the diverted funds would be used to complete the Midland Main Line (MML) electrification, and it is likely that this would be undertaken in seven piecemeal stages.

Gospel Oak to Barking – 5/18

The Gospel Oak to Barking line has existed in its current form since 1981 (see Chapter 14) and is made up of various lines, namely:

- Tottenham and Hampstead Junction Railway (opened in 1868 by the Great Eastern Railway but subsequently becoming a joint line with the Midland Railway) and
- a link from South Tottenham to Woodgrange Park, Barking built by the Tottenham and Forest Gate Railway (joint between the Midland and London, Tilbury & Southend railways) to serve a major housing development (1894).

Over the years, the passenger service between Upper Holloway, South Tottenham and Woodgrange Park has been linked to a number of termini at both ends of the route. These included St Pancras, Moorgate, East Ham and Barking. A service to Gospel Oak was added in 1888 but abandoned in 1926 due to poor patronage, and subsequently the service to East Ham was closed in 1958, all trains then terminating at Barking. Beeching actually proposed the withdrawal of the local service in 1963, but with the aid of concerted local community pressure, the line survived, albeit in a deteriorating state of reliability and repair. By 1980, services had been cut back to an hourly service between Kentish Town and Barking.

However, with a changing demographic and the enhanced commercial activities within the capital since the 1980s, the route became a busy commuter line and, as a result, the quality of the service was improved from 1981. Electrification and upgrades in connection with the Bedpan scheme, and later Thameslink, displaced the hourly service from Kentish Town and the service was diverted to run from Gospel Oak. The revised train plan had a two train per hour service on this now 22km (14 miles) long route.

Often referred to as the GOBLIN Route, the line also carries heavy freight services that have significantly increased following the openings of the Thames Gateway port and the major growth in activity at Tilbury.

At privatisation, and for most of the period thereafter, this twelve-station line was operated under the Silverlink brand on Railtrack, later Network Rail, infrastructure. Despite the service quality improvements mentioned earlier, the line still suffered from limited investment in both stations and trains.

Approval was given in July 2013 to electrify the route with 25kV AC NR Series 2 overhead line equipment.

The work involved substantial civil engineering works. However, the project was beset by a number of problems concerning the design of the enhancements, and the inherent delays were compounded by the late delivery of materials.

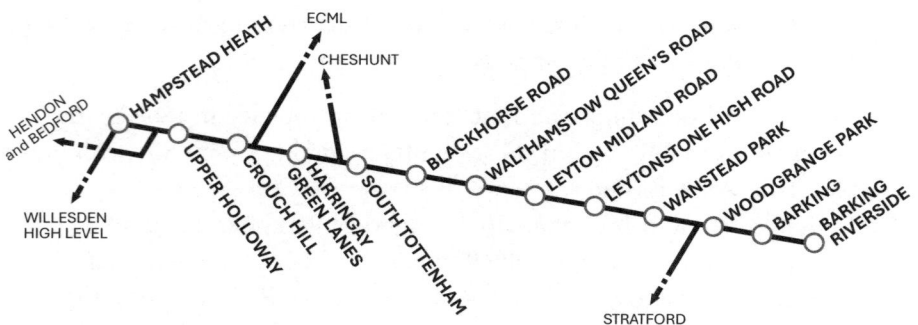

Gospel Oak to Barking

Originally planned for commissioning in June 2017, the infrastructure work was eventually completed in May 2018. However, deliveries of the new Class 710 trains were delayed – these being finally introduced from May to August 2019.

A 4.5km (3 miles) electrified extension to Barking Riverside from the Tilbury Loop line between Barking and Dagenham Dock stations was commissioned in June 2021 six months ahead of plan. The new terminus station is located in the town square at the heart of the Barking Riverside development and has step-free access.

The Chase Line (Walsall to Rugeley, Trent Valley) – 4/19

The electrification of this route was included in the original 1956–67 electrification of the West Coast Main Line programme but was dropped in the reauthorisation review in the 1960s. In July 2012, the government decided that this route should be wired, and that work would start in 2014 at a cost of £36M (£50M at 2025 prices). In anticipation of this electrification, the route was resignalled, with the closure of three manual signal boxes in August 2013. Control was transferred to the Saltley signalling centre. Again, the NR Series 2 OLE system was utilised.

The scheme should have been completed by December 2017 but, due to the presence of old mine workings, which had not been foreseen, there was an impact on the foundations for the overhead line masts. The work was delayed by twelve months and electric services finally commenced in May 2019. The general line speed limit was uplifted from 72kmh (45mph) to 97kmh (60mph). Bloxwich level crossing was closed, and a severe local permanent speed restriction was raised from 40kmh to 80kmh (25 to 50mph). The resulting service provided shorter journey times with more trains (2tph) and more seats.

An interesting event occurred in that the local MP tried to impose speed limits on the new trains in order to minimise the impact of the noise of the new service on local residents! Fortunately, this initiative of his was not successful.

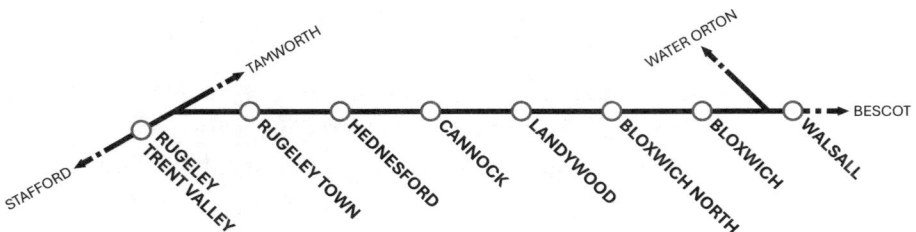

The Chase Line – Walsall to Rugeley, Trent Valley

Great Western Main Line – 6/20

Original Project Scope

The decision to electrify the Great Western Main Line was a political one taken in 2009. The government of the day were keen to demonstrate to the people of South Wales that they had their interests in mind and, after the briefest consultation with Network Rail, announced that the route from London to Swansea, including both routes to Bristol, would be electrified at an early date. The net result of this was that a project team was put together in great haste and the project remit had to be developed in retrospect.

As previously noted, electrification between Airport Junction and Maidenhead had proceeded under the auspices of the Crossrail Project, with this scheme subsequently extended to include electrification to Reading. Following Lord Adonis's announcement in July 2009, electrification work on Great Western commenced in June 2010 and was due to be completed in 2016–17.

The original scope of the £874M (£1.35M at 2025 prices) project included electrification of the following route sections:

- Maidenhead (later to become Reading) to Bristol Temple Meads via Swindon, Chippenham and Bath Spa
- the South Wales Main Line from Wootton Bassett Junction to Swansea via Bristol Parkway, the Severn Tunnel and Cardiff
- Bristol Parkway to Bristol Temple Meads
- Didcot to Oxford, and
- Reading to Newbury.

Subsequently, the Windsor, Marlow and Henley branches were also added to the project, although electrification of the last length to Marlow was later cut back due to the lack of space to reverse a four-car EMU at Bourne End.

Initially, Network Rail identified 113 bridges, tunnels and aqueduct structures that required overhead clearances to be increased to accommodate the overhead line equipment. Subsequently, it was found necessary to increase the number of structures that needed to be dealt with by more than twenty. In addition, the work required at many of the sites was far more extensive than had been anticipated.

It was also originally thought that the 7km (4.35 miles) long Severn Tunnel would be relatively easy to electrify as it had been built with generous clearances to aid its ventilation. However, no account was taken of both the corrosive atmospheric and damp conditions under the River Severn, later resulting in much of the original OLE that was installed having to be taken out and replaced with equipment having a higher degree of corrosion resistance. As a result, the

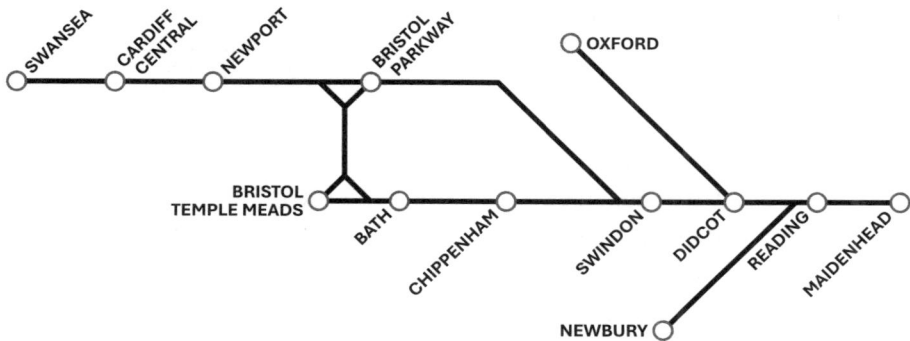

Great Western Main Line Electrification

Severn Tunnel was the very last section of the route to be commissioned into operational use.

Design and Construction Issues

From the very start, issues associated with the design of the OLE had a major impact on the project. To comply with the 2008 Technical Standard for Interoperability (TSI) for the Energy Subsystem of the trans-European high-speed rail system, Network Rail decided, in the following year, to redesign its electrification equipment. In addition to meeting the TSI requirements, it was proposed that the new design should be an improvement on existing designs by eradicating known failure modes, improving reliability and reducing the requirements for maintenance. It was also hoped that construction would be simplified.

Network Rail set up two work streams, one to create a higher speed design capable of operating at 125mph and above to 140mph (Series 1), the other to develop the legacy Mk 3 design range initially to meet electrification requirements for compliant lower speed operation up to 100mph (Series 2).

The work stream for the design of the TSI compliant OLE for GWML was subject to a tendering exercise carried out in late 2011. This design contract (that is, for Series 1) was awarded to Furrer+Frey with Network Rail retaining joint intellectual property rights (IPR) for the design.

The Series 1 design contains the following principal changes from Network Rail legacy systems:

- reduced structure spacing
- mechanically independent support and registration for each track
- much higher catenary and contact wire tensions and the use of different conductors
- separate tensioning arrangements for catenary and contact wire

- use of a spring tensioning device situated over the track (removing balance weights and wiring crossing other tracks for anchoring purposes)
- tangential wiring arrangements
- wire runs restricted from crossing other tracks (to limit equipment damaged in a de-wirement)
- reduced component count
- construction by high output methods.

Further work took place to integrate the two design ranges to remove duplication and provide a complete design range certificated to meet the future requirements of the integrated high-speed and conventional rail TSIs. It was the intention that this work would provide a complete suite of designs for Network Rail routes, to be known as the UK Master Series.

Due to the lack of experienced design staff, the design approval work proceeded slowly. The net result was that the detailed design of the GWEP system was significantly delayed. Consequently, the GWEP basic designs were still incomplete five years after authorisation of the scheme.

There was considerable debate over the parameters for the revised design and it was eventually decided that it should be fit for 225kph (140mph) operation with two pantographs on a nine-coach train raised, or 175kph (110mph) operation by a twelve-car EMU with three pantographs raised.

To minimise disruption to normal services and speed up installation it was envisaged that a special High Output Plant System (HOPS) would be devised to work with the adjacent running line open to normal traffic. The HOPS equipment was designed and built by Windhoff and was said to be capable of installing 1.5 kilometres (0.93 miles) of electrification per eight-hour shift. A purpose-built High Output Depot was established at Swindon to service and supply the HOPS trains.

However, this equipment was procured before the decision concerning OLE deflection criteria were decided – these criteria being a source of much technical debate. Deflection criteria is defined as the permissible degree of deflection of an OLE structure and its foundation that may contribute to the lateral displacement of the contact wire, taking cognisance of wind loading conditions.

The risk-averse decisions made on this subject resulted in stiffer structures, larger foundations and the allocation of more portal structures than had been originally envisaged. In addition, because of the unknown location of signalling cables along much of the route, supporting structures had to be sat back further from the track. For Twin Track Cantilever (TTC) structures in particular, this meant that much longer booms were required, which triggered the need for even stiffer supporting structures and increased foundation support. The HOPS units were not designed to handle these heavier and larger OLE components, so it is not surprising that the HOPS units could not meet the forecast rate of construction. Ironically, the project eventually settled on using the more generous deflection specifications used on the ECML and WCML routes.

By 2014 the HOPS equipment was operational, but productivity was much lower than planned due to:

- design shortcomings and teething problems with the HOPS equipment
- difficulties locating buried operational S&T cabling alongside the railway tracks where the OLE masts were to be located
- buried rocks and boulders, particularly in the fill that made up embankments.

Originally it was assumed that 80 per cent of foundations would be piled with 5.5m piles, but due to the changed circumstances described, only about 40 per cent were piled and many of these required piles longer than 5.5m.

On top of all this, because the OLE technical design was new and unproven, there was a need for a special test site to be established on the route to prove that the theoretical predictions in the OLE modelling could be realised in practice. This inevitably further delayed matters.

Programme Delays

The foregoing chain of events resulted in major delays to the start of the construction process. Furthermore, within a year it was clear that rate of installation of the OLE would not be achieved within the planned timescales. In addition, as already mentioned, it was also found that the bridgework and track lowering necessary to achieve the revised design electrical clearances, partly as a result of the adherence to the TSI, had been significantly under-estimated.

In 2015, Network Rail estimated that the project was running a year behind schedule, with completion during 2017, and that the budget outturn would need to be £1.74Bn. Following the May 2015 election, the new Secretary of State for Transport, Patrick McLoughlin, asked the new Chairman of Network Rail, Sir Peter Hendy, to review not only the Great Western Main Line Project, but all Network Rail's capital investment programmes being undertaken in CP5 (2014–19). He was to report back in the autumn of 2015.

As far as the Great Western Main Line works were concerned, Sir Peter Hendy's report concluded that the outturn would be £2.8Bn, more than a billion pounds more than estimated the year before. As a result, it was decided that the core routes would go ahead as planned but with a revised completion date of

March 2019. However, commissioning of the Cardiff to Swansea section would be deferred to CP6 (2019 and 2024).

After some further reprogramming work, revised dates for the completion of electrification work were published in early 2016 as follows:

- Maidenhead to Cardiff via Wootton Bassett and Bristol Parkway: December 2018
- Reading to Newbury: December 2018
- Didcot to Oxford: June 2019
- Wootton Bassett to Bristol Temple Meads via Chippenham and Bath: between February 2019 and April 2020
- Bristol Temple Meads to Bristol Parkway: between February 2019 and April 2020.

However, as a result of the costs of the project continuing to escalate plus further programme slippage, the government announced in November 2016 that electrification of the following sections were to be deferred indefinitely:

- the branch lines to Henley, Windsor and Bourne End/Marlow
- Didcot Parkway to Oxford (which was also planned to form a part of the Electric Spine)
- Bristol Parkway to Bristol Temple Meads
- Thingley Junction (just west of Chippenham) to Bath Spa and Bristol Temple Meads.

Furthermore, after what many consider was a less than conclusive debate, in July 2017 it was announced that the Cardiff to Swansea electrification section of the project had been cancelled.

All these cancellations resulted in the requirement for far less 'straight' electric EMUs and consequently many more bi-mode trains. Unsurprisingly, the running costs of these bi-mode trains would be far in excess of the original proposals for the project. In addition, even allowing for the cut-backs in scope, the costs had ballooned to more than three times the 2013 £874M estimate by 2018, and completion deadlines were repeatedly being extended.

Severn Tunnel and Patchway Tunnels

As previously noted, electrifying the 7km (4.35 miles) long Severn Tunnel proved to be particularly problematic. The tunnel suffers from significant water ingress through the lining and, in far more significant quantities, from the Great Spring on the Monmouthshire side. Generally, 14 million gallons of water are pumped from the tunnel every day from the Sudbrook Pumping Station near the Monmouthshire bank. The water ingress through the lining pours rather than seeps through the brickwork in many places and, in the worst areas, down pipes are installed to divert this brackish water into the tunnel drainage system. The tunnel has a forced ventilation scheme, with salt-laden air pumped down a shaft at Sudbrook, clearing the tunnel of diesel fumes that are being expelled from both portals. The salt-laden air forced into the tunnel mixes with the diesel fumes together with the residues of coal and ash dust, left by decades of steam train operations, and the compounds from the water ingress to form a very corrosive environment.

The installation of 26 single track km (16 miles) of OLE taking in the Severn and Patchway tunnels was undertaken in a six-week closure from 12 September 2016. Track lowering was necessary in the single-bore Patchway Tunnels where clearances were extremely tight. This also involved the installation of rock-anchors in the weakest areas of the tunnels.

Due to the stability issues in the Patchway tunnels (between Bristol Parkway and the Severn Tunnel), a comprehensive systemised set of measures were implemented to undertake the works safely (see Appendix 6). The system was initially developed for the GWEP track lowering at Alderton Tunnel on the Badminton Line, and further improved for the track lowerings subsequently undertaken in Box Tunnel (between Chippenham and Bath) and through the Bath, Sydney Gardens site.

Further enhancements to the monitoring system were needed to ensure the safety of the more fragile Patchway 'Up' and 'Down' line tunnels. This involved the installation of 'tilt' and 'shape array' sensors to detect movement of the tunnels' linings on a 'real-time' basis. Risk assessments were prepared, and an action plan put in place detailing actions that should be taken in case movements of the lining exceeded those predicted by the structural analysis that had been undertaken. This innovative system, including the technology utilised, has the potential to benefit future electrification schemes where track lowering is necessary.

Reverting to the Severn Tunnel, since it was built in 1886, the track has been subject to the regular replacement of rail, sleepers and ballast due to a com-bination of the poor atmospheric conditions corroding the rail and fastenings and the degradation of sleepers and ballast due to the limited ballast depth. The overhead clearances in the Severn Tunnel were quite generous in order to ensure the effectiveness of the ventilation system. In the years prior to elec-trification, some minor track lowering took place when track was renewed in order to obviate any clearance issues, so no significant further track work was required during the blockade to enhance clearances for the OLE.

The OLE comprised a conductor bar holding a copper contact cable on its underside retained by means of high-grade stainless-steel fixtures. This form of

OLE is also used in other tunnels that were electrified under the Great Western Electrification scheme, as it is more compact than conventional overhead wires and it was anticipated that it would require less maintenance.

The tunnel electrification involved installing circa 850 anchoring points, 1,700 vertical drop tubes and 14km of aluminium conductor bar, with the copper contact wire attached by 7,000 high-grade stainless-steel fixtures.

As already noted, there has been a long history of the effects of the corrosive atmosphere in the Severn Tunnel. Galvanic corrosion of the stainless steel, aluminium and copper components had been anticipated but, in the event, the level of degradation was far greater than had been expected.

The cause was found to be a form of anaerobic sulphur-reducing bacteria. Normally, the formation of oxide on the surface of aluminium protects the metal against further deterioration. However, the bacteria were found to be consuming this protective layer, exposing bare metal and allowing it to oxidise again and cause further corrosion. This cyclic action eventually totally consumed the metal. Corrosion was particularly acute where there was bi-metal contact.

Severe corrosion of the OLE became apparent within a short period after the installation. Among other things, the fixtures holding the copper contact wire failed so that the wire sagged below the conductor bar. The OLE was therefore not energised for service use. As a result, design modifications to the OLE were put in hand. First, more drip trays were fitted to prevent salt-water ingress from the Severn Estuary directly contaminating the OLE components. Network Rail's OLE contractor, Furrer+Frey, amended their design, replaced the normal copper conductor with aluminium and changed the composition of other components to reduce galvanic effects.

A further major closure of the tunnel was needed for these rectification measures to be implemented. The section of line between Patchway and Severn Tunnel Junction was therefore closed for three weeks, commencing on 29 June 2018, to undertake the modifications necessary, including replacement of the copper contact wire with aluminium. Earthing components were also replaced or sheathed in plastic to prevent future deterioration.

The whole system was then energised and subjected to a long-term test to ascertain its ability to stand up to the atmospheric conditions. Unsurprisingly, as already indicated, this was the last section of the Great Western Main Line Electrification to be commissioned. Some further track lowering was also undertaken in the Patchway Tunnels during the 2018 blockade, again requiring special measures to be implemented due to potential instability issues. As regards additional costs to re-engineer the Severn Tunnel OLE, Network Rail spent a further £20M on works to prepare the Severn Tunnel for the start of electric operation over a three-year period.

Cardiff Intersection Bridges

A further location that also presented a huge technical challenge was the set of Intersection Bridges on the approach to Cardiff Central Station. One bridge carries the two-track Valley Lines over the main line as they climb from Cardiff Central to Queen Street Station. The other bridge supports the single-track Cardiff Bay to Cardiff Queen Street Valley Lines branch.

Originally, it was proposed to undertake a track lowering but because there was also a large culvert, known as the Bute Feeder Canal, passing below the main line at this location, track lowering would be extremely difficult to achieve from a technical point of view. The cost would also have been extremely high for, to effect the required 331mm lowering, the Bute Feeder culvert would have required extensive reconstruction.

Network Rail therefore generated several potential solutions including:

- Option 1 – Bridge reconstruction
- Option 2 – Culvert reconstruction with syphon and slab track
- Option 3 – Earthed Section
- Option 4 – Contenary* wires having reduced OLE clearances with the aid of effective insulated coatings and surge voltage protection.

Option 1 was discounted as it was too expensive, and the implementation would be too disruptive to train services.

Option 2 would be an expensive and onerous choice requiring the feeder canal to be closed, which, as a critical water course, would be difficult to arrange; also the restrictive site would make piling difficult.

Option 3 had the severe disadvantage that a train may become stranded on the low-speed section of track beneath the bridges; in addition, the close proximity of signals would require severe operational restrictions.

Option 4 was unproven but, by accepting shallow ballast depth, a higher track fixity and providing effective insulation, it appeared to offer the most practical and cost-effective solution.

With the benefit of some positive initial information, it was decided to proceed with Option 4. This novel solution had the clear advantage that it would avoid the costs of having to make significant modifications to either the bridge or culvert; nor would it be necessary to lower the track to any significant extent.

Insulated coatings had been used successfully in the past on bridges on existing electrified lines with known clearance issues as a 'patch' remedy. However, the use of an insulated covering had not been used as part of a

---

* A solid conductor wire that combines the catenary and contact wires, which is used in confined spaces.

long-term design proposal for new electrification schemes. Under-bridge arms would also need to have insulation applied.

Network Rail's consultants, Andromeda Engineering, recommended the use of the 100R insulation product marketed by GLS Coatings Ltd. Voltage surges in the OLE would be mitigated using Siemens Surge Arrestors (see Appendix 3).

Further development of this solution proposed utilising the following:

- an applied GLS 100R insulating layer to the bridge soffit
- insulated OLE under-bridge arms
- surge arresters in circuit with the overhead line system
- twin-wire to limit wire lift
- shallow ballast depth with glanding rails to improve 'fixity' and
- data logging monitoring equipment.

Some key information about how the successful solution was developed is in Appendix 3.

The development of this radical new solution cost less than £1M to implement, including design costs, saving more than £10M compared to the other proposed options. This was the first such installation in the UK and, clearly, the solution offers significant benefits both in terms of the cost and speed of electrification works. Potentially, it may obviate the need for track lowering or for the reconstruction of buildings, bridges or tunnels at many wiring sites in the future, thus significantly reducing the cost of electrification schemes. Site-specific derogation was, however, required for the Cardiff Intersection Bridges and a more generic acceptance in standards would be needed before widespread use of this solution could be considered.

The Cardiff Intersection Bridges and the Severn Tunnel were the last two major hurdles to be overcome before the completion of the scaled back Great Western Main Line electrification could be achieved. Eventually, full electric services between Cardiff and London Paddington commenced on 5 June 2020.

In summary, the actual phased commissioning dates of the Great Western Electrification Project were as follows:

- Reading to Newbury: 31 December 2018
- Reading to Didcot Parkway: 27 December 2017
- Didcot Parkway to Swindon: 1 November 2018
- Swindon to Bristol Parkway: 18 December 2018
- Wootton Bassett to Chippenham: 21 April 2019
- Severn Tunnel Junction to Cardiff Central: 5 January 2020
- Bristol Parkway to Severn Tunnel Junction: 5 June 2020

Conclusion

The outturn costs for the project came in at £2.8Bn, more than three times the original estimate, and for a significantly reduced scope. Key lessons that should be learned from a political, technical, project control and management point of view follow:

- While it is important to stimulate a positive political will at all levels, projects shouldn't be promoted and authorised on a 'last minute' political whim.
- High-handed dictats such as the directive to implement the 2008 TSI should have been robustly challenged by Network Rail with a derogation sought from the DfT.
- Scope and specifications should be based on appropriate, proven materials and technology. Projects should not be used as a 'test bed', as GWEP was, to develop a new form of OLE.
- GWEP's original assumptions and costings were flimsy, and there was little time to consider alternatives. It is important for such schemes that viable options are generated and costed, taking cognisance of certified contemporary unit costs and local conditions.
- The inaccurate initial GWEP costings skewed the economics of the scheme. It is essential that the costings and economics for all project options are scrutinised to ensure the best payback is established.
- In the rush to get the GWEP scheme going, there was little time for option development. Had there been more time, proposals and assumptions for scheme options should have been rigorously examined, challenged and compared in order to select the preferred scheme to take forward for authorisation.
- Once authorised, the GWEP project controls were too loosely managed with many decisions, which significantly impacted on costs, effectively decided at an inappropriate level. It is important that firm project controls are implemented, presided over by an experienced project director with the knowledge and ability to maintain a firm grip on the overall scheme and his or her organisation.
- Technically, decisions on specifications for GWEP were not challenged as robustly as they should have been, again with significant impacts on costs. Warranting a competent technical approach is key to the success of any major project.
- The form of contract used on GWEP, and the management of the same, resulted in too much of the budget being paid to contractors in the early phases of the project. It is essential to ensure an appropriate type of

contract is chosen and that the contractors are diligently managed in a firm but fair manner.

Many of the lessons learned were put into effect following the Hendy Review in 2016. This enabled the final cost to be kept within the revised budget set in the review.

On the 'up side', it can be seen that a number of cost-saving innovations were developed by the project (for example, the comprehensive structural monitoring systems developed for track lowering in tunnels and the insulating coating and surge arrestors adopted for the Cardiff Intersection Bridges). These advances should benefit future electrification schemes.

North West England – Ongoing
Liverpool, Manchester, Blackpool and the Lake District

The National Infrastructure Commission was asked to advise on strategic improvements to transport connectivity in the North in October 2015. The report published on 15 May 2016, stated that 'the North needs immediate and very significant investment for action now and a plan for longer term transformation to reduce journey times, increase capacity and improve reliability'. Among other things, the report called for the North TransPennine rail route to be electrified and upgraded.

A new transport body, Transport for the North (TfN), was formed at about this time, comprising twenty local transport authorities, Network Rail, Highways England and HS2 together with business leaders. Transport for the North formally became a statutory sub-national transport body in April 2018.

Transport for the North has championed the Northern Powerhouse Rail programme and has a vision that combines both the existing and future rail network of the north. George Osborne was also an active supporter of an east–west high-speed line (dubbed HS3), and in his time as Chancellor, he allocated £60M (£79M at 2025 prices) of funding to enable plans for a route to be generated by 2017.

In October 2017, a further £300M (£390M at 2025 prices) of funding was authorised by Chancellor Philip Hammond, in particular, to identify and plan for junctions between Northern Powerhouse Rail and HS2 to enable east–west services to use HS2, the proposed HS3 infrastructure as well as upgraded conventional lines.

Transport for the North published their draft thirty-page Strategic Transport Plan on 16 January 2018. This outlined the proposed staged rail developments for northern England, including the following:

- a new twin-track railway linking Liverpool to HS2 and beyond to Manchester via Warrington and Manchester Airport
- a new TransPennine rail line linking Manchester and Leeds via Bradford including a Bradford city centre station
- capacity upgrades at Manchester Piccadilly Station
- significant upgrades along the corridor of the existing Hope Valley line between Manchester and Sheffield
- rail links to integrate HS2 with Sheffield and Leeds and upgrading of the line from Sheffield
- upgrading the existing line from Leeds to Hull (via Selby) and Sheffield to Hull (via Doncaster)
- upgrading the rail infrastructure between Leeds and Newcastle via a HS2 junction and the East Coast Main Line
- developing a Hub concept for Northern Powerhouse Stations.

Kicking-off the Northern Powerhouse Rail Project in July 2019, Prime Minister Boris Johnson pledged funding for the Leeds to Manchester route section. Following this announcement, Network Rail started a public consultation process concerning their plans for electrification and four-tracking of the existing slow and capacity constrained section of the route between Huddersfield and Dewsbury. Grade separation of the junction at Ravensthorpe is also proposed.

The initial authorisation of the electrification of the Liverpool to Manchester, Deansgate via Newton le Willows – the Northern or Chat Moss Route – was received in July 2009 and preliminary work quickly got underway with a planned completion date of 2013. The scheme was enhanced in December 2009 when the then Transport Secretary Lord Adonis added electrification of the line between Huyton and Springs Branch Junction, electrifying the entire Liverpool to Wigan route. Furthermore, with the later announcement in July 2012 of the Manchester to Preston and Blackpool electrification projects, electric train operations from Liverpool to Preston and Blackpool North could be realised.

As previously noted, a number of delays, including that caused by the further review of the scheme following the change of government, pushed back the original completion dates. The Manchester Castlefield Junction to Newton-le-Willows section was completed in December 2013. This enabled Manchester–Scotland services to be operated by electric traction. Class 350 electric multiple units generally operate the passenger services between Manchester Airport and Glasgow Central and/or Edinburgh Waverley Stations, utilising the Chat Moss route and joining the West Coast Main Line at Golbourne Junction. The final section to Liverpool was completed with the regular electric service starting on 5 March 2015.

The refurbished Class 319 (ex-Thameslink) units generally used on the Liverpool Lime Street to Manchester Victoria service have reduced the journey time between the two cities to 30 minutes, saving 15 minutes compared to the previous diesel schedules. The newly electrified Northern route also formed an alternative route for freight trains from the West Coast Main Line to Liverpool and an electrified diversionary route between Crewe and Golbourne Junction.

A key element of the Northern Hub project involved diverting Liverpool to Leeds TransPennine services from the Southern Route, via Warrington Central and Manchester Piccadilly, to the faster Northern Route via Newton-le-Willows and Manchester Victoria. Important elements of the plan were the removal of infrastructure bottlenecks in Manchester to enhance rail services and reduce journey times between major cities and towns in the North of England. The works also involved a major programme of electrification.

Victoria Station was refurbished in October 2015 to become the key rail interchange point between Liverpool and Leeds. The construction of the Ordsall Chord, linking Manchester Victoria and Oxford Road Stations, enabled rail services from the North East to run directly to Manchester Airport via Manchester Victoria and Piccadilly without reversing. The new chord line also facilitated improved connections between through trains and local services. The Ordsall Chord was commissioned on 10 December 2017 and the Northern Hub Project was completed in 2018.

The planned completion of the electrification of the line between Huyton and Spring Branch Junction was December 2014, but delays, similar to those described earlier, again postponed the commencement of the Liverpool to Wigan services until March 2015.

The Manchester to Preston (via Bolton and Euxton Junction) and Preston to Blackpool North schemes were linked, it being proposed to run through services between these locations. It was originally planned for the Preston to Blackpool electrification to precede the Manchester to Preston section, but as the Preston to Blackpool line required remodelling and re-signalling works, electrification work on the Manchester to Preston route started first.

Electrification work commenced in May 2015 involving extensive bridge-works and the re-boring of Farnworth Tunnel to provide the necessary OLE clearances. Poor geological conditions delayed the tunnelling works by a month with breakthrough achieved in late October 2015. The first train ran through the new twin-track tunnel on 14 December 2015. During April 2016, the Orlando Street Bridge in Bolton was replaced, Soho Street bridge was demolished and track lowering undertaken in the two Bullfield Tunnels.

In a further fifteen-day blockade from 12 to 27 August 2017, in addition to the installation of OLE, re-signalling works were completed. At Bolton Interchange

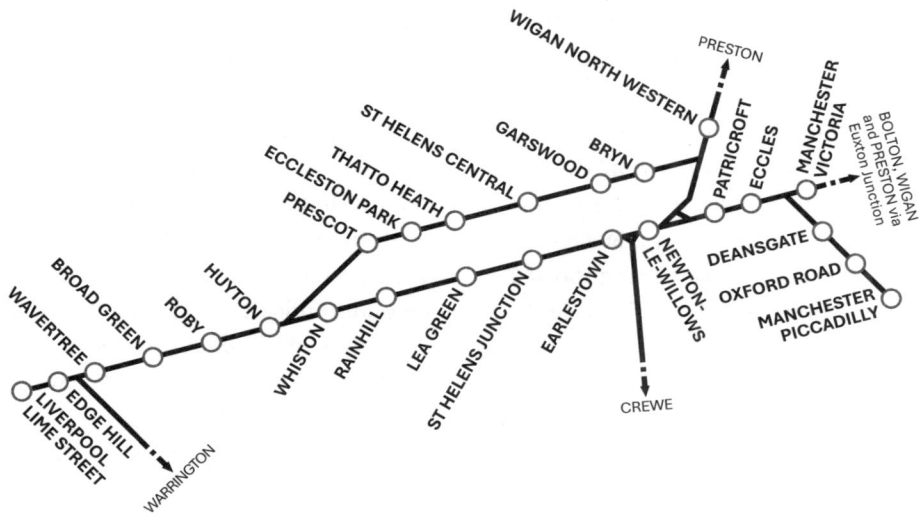

Liverpool to Manchester/Wigan

Station, a new footbridge was erected and a platform line reinstated. However, the Manchester route remained closed after the blockade due to an embankment collapse and bridge damage at Moses Gate caused by a burst water main.

The electrification works slipped from the planned completion date of December 2016, initially stated as twelve months' delay but subsequently flagged up as an eighteen-month slippage. A further delay was announced in January 2018 due to difficult ground conditions, which meant that, among other things, 200 OLE masts could not be installed as planned.

On the Preston to Blackpool North route, difficulties were also encountered in undertaking the aforementioned remodelling and re-signalling work, causing the planned completion of this section to slip. Eventually, the full electric service between Manchester and Blackpool North, operated by refurbished Class 319 units, commenced on 11 February 2019.

As an 'add-on' to the Manchester to Preston scheme, the Transport Secretary Patrick McLoughlin announced in December 2013 that the line between Wigan North Western station and Bolton's Lostock Junction would also be upgraded and electrified. The work was costed at £37M for completion to be 'by 2017'. However, in September 2016, the scheme was deferred, with only the outline options scoped by this date. Further work on the project was then deferred. In December 2020, the scheme was described as 'hanging in the balance' after the government had cut £1bn from Network Rail's enhancement budget.

It had long been the aspiration back to British Rail days to improve services on the North TransPennine route between Liverpool, Manchester and York via

Huddersfield and Leeds. The TransPennine Express brand came into being with the introduction of British Rail Regional Railway Sprinter trains in the 1980s, which delivered modest journey time improvements. The brand name for the TransPennine routes was retained when the Regional Railways North East Franchise was privatised in 1997.

With the objective of separating the local and longer distance services, the Strategic Rail Authority decided to reorganise the North West and North East franchise, including the creation of a new separate franchise, under the TransPennine Express brand, for the long-distance regional services. It was the intention for TransPennine franchisee to operate Inter-City style services, although the Siemens Desiro Class 185 units, ordered by the new franchisee in 2003, were more akin to an outer-suburban rather than an Inter-City rolling stock.

Along with the new trains, there were aspirations to reduce journey times, particularly between Manchester and Leeds, and the scope for line-speed improvement works were investigated. Steps to deliver significant journey time improvements commenced in 2009, with the development of the Northern Hub Project by the re-routing of North TransPennine Express services through Manchester Victoria and the Northern route to Liverpool. In 2014, proposals to improve linespeeds between Manchester and Leeds to reduce journey times by 15 minutes were included in Network Rail's CP5 programme.

Further infrastructure enhancements under development at this time included electrification of the North TransPennine route, and in 2011 limited funding was authorised to commence preliminary electrification work on the route between Manchester and Leeds.

In 2013, the Department for Transport authorised the extension of the electrification to York. It was proposed to electrify the line between Selby and Hull to enable electric services to be further extended. Subsequently, Hull Trains proposed electrification of the line between Temple Hirst Junction, on the East Coast Main Line, to Hull line. This Open Access Operator promoted the extension and planned to seek private finance to fund the works. On 20 March 2014, Transport Secretary Patrick McLoughlin made £2.4M available to move the project to the GRIP 3 'Options Selection' stage.

However, on 25 June 2015 the Transport Minister, Chris Grayling, announced that, along with a number of other electrification projects, the TransPennine route between Manchester and York would be 'paused' because of rising costs. Grayling went on to voice his support for bi-mode trains and argued that electrification of the entire TransPennine route may be 'too difficult'. Grayling later chose the TransPennine Route to be the UK's first digitally controlled intercity route but remained silent as to whether electrification would still be part of the upgrade. Following on in November 2016, unsurprisingly, the extension to

Hull was also dropped. These proposals were largely overtaken with the publication of the *Integrated Rail Plan for the North and Midlands* in 2021 and further details are shown in the TransPennine section below.

The £16M (£22M at 2025 prices) electrification of the Windermere branch was announced by the Department for Transport in August 2013. The scheme would enable the operation of through electric services from major northern cities and London Euston. Funding approval followed in 2014 with the electrification work planned to be undertaken from 2019 to 2024 (CP6).

Apart from the electrification of Platform 3 at Oxenholme Lake District Station, no physical work had been undertaken by 20 July 2017 when it was announced that electrification of the Windermere branch had been cancelled. The use of battery bi-mode units was being considered in 2022, utilising adapted Class 331 Civity units.

Bolton to Wigan North Western – 2025

Electrification of the Bolton to Wigan North Western line was first announced in 2013 but was deferred by 2017.

On 1 September 2021, the Rail Minister, Chris Heaton-Harris, announced the initiation of a major upgrade to the railway line between Wigan East and Lostock Junction near Bolton. He promised that the project would focus on providing passengers with a greener and more reliable railway.

The work, budgeted at £78M, includes the electrification of the twelve kilometres of double-track railway, requiring the erection of 450 UKMS100 OLE masts and the modification of seventeen bridges and two-level crossings. In addition, platforms will be extended at Hindley, Westhoughton and Ince stations to cater for six-car trains. The project is targeted for completion in 2025.

The diesel trains operating this route were based at depots more than 30 kilometres distant. The electric trains would be based locally at the recently upgraded train maintenance facility at the Wigan Springs Branch Depot, thus reducing empty stock moves. In addition, this electrification would permit the expansion of through electric train services to destinations beyond the extent of the line, as well as facilitating another electrified diversionary route option for the West Coast Main Line.

Although only short in length, the project represents a useful infill scheme. It is noteworthy that although the cost per single track kilometre is high, due in part to the risks associated with old mine workings, the scheme was authorised on the basis of a strong business case.

From a contractor perspective however, all did not go to plan, as the main contractor, the Buckingham Group, fell into administration on 23 July

2023, and the project was paused. Remarkably, within a week, Network Rail managed to appoint J Murphy and Son Ltd as its new contractor and work recommenced.

Regular blockades were imposed on rail services during 2023/4 to delivering a range of major improvements, including electrification work, level crossing renewals and signalling upgrades. In addition, longer blockades were necessary for bridge works, including a six-day closure to renew the road over rail bridge at Hindley.

The wires were energised on 1 January 2025 and, going forward, the newly upgraded railway should offer customers 'greener' and more reliable journey opportunities.

TransPennine – In Hand 2025

The main TransPennine route runs from Manchester across the Pennines to Huddersfield and Leeds. Historically, the passenger services went from Manchester Piccadilly through Guide Bridge to Stalybridge up to Diggle under Marsden Moor and down to Huddersfield. From there it ran along the Calder Valley to just south of Dewsbury, through there and up to Batley and on to Leeds. From Leeds, services ran to Hull, York and on to either Scarborough, Middlesbrough or Newcastle. With the advent of diesel multiple unit workings and then, subsequently, the opening of Manchester Airport station in 1993, the majority of services served the airport by reversing in Manchester Piccadilly station but one train an hour ran to and from Liverpool.

With the opening of the Ordsall Chord the service pattern was amended so that virtually all the trains ran through Manchester Victoria station and, other than the ones to and from Liverpool, used the Ordsall Chord to access Manchester Piccadilly and thence to run on to the airport. This had the big advantage of removing the many conflicting movements at Slade Lane Junction and across the throat of Piccadilly station.

The electrification of the route from Manchester to Leeds and onwards has been an aspiration for many years. Network Rail established a project team to develop this scheme and, after five years' work, they had still not reached a decision on the precise content of such a project. One of the biggest contentions was the extent to which the route should be wired because with bi-mode passenger trains it would be possible to have significant stretches unwired where the provision of electrification clearances were going to be very expensive. The downside of such a proposition is that freight trains would not be electrically hauled and there are significant flows of freight over the route, which, in those

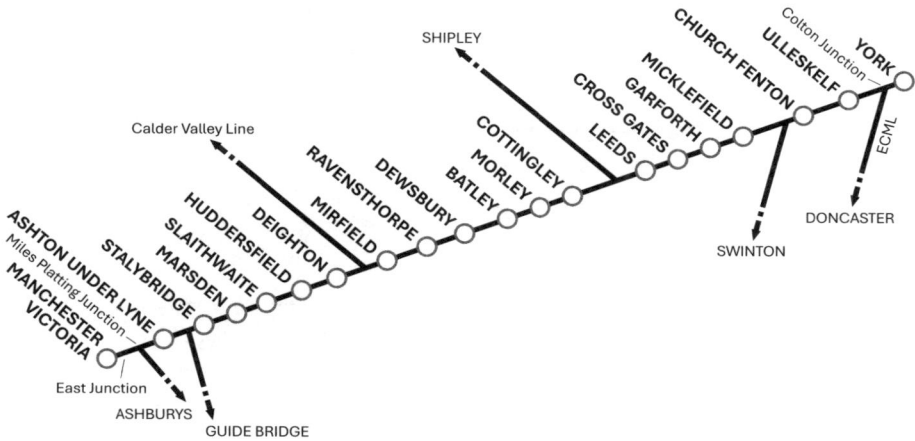

TransPennine Upgrade

circumstances, would have to remain diesel hauled thus losing an opportunity to make a step change in the volume and distance of electric freight operation. Because there are forecasts of significant increases in both passenger and freight traffic on this route, the project team examined the extent to which grade separation of junctions could be provided along with re-quadrification of the Marsden to Ravensthorpe section.

The use of the Ordsall Chord means that there are additional trains placed on the Manchester, Castlefield corridor. Prior to the 2020 Covid pandemic reduction in the number of passenger trains being run, this section of route had become notorious for creating large numbers of delays daily and there were moves afoot to review service specification levels to resolve this problem.

However, the whole scheme was effectively re-evaluated under the *Integrated Rail Plan for the North and Midlands* (IRP) published in November 2021. Under the plan, the proposed new high-speed route was scaled back and instead the revised Northern Powerhouse Rail route incorporated the upgrading of the existing Transpennine route, which would now be fully electrified between Manchester and York and also digitally signalled. In addition, the route was to be cleared for W12 loading gauge to accommodate 2.9m × 2.6m shipping container trains.

During the autumn of 2022, Network Rail started the erection of masts in the section from Colton Junction towards Leeds as far as Church Fenton and from Victoria East Junction to Miles Platting and on to Stalybridge. The work also involves the rebuilding or elimination of many of the bridges on the route and improvements or closure of level crossings. Huddersfield Station, a Grade 1 listed building, is to have a special bespoke design of OLE and was to

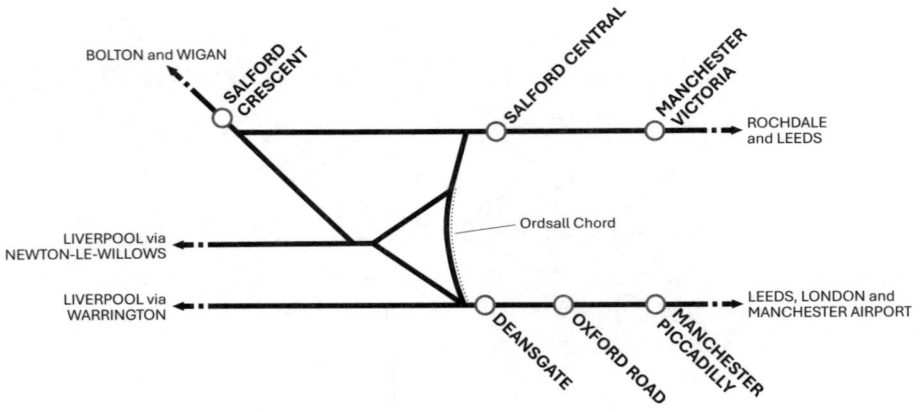

Ordsall Chord

be subject to two 32-day closures during 2024/5 to enable the upgrading work to be undertaken.

Commissioning of the Church Fenton and Colton Junction electrified section is scheduled for Spring 2025, with energisation between Church Fenton and Cross Gates planned for Easter 2027. The final section around Neville Hill is to be commissioned in late 2028.

Chapter 13

Post-privatisation Schemes in Scotland since 2005

The responsibility for Scotland's railways was devolved to the Scottish government from the Department for Transport in 2005. Since that time, Holyrood has consistently supported railway electrification, providing it can be delivered at an affordable cost. The Scottish government has been convinced of the economic, social and environmental benefits that a rolling programme of electrification will bring.

In contrast to the rest of the UK where confidence in electrification has suffered, the Scottish government, through Transport Scotland, has maintained a consistent policy and programme of electrification work. It is easy to see why this consistency has been maintained, since the cost of the 25kV electrification programme per STK north of the border has generally out turned at under £1.5M, compared to the GWML, which cost double this figure. Some schemes, for example the Cumbernauld to Springburn electrification, came in as low as £1.2M per single-track kilometre, which compares well with the best rates achieved in Germany and Denmark in recent years.

Four things stand out from the success of these Scottish projects. First, the body responsible, Transport Scotland, is a strong, informed client; secondly, the Network Rail strategy and investment team agreed with the client a development programme for future electrification then co-ordinated this work; thirdly, the project team has taken an integrated approach with stakeholders; fourthly; the projects interlink to provide a modest but consistent year-by-year rolling programme ensuring that knowledge and skills are retained.

Where things have not gone so well, for example the Edinburgh to Glasgow electrification and route improvement programme (EGIP), the approach has been to get all the interested parties around the table and fix the problems. In contrast, the Westminster government's approach for England and Wales has been to simply abandon electrification schemes when costs are over-running and claim that less-powerful bi-mode trains are the answer.

Electrification Schemes

Since 2005, 25kV AC electrification has proceeded in Scotland with multiple schemes in the Central Belt. These are:

- Hamilton to Larkhall
- West Coast Main Line Upgrade
- Airdrie to Bathgate
- Paisley Canal Line
- Rutherglen and Coatbridge (R&C line)
- Edinburgh Glasgow Improvement Programme (EGIP)
- Muirhouse Central to Barrhead and Busby Junction to East Kilbride.

An outline description of these projects (plus schemes added later) follows.

Hamilton to Larkhall – 12/05

The 4.7km (3 miles) branch from Haughhead Junction on the Hamilton Circle Line to Larkhall was reopened on 12 December 2005 providing a through passenger service from Larkhall to Glasgow via Hamilton. Signalling is controlled from Motherwell Signalling Centre utilising two and three aspect colour light signals and a passing loop is provided at Allanton. The line was electrified from the start using Mk 3 OLE. A new electrification team was set up for the implementation of this project, the first post-privatisation electrification scheme delivered in Scotland.

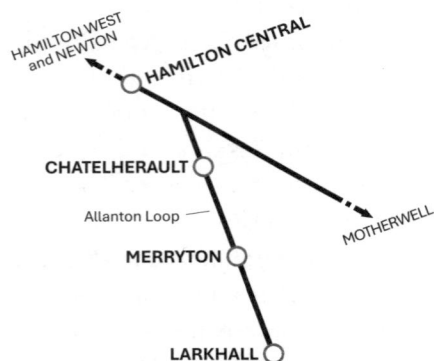

Hamilton to Larkhall

West Coast Main Line Upgrade – 12/08

From the Hamilton and Larkhall scheme, the same electrification team moved on to the Scottish section of the West Coast Main Line as part of the major upgrade of the route completed in December 2008. No further details are included here as this was an enhancement rather than the electrification of a new route, but it does show the advantage of having a rolling programme to maintain the skills within a team. On completion of these works, the team were deployed on the Airdrie to Bathgate electrification scheme.

Airdrie to Bathgate – 12/10

This project originated from the Central Scotland Transport Corridor Study (CSTCS) undertaken for the Scottish Executive in 2001/2. Scottish Ministers decided in January 2003 to accept the recommendations of the CSTCS and, as a result, funding was made available later in 2003 to West Lothian Council to fund a re-opening study involving Network Rail for the Airdrie to Bathgate Line that included electrification. The study commenced in December 2003 and was completed towards the end of 2004. The study informed the decision to initiate the Airdrie to Bathgate Rail Link Improvement Project to reopen 22km (14 miles) of closed route from Drumgelloch to Bathgate as an electrified, double-track railway. At a cost of £300M (£450M at 2025 prices), the route reopened on 12 December 2010 with the full fifteen-minute frequency passenger service commencing in May 2011.

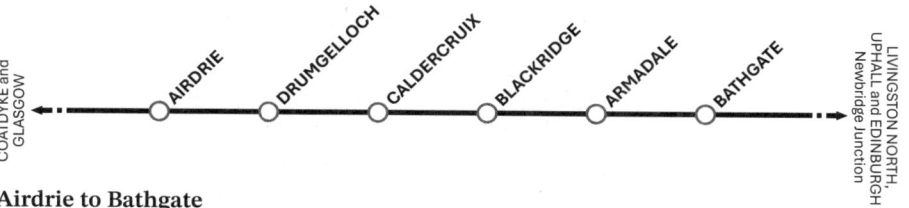

Airdrie to Bathgate

Existing stations at Airdrie, Livingston North and Uphall were enhanced, and new stations were provided at Caldercruix, Blackridge and Armadale. The line was electrified utilising Mk 3 OLE. The scheme also included remodelling at Bathgate itself and running the wires to Newbridge Junction and on as far as Haymarket East Junction to join the electrified route from Carstairs.

It is worth noting that this scheme was Britain's first significant electrification since the 1994 Heathrow and Leeds North West electrification projects. It was delivered to time and budget.

Paisley Canal Line – 12/12

On completion of the Airdrie to Bathgate Project, the more detailed electrification development study for this route was nearing completion. The First ScotRail/Network Rail Alliance had by this time identified that the Paisley Canal Line should be electrified to improve the performance of the Ayrshire and Inverclyde routes and support the successful delivery of the Transport Scotland funded PCI (Paisley Corridor Improvements) enhancement project. This included three-tracking between Shields Junction and Arkleston Junction to increase line capacity for additional passenger services. The new timetable being introduced in December 2012 was modelled and the Paisley Canal DMU service was not compatible with the new timetables. First ScotRail demonstrated that the route could be electrified and operated by EMUs without requiring any additional rolling stock. After various iterations of the electrification proposals, the Network Rail project team started work on the development of the Paisley Canal line final electrification specification during 2011. This line had been closed to passenger traffic in 1983 and to through freight traffic in 1984, although freight traffic to a fuel depot at Hawkhead continued. In 1990, passenger services were restored from a new Paisley Canal Station to Glasgow Central. The scheme included five intermediate stations.

The first 5km of the 12.2km (8 miles) route between Glasgow Central and Corkerhill Depot was already electrified so that no additional HV switchgear was required.

However, for a conventional electrification scheme, nine of the twelve overbridges on the line needed electrification clearance work, including three bridges next to stations where track lowering and platform re-construction would be required. The estimated cost was twice that which could be justified, so an innovative approach had to be adopted if the line was to be energised.

A minimum-cost electrification scheme was therefore developed by specifying the lowest possible wire height for EMU operation. A similar approach

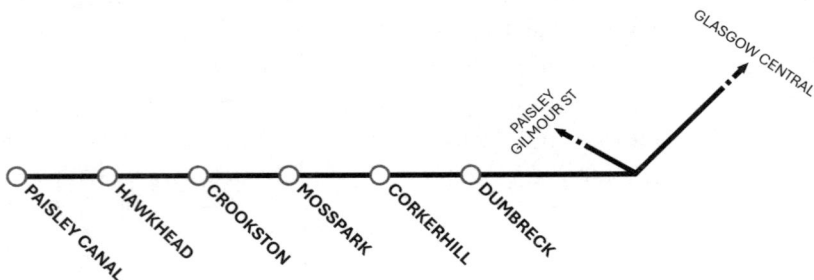

Paisley Canal Line

had been used to electrify parts of Thameslink in the 1980s. However, there was an additional requirement for the Paisley Canal scheme because the route had to retain its current W7 structure gauge to allow for a possible resumption of freight traffic to the Hawkhead Oil terminal.

Accordingly, a new method of working was developed to allow EMUs to operate with the OLE power on (daytime) and freight or engineering trains with power off (night-time). For the single-track Paisley Canal branch line, this was to introduce an 'Authority Key' system. This unique method of working required a freight or engineering train to come to a stand at a 'stop board' to obtain the 'Authority Key' before proceeding.

To provide the required electrical clearance for EMUs and W7 gauge to permit the passage of Class 66 locomotives (with the power de-energised), it was computed that a minimum wire height of 4.030 metres was possible. As a result, the minimum required bridge soffit clearance could be reduced from 4.440m to 4.305m and thus the number of bridges that required clearance works was reduced from nine to five. Only one of these structures was near a station. It was also found possible to simply provide a 70m long neutral section at one other bridge that was well clear of signals or stations, so that, in the event, only four bridges required track lowering by up to 160mm. Thus, the only major civil engineering work was the reconstruction of just one station platform at Hawkhead.

A fixed-price design and construct contract was let to Babcock in June 2012 for all the electrification works on the branch with a six-month programme starting the following month. The £10.64M job was undertaken during possessions at weekends and after 8pm on Mondays to Thursdays. Utilising the Medium Output Ballast Cleaner, track lowering through three of the four underbridges was undertaken over two weekends in early October. Hawkhead station platform was rebuilt, and the track lowered through the platform and the adjacent overbridge in a nine-day blockade. Of special note was the speed and efficiency in which the work was undertaken – it took only forty-four days from driving the first mast pile to the energising of the OLE!

The scheme utilised a KIROW 25 tonne capacity rail crane working with four modified SALMON bogie wagons to install the steel mast piles. The methodology involved the KIROW crane and two wagons to run in the possession at a maximum of 25mph between the construction depot (a disused oil siding yard near Hawkhead) and the installation site. The KIROW crane had a pile handler attachment together with two different pile driving attachments, all of which were powered from a diesel-powered module mounted on one of the wagons. The other two SALMON wagons would be at Hawkhead being loaded with steel pile tubes for the next shift. During the nine-day blockade every day the crane worked two 12-hour shifts with a wagon marshalled on either side of the crane. This novel method of working supported the overall project programme and allowed the energisation date to be achieved.

The first electric passenger service operated during the morning of Monday 19 November 2012. Between then and Monday 10 December 2012, the half-hourly service was operated with a mix of DMU and EMU rolling stock. The full EMU introduction facilitated the extension of DOO (Driver Only Operation) on this line. The release of three DMU sets provided resources for the planned re-opening of the Borders Railway in 2015.

An efficiency of £1.085M was achieved on the original financial authority, which meant that, commendably, the scheme was delivered for £9.6M. Also of note is that this was the last project in Scotland to use the Mk 3 OLE design.

Rutherglen and Coatbridge (R&C Line) – 12/14

The Rutherglen and Coatbridge Railway was opened by the Caledonian Railway in the mid-1860s linking the West Coast Main Line in the Rutherglen area with the Scottish Central Line at Coatbridge, providing a through route to Stirling and Perth. Freight and regular passenger services operated on the line until November 1966 when passenger services running to Coatbridge Central succumbed to the Beeching closures. Scheduled passenger trains between Glasgow Central and Perth returned to the route briefly from 1972 to 1974, but, apart from this, the line was only used by freight and diverted passenger services until 1993. On 4 October of that year, British Rail reopened the route to scheduled passenger services and reinstated many of the intermediate stations. It was decided to terminate these services at the recently completed new station at Whifflet rather than the original terminating station at Coatbridge Central, as this avoided using the single connection at Coatbridge Junction, since redoubling was not considered to be economically viable and because turning back trains in Coatbridge Station was now operationally unacceptable.

The business case was put forward based on two criteria. The first was that by electrifying the R&C First ScotRail could implement a re-cast of the Lanarkshire suburban network without requiring to lease any additional rolling stock as well as releasing DMU stock for use elsewhere in Scotland. The second allowed the R&C to provide an electrified diversionary route for long distance West Coast Main Line passenger and freight operations, together with an enhanced passenger service between Whifflet and central Glasgow and beyond to Dalmuir. Network Rail had previously carried out two electrification studies that supported the proposals to electrify the route in 2006 and

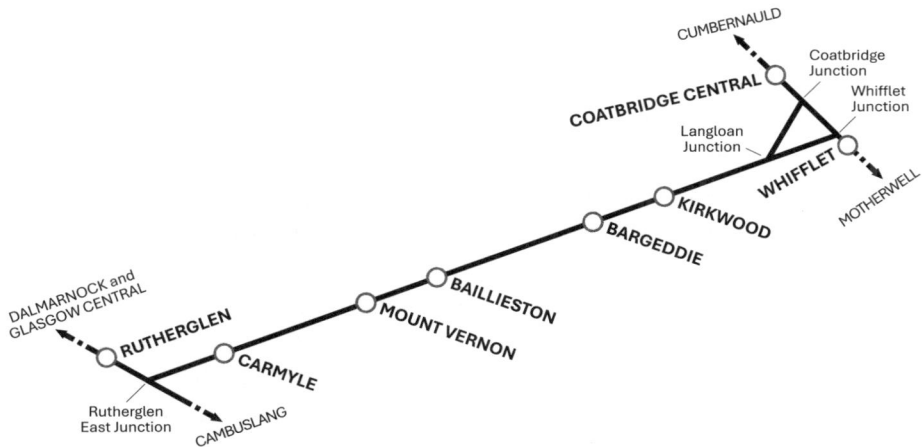

CUMBERNAULD

Coatbridge
Junction

Whifflet
Junction

COATBRIDGE CENTRAL

Langloan
Junction

WHIFFLET

KIRKWOOD

BARGEDDIE

MOTHERWELL

DALMARNOCK and
GLASGOW CENTRAL

RUTHERGLEN

BAILLIESTON

MOUNT VERNON

CARMYLE

Rutherglen
East Junction

CAMBUSLANG

Rutherglen and Coatbridge

coined the acronym RACE (Rutherglen and Coatbridge Electrification). Subsequently, Scottish Ministers' High Level Output Strategy included the scheme and scheduled it for delivery by 2018/19.

On 30 May 2013, the Scottish government announced that the electrification of the Rutherglen and Coatbridge (Whifflet) Line would be brought forward so that electric services could be in operation for the July 2014 Commonwealth Games being held in Glasgow. An Alliance Framework Agreement between Network Rail and First ScotRail was put in place to sponsor the electrification of the required 25.5 single-track kilometres (16 miles). Unfortunately, it did not prove possible to meet this tight deadline, the works being completed by Carillion on 28 September 2014. The new electric services were incorporated into the Argyle Line and commenced with the December timetable change. The scheme included the electrification of the double track Coatbridge Curve, from Langloan Junction to Coatbridge Junction.

The construction activities required the diversion of the late evening and night-time Hunterston to Longannet Power Station coal trains on three nights each week from January to May 2014. These diversions had to be carefully planned to avoid disrupting the Springburn to Cumbernauld electrification scheme construction activities. The loaded and empty coal trains had to be routed via the City Union, Springburn, and the E&G to Greenhill, which necessitated additional inspection staff to monitor the City Union Line structures, and additional maintenance payments to NR maintenance due to the increased tonnages on the diversionary routes. These diversionary routes had not seen this level of freight tonnage since 1990. A banking locomotive also had to be hired to provide back-up assistance for loaded trains suffering low adhesion challenges on the steeply graded section from Bellgrove to Springburn. The Class 66 locomotive rostered for this duty was stabled in Eastfield Loop each evening ready to assist if required. This proved to be a worthwhile provision as the 'banker' was called out on several occasions.

A number of overline structures required alterations to the parapets and wingwalls to provide separation from the live OLE. There was only one public highway bridge that had to be reconstructed, which was adjacent to Bargeddie Station, and collaborative working with North Lanarkshire Council led to a road closure and traffic management plan that minimised the disruption to road users.

To meet the Technical Specification for Interoperability (TSI), Series 2 OLE was specified. This system has the advantage in that the cantilevers are delivered to site as one complete factory-made component, which has many advantages in terms of quality, logistics and ease of construction, although some consider the design to be over-engineered.

This was the first electrification scheme in Scotland that had to meet the TSI electrification requirements. This was a learning curve for all concerned and resulted in the NR sponsor and project manager having to seek additional funding authority to complete all the authorisation requirements. The amended scheme financial authority was £39M (£50M at 2025 prices).

Most of the mast foundations were formed with driven 610mm diameter steel piles but where ground conditions were poor, mass concrete footings were used.

As vandalism and theft are key concerns along the route, the existing power and signalling cables had been buried and proved difficult to locate. This delayed the mast erection. In addition, the proposed position of the cable-duct also had to be amended to avoid obstructions that were unearthed, causing further delays.

Money was, however, saved by utilising two Track Section Cabinets (TSCs) that had become redundant elsewhere on the Scottish network, which were installed at Langloan Junction and Eglinton Street in Glasgow. Booster trans-formers, made redundant in the Rugby area, were also purloined, not only saving cash but also providing a more sustainable recycling of this vital equipment, which was required to assist with management of the immunisation interfaces. In addition, a redundant harmonic damper was acquired from Bourne End, adding further to the savings. This piece of equipment was necessary to smooth interferences in the OLE supply patterns caused primarily by the Pendolino trains that were planned to use the line as a diversionary route, once electrified.

This new electrified line is controlled from the existing control room at Cathcart with new communications links provided to interface with the SCADA systems and control displays.

It is worth specially noting that to overcome the skills shortage prevalent during this period, the main contractor, Carillion, set up training facilities for new entrants to the rail industry so that they could achieve the recognised OLEC training qualifications.

The modified timetable facilitated by the R&C electrification scheme led to an increase in footfall and revenue on parts of the Strathclyde electrified network due to the increased journey opportunities that were possible with the new timetable.

During 2017 the R&C was used by electrified inter-modal services to and from Mossend and Coatbridge that were diverted due to junction renewals in the Mossend area. These included several double-headed Class 86 and Class 90 hauled services and the electrified infrastructure performed faultlessly with the high current demand from these heavy trains.

Edinburgh Glasgow Improvement Programme (EGIP) – (Various Dates – See Each Element – Completed 5/21)

In the early part of the twenty-first century, the Labour–Liberal Democrat coalition government in Scotland was very focused on improving the link between Scotland's two principal cities. The 'Edinburgh Glasgow Improvement Programme', announced in 2006, envisaged a six trains per hour service between Edinburgh and Glasgow on the direct route via Falkirk High station with a 37 minute journey time. The initiative was continued by the subsequent Scottish National Party government, but the scope was cut back following a review by the consultants, Jacobs, in 2012. The new proposals abandoned the expensive provision of a grade-separated junction at Greenhill and a new chord at Dalmeny. In addition, means of extending the platforms at Glasgow Queen Street station became possible so that longer, higher capacity eight-car trains made a four trains per hour service feasible.

The introduction of electric trains would reduce journey time on the fastest trains from 54 to 42 minutes. The lengthening of platforms at Queen Street allowed for a significant increase in capacity on the peak services by using eight-car sets in the peak periods. The increase in train lengths from the previous six cars to eight cars increased capacity of the route by 30 per cent.

The original estimated cost for the scheme was £742M (£1050M at 2025 prices), with the new, longer electric trains running by 2017 and the full completion of the scheme in 2019.

In collaboration with Network Rail and funded through Transport Scotland, the scheme was developed into a package of enhancements (known as the EGIP

Scheme), which included major improvements to the facilities at Edinburgh Waverley station, electrification of the direct line from there to Glasgow Queen Street together with the complete rebuilding of the latter. The electrification programme also included the diversionary route through Falkirk Grahamston station and Cumbernauld and totalled 150km (94 route miles). Associated with these works was a further 50km (32 miles) extension to Dunblane and Alloa.

At an early stage of the EGIP development it was agreed that due to the large number of overline structures, several on busy public highways, route clearance works should be undertaken to align with the OLE construction programme to avoid delays. At some locations the responsible local authority for the highway infrastructure required long periods of notice due to the need to arrange reliable diversionary routes, particularly where emergency services and local bus routes were impacted. The contractor for much of the route clearance work was BAM. A number of the structures scheduled for reconstruction received local authority grant funding contributions to remove axle weight restrictions, widen carriageways and upgrade the pedestrian footways.

The intention was for the electrification of the line to Cumbernauld to be completed by July 2014, to provide acceptable access to the Commonwealth Games to be held in the summer of that year. Subsequently, the construction of the Edinburgh Gateway station would provide an interchange with the Edinburgh tram network as well as with the train services to Fife.

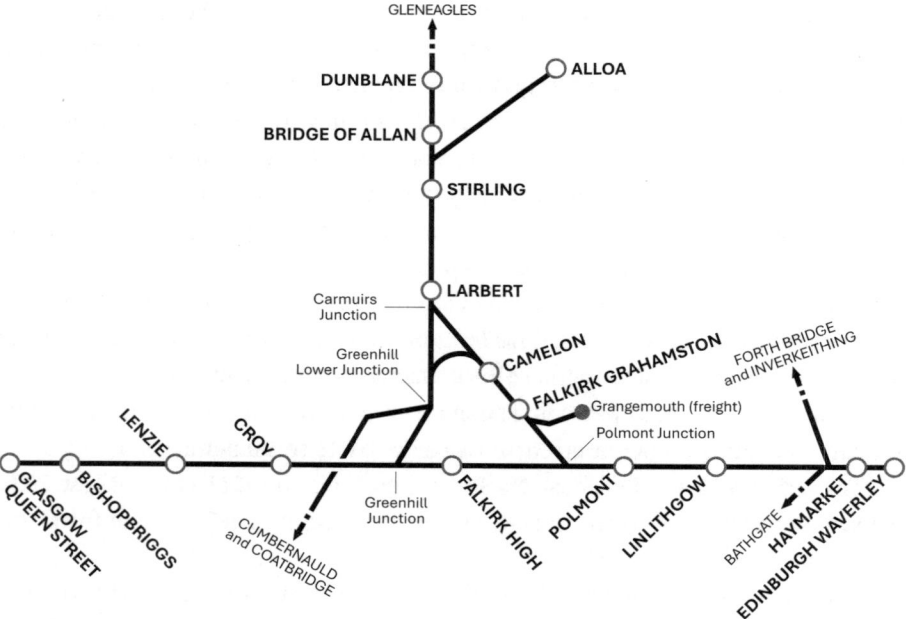

Edinburgh Glasgow Improvement Programme (EGIP)

The initial EGIP works undertaken included the £25M (£35M at 2025 prices) rebuild of the Grade A listed Haymarket Station, which was completed in 2013. Electrification of the 26km (16 miles) Cumbernauld Line followed with work taking place from June 2013 to May 2014.

The contract to deliver more than 50km (32 miles) of railway electrification for the Cumbernauld scheme was delivered by Carillion, working in partnership with SPL Powerlines. The work involved installing 1,000 foundations for the Series 2 OLE and 63,000m of wiring. The use of Tensorex anchors eliminated the need for anchoring wires and heavy balance weights.

The works included track, signalling and telecommunications renewals, together with layout remodelling at Springburn. Platform extensions were also carried out at Stepps, Greenfauld and Cumbernauld.

In addition, the scheme included the infill electrification of the remaining section of the Cumbernauld to Motherwell line from Gartsherrie South Junction (north of Coatbridge Central), including the wiring of the triangle. A new modular station building was provided at Cumbernauld as part of the scheme. This infill electrification allowed First ScotRail to introduce EMU stock on the Motherwell to Cumbernauld service, releasing DMU stock for other purposes. This service change did not require any additional EMU stock.

A novel innovation was the installation of fixed earthing devices, which allowed for remote operation and control of earthing requirements. This initiative brought about safety and efficiency improvements, obviating the need for infrastructure staff to carry out isolation and earthing tasks trackside.

The new three trains per hour electric services from Cumbernauld commenced on 19 May 2014, two of them serving as an extension to the North Clyde Line Springburn Branch trains with the third being an extension of an existing Springburn Dalmuir service. Initially, services to Glasgow Queen Street Low Level necessitated a reversal at Springburn Station. Effectively, this was the first stage of the EGIP scheme to be energised, commissioned and placed into service. Non-critical infrastructure and snagging work continued along the route until December 2014.

The wires to Cumbernauld were subsequently extended to Greenhill Lower Junction so as to connect with Falkirk Grahamston and provide a secondary route to Edinburgh as part of the Stirling, Alloa and Dunblane project. Half-hourly electric services replaced the hourly DMU service and also took over the existing EMU service between Springburn and Cumbernauld. These new electric services were diverted to Glasgow Queen Street High Level from the Low-Level station.

The route was energised with the first train running the full length of the route under test on 31 October 2017. A new half-hourly Edinburgh–Falkirk

Grahamston–Cumbernauld–Glasgow Queen Street electric service was introduced on 9 December 2018.

To operate the Central Belt suburban and inter-urban services, a total of 70
Hitachi Class 385 units were ordered in 2015, 46 three-car and 24 four-car sets.
The new fleet of trains was expected to be deployed in the autumn of 2017, but
driver visibility and software issues delayed their introduction until July 2018.
In the interim, Scotrail procured ten Class 365 units as a stop-gap measure. By
December 2019, all 70 of the Class 385 trainsets had been delivered.

Stirling, Alloa and Dunblane – 12/18

Work on this 50km (32 miles) scheme to electrify the lines from Greenhill, Larbert
and Polmont junctions to Stirling, Dunblane and Alloa commenced in October
2016 as an 'add-on' to the EGIP programme. This work started with vegetation
clearance, followed by the piling of the foundations for the masts supporting the
OLE. Modification or reconstruction of bridges along the route was also undertaken to provide adequate clearances for the Series 2 OLE. Much of the work was
undertaken in a four-week blockade from 17 March to 16 April 2017.

Significant works were undertaken at Stirling Station to improve accessibility for customers, including the installation of lifts to provide step-free access
to all platforms.

The first section of the Stirling–Dunblane–Alloa (SDA) electrification project
was completed in May 2018 covering 26 STKs (16 miles) between Greenhill,
Larbert and Polmont junctions and included Falkirk Grahamston and Camelon
stations. The scheme involved the installation of more than 750 piled foundations, erecting more than 500 masts and running out 120km of wires.

Work continued in the second half of the year and on the night of 27/28
November 2018 the first train successfully completed test runs at line speed along
the newly electrified lines to Dunblane and Alloa. Testing was completed on
7 December 2018 with electric passenger services introduced from 9 December
2018.

In order to achieve the December timetable introduction date, it was identified in August/September 2018 that a nine-day access period would be necessary
for the line from the Carmuirs triangle north to Dunblane, and east to Alloa
and Longannet Power Station. This resulted in very detailed and occasionally
heated discussions with all the passenger and freight operators affected by the
long possession period.

The 9-day disruptive possession was agreed for October 2018 and a member
of the Network Rail sponsor team had to arrange the service alterations and

route conductors for the diverted services. The freight operators affected were DB Cargo (oil and gas pipes and calcium carbonate), DRS (inter-modal and nuclear services) and Colas Rail Freight (cement). The passenger operators affected were ScotRail, LNER and Caledonian Sleeper. Without the assistance of West Coast Railway Company who provided route conductor driver resources, the disruptive possession would not have worked. The nine-day possession was a success, with the system-proving, energisation and entry into service milestones all achieved.

In total, along the 50km (32 miles) of the route, Network Rail installed more than 2,000 structures carrying the overhead line wire.

The electrification of the route between Glasgow Queen Street and Falkirk Grahamston via Cumbernauld provides a diversionary route for electric trains using the main Edinburgh–Glasgow via Falkirk High route. It also delivered the additional benefit of facilitating the electrification of the short freight-only line from Grangemouth Junction to Grangemouth (Fouldubs Junction). This electrification received a separate CP5 enhancement funding contribution because of the extension to the Scottish rail freight network. The electrified network at Fouldubs Junction includes the freight loops, part of the Docks Branch and the line connecting with the Inter-Modal freight terminal. The latter has the OLE terminated at structures that prevent the container/swap body reach stackers from encroaching on the live OLE. Special operating instructions were produced to maintain safety for the rail and road operations within the terminal. There are electric hauled services to and from Daventry International Freight Terminal that operate on a daily basis, which reduces the amount of diesel haulage in Scotland.

The Forth Port Authority runs the expanding Grangemouth port, which has a container terminal and an oil terminal. The latter supports the adjacent INEOS oil refinery. In 2020 it was announced that rail handling facilities at this Central Scotland port terminal, costing £3M, would be greatly extended by doubling the number of reception sidings and extending their capacity to accommodate 775-metre trains when completed in 2021. This investment was completed as planned by Forth Ports Authority. In 2024 it was announced that the oil terminal is to close. However, plans are afoot to restore passenger services on this line.

Edinburgh to Glasgow Direct Line (E&G) – 12/17

The work to electrify the direct route between Edinburgh and Glasgow via Falkirk High commenced in 2013. Associated with this scheme was the redevelopment of Edinburgh's Haymarket station previously mentioned and the construction of the new £41M (£55M at 2025 prices) Edinburgh Gateway station and tram-train interchange in 2016. Major works were also required at Glasgow Queen Street Station where platforms needed to be extended to accommodate eight-car trains, and the station facilities completely remodelled.

The project required a four-year programme to reconstruct 60 overbridges to provide adequate clearance for the OLE. In addition, more than 100 bridge parapets were raised to comply with safety standards to ensure that the public is kept a safe distance from the overhead equipment.

Two major blockades of the route were required. The first in 2015 closed the 335m (372 yds) long Winchburgh tunnel in West Lothian for six weeks to enable the installation of slab track to be undertaken. Even more substantial works were undertaken in the 979m (1006 yds) Cowlairs tunnel on the approach to Glasgow Queen Street. Here, Queen Street High Level was completely closed for a twenty-week period from March to August 2016. This enabled 1800m of permanent way in Cowlairs tunnel to be lowered and reconstructed in slab track form to accommodate the installation of the new OLE.

Platform extensions were constructed at five other stations, at Linlithgow, Polmont, Falkirk High, Croy and Platform 12 at Edinburgh Waverley. The 75m extension of Platform 12 at Waverley Station has created a 204m (670ft) long platform and involved the demolition of redundant station buildings. Substantial trackwork remodelling was undertaken at this station in advance of the erection of the OLE. The project also included resignalling between Haymarket and Inverkeithing.

A new LMD (Light Maintenance Depot) was constructed in the former Up Yard area at Millerhill. Part of this utilises the BR era Fuelling and Stabling Point and associated train crew/administration building. The latter was operated by DB Schenker who had to be re-located to another part of Millerhill Yard to allow the demolition and construction works to proceed. The new LMD was required to service the new Class 385 electric train fleet. This included the installation of a new train wash facility and the erection of high security fencing, together with staff accommodation.

The existing LMD at Eastfield in the Springburn area of Glasgow, which had been reopened by National Express ScotRail in 2005, was identified as requiring expansion and further enhancement to support the new Class 385 EMU fleet. This work included significant cutting slope excavations and piling works to extend the reversing headshunt at the Bishopbriggs end, and the installation of additional stabling sidings. Other facilities provided included staff access platforms, staff walkways, electrification compliant yard lighting systems, additional CET (Controlled Emission Toilet) servicing equipment and staff accommodation. In addition, the 'wiring' of the whole LMD track layout was carried out. OLE distribution arrangements on the main line had to be modified to accommodate the newly electrified LMD site.

On top of this, 15km (10 miles) of walls and fencing were raised or rebuilt at a cost of £3.3M in order to comply with the 1.8 metre safety standard required for the operation of an electrified railway.

The erection of the Series 2 OLE required the installation of 3,000 mainly piled foundations to provide the bases for the overhead masts, which support the OLE.

One factor, which was outside the control of the project team, had a significant negative effect on the delivery of the project. From early 2015, UK railway projects had to comply with European Standard EN TSI. (This refers to British Standard BS EN 50122, which protects people against electric shocks.) Annex G of this standard permitted 25kV equipment to be outside a 2.75m (9ft) radius from the platform edge. In the rest of Europe, it had to be outside a 3.5m (11ft 6in) radius. A British Standards Committee had ruled in 2013 that Annex G should not be used. So, from 2015, unless justified by risk assessment, 25kV equipment has to be outside the normal European 3.5m (11ft 6in) radius. This requirement was to apply to all new schemes, including those under construction. On 25 August 2016, Network Rail announced that a re-evaluation of the EGIP OLE design clearances would be necessary that both delayed implementation and inflated the cost of the scheme.

By this time the project was already behind schedule by seven months with costs over-running due to systemic project planning and delivery issues including revisions to designs, changes to access and unexpected site conditions. The OLE clearance problems added further to these escalations. However, an electrified service between the two cities was introduced in December 2017, initially with seven-car trains pending the completion of the Queen Street Station platform lengthening works. The cost overrun was in the order of £110M.

The key requirement to extend the platforms and reconstruct Glasgow Queen Street Station was undertaken as a separate work stream. In 2017 Network Rail secured a Transport and Works Scotland (TAWS) Order, giving the organisation powers to compulsorily purchase the eight-storey Consort House and the Millennium Hotel extension, which had been erected between 1969 and 1973 as part of an earlier station redevelopment scheme.

Balfour Beatty was awarded a £16M enabling-works contract for the station's redevelopment in April 2017. Shortly after, this company was also awarded the £63M 'target cost' contract to demolish buildings and build the new station. Arup had the responsibility for the scheme's design, with BDP architects acting as architectural sub-consultants.

Demolition work was extensive, including taking down the multi-storey Consort House and part of the Millennium Hotel extension. These works, completed in October 2017, involved the removal from site of 14,000 tonnes of material, most of which was crushed and recycled.

Cladding, roofing, curtain walling together with the striking glass façade was completed in September 2019, with the finished works nearly doubling the size of the station concourse.

October 2019 saw the overall completion of the platform extensions and OLE works, enabling Platforms 2, 3, 4 and 5 to be able to accommodate eight-car electric trains. In total, five platforms in the station have capacity for trains of this length as Platform 7 could already accommodate them.

Work at Glasgow Queen Street Station was on track to be completed by 1 April 2020, but all work was stopped on 17 March 2020 due to COVID-19. A recovery plan was subsequently implemented with on-site work resuming in June 2020. The works were finally completed in June 2021. In total the station works were budgeted at £120M in 2014 (£160M at 2025 prices).

Prior to the EGIP programme, the Edinburgh to Glasgow direct line carried four DMU trains an hour in each direction between the two cities with a minimum journey time of 49 minutes. Each six-car train had 396 seats. From December 2019 four eight-car EMU trains per hour with a capacity of 546 seats covered the distance in 42 minutes, giving not only a significantly faster journey but also an increase in capacity of 38 per cent.

Holytown Junction to Midcalder Junction: The Shotts Line

The Shotts Line also links Edinburgh and Glasgow, a mixed traffic railway that prior to the pandemic had a stopping and semi-fast two trains per hour passenger service frequency, and several inter-modal and heavy haul freight services. The route is also used for passenger and freight diversionary purposes. Journey times between the two cities take approximately 50 per cent longer than the direct route via Falkirk High. From Edinburgh, the route shares the Edinburgh branch of the West Coast Main Line as far as Midcalder Junction, where it branches off towards Livingston, then running via Shotts, Holytown and Bellshill before joining the main West Coast Main Line to Glasgow Central at Uddingston.

The sections of the route between Edinburgh Waverley to Midcalder Junction and from Glasgow Central to Holytown had previously been electrified at 25kV AC, which resulted in the project consisting of the electrification of 75 STKs (46 miles) between Midcalder Junction and Holytown. Work commenced on this £160M scheme in 2013 when the initial financial authority was obtained to allow the Route Clearance package of works to commence. This was in the form

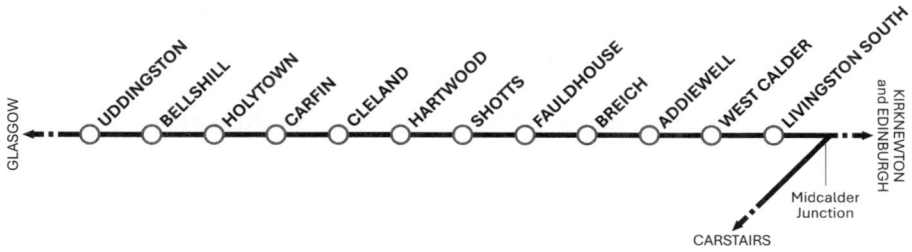

The Shotts Line

of an ECI (early contractor involvement) contract with BAM. This led to BAM letting detail design contracts to three design houses (Tony Gee and Partners LLP, URS and Mott MacDonald) to meet the timescales for delivery of the route clearance works. As detailed development proceeded it was realised that a number of structures proposed in the feasibility studies for complete removal would have to be retained. This, coupled with the complexity of some of the overline structure reconstructions, resulted in the project cost increasing. There were more than 29 overline structures that required intrusive works, including several that required demolition and full reconstruction. Six of these were all on strategic public highways, which required significant road closures and diversionary routes to be agreed with the two local authorities involved.

One structure at Shotts required a road closure of nine and a half months that necessitated a 6km (3½ mile) diversion for road vehicles. Another, in West Lothian, required a sixteen-week closure of the A71 road, a strategic east–west road including several bus routes.

The route clearance work included the provision of a new footbridge and 'Accessible' lifts at West Calder Station, together with the careful dismantling of the listed footbridge and transport to a registered railway museum. The Victorian footbridge at Addiewell Station was replaced with a standard footbridge that could be fitted with Accessible lift towers by a separate Accessibility project. While the utility service diversions were complex, the route clearance works were successfully completed in early 2018.

The electrification element of the scheme involved the erection of 1,400 steel masts carrying the overhead wires, installation of three TSCs (Track Sectioning Cabins), SCADA systems linked to Cathcart ECR and alterations to the existing electrified network and stations at Holytown and Midcalder Junctions, Livington South and Breich. The latter station that had been listed for closure prior to electrification was retained, rebuilt and provided with an hourly service utilising the Shotts Line stopping services.

Platform alteration works were undertaken at the following stations to accommodate the various rolling stock types that would run on the route once electrified: Carfin, Hartwood, Shotts, Fauldhouse and Addiewell. The electrification element contract was let to the Carillion Powerlines consortium (CPL). When Carillion plc went into administration in January 2018 the scheme experienced some unplanned challenges, however the CPL equity was acquired by Powerlines in February 2018, which ensured any programme delays were minimised.

The installation of the OLE utilised the Italian-built Alstom wiring train that was based at the main construction depot adjacent to Cleland Station and at the subsidiary depot adjacent to West Calder Station. The wiring train allowed the contact system to be installed under tension thus effecting a significant reduction in installation times and labour resources.

The route was successfully energised for system proving test trains on Sunday 24 February 2019. (These trains utilised Class 86 electric locomotives and rolling stock hired from the Freightliner Group.) Energisation for passenger and freight operation was achieved on 31 March 2019. This complied with ORR CP5 enhancement milestone for the scheme.

Due to EMU rolling stock and train crew availability challenges, a limited ScotRail electric service did not commence until later in April 2019 with the full ScotRail passenger service commencing at the timetable change in May 2019. The first electric passenger services to run the route in both directions were planned diversions of the Caledonian Sleeper Lowlander services and associated ECS workings between Polmadie and Edinburgh Waverley, all utilising Class 92 locomotives.

The project was completed to time and budget, creating a fifth electrified route between Scotland's largest cities.

The route clearance works facilitated a separate Transport Scotland accessibility project to provide accessible lifts at the following stations: Cleland, Fauldhouse and Addiewell. This was a further indication of the excellent collaborative working arrangements between Transport Scotland, Network Rail and ScotRail.

Summary

The Edinburgh Glasgow Improvement Programme (EGIP) has delivered a cleaner, greener and quieter railway with lower carbon emissions. Passengers have benefited from more frequent services and faster journey times in modern, attractive and more energy-efficient trains.

Over the period 2014 to 2019, Network Rail, on behalf of the Scottish government, completed a rolling programme of electrification of 325km (202 miles) of Scotland's central network between Edinburgh and Glasgow. This increased

Scotland's electrified routes from 711km (441 miles) to 1037km (643 miles), representing 37.35 per cent of the total 2776km (1721.5 miles) network.

As a result, in 2022, following a fifteen-year rolling programme, electric trains now carry 76 per cent of rail passenger traffic and 45 per cent of freight haulage in Scotland, more than triple what it was in 2008.

Electrification schemes currently active in 2024/5

Muirhouse Central Junction to Barrhead and Busby Junction (Pollokshaws West) to East Kilbride

Advanced works involved route clearance works between Muirhouse Central Junction and Barrhead, including lineside vegetation clearance and the provision of enhanced boundary fencing. These works were followed by the installation of pile tubes or cast foundations and the erection of OLE masts. A significant public highway bridge reconstruction was undertaken in the Strathbungo area of Glasgow.

The project team also identified the need for additional electrical capacity on the existing South Clyde electrified network. Collaborative working with National Grid and the Transmission Network Operator (TNO – Scottish Power) from 2019 to 2021 involving detailed electrical modelling enabled the optimum grid supply points, TSCs, Neutral Sections and Motorised Line Switch locations

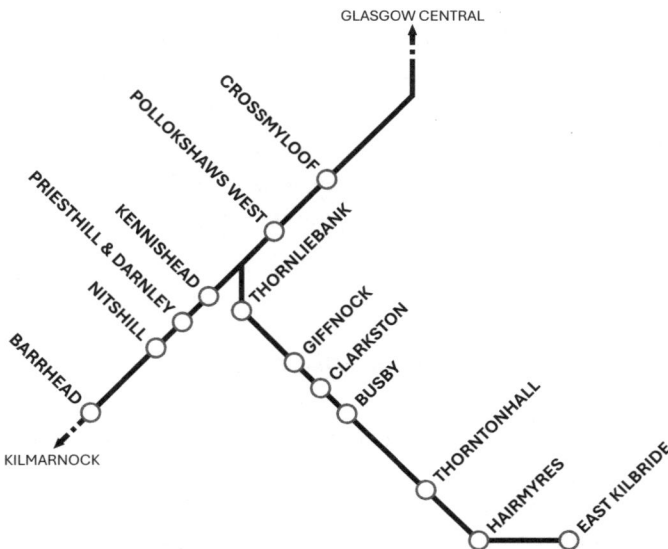

Barrhead, Busby and East Kilbride

to be identified. In addition, where necessary, alterations to the existing OLE distribution arrangements were also pinpointed. Following this advance work, financial authority from Transport Scotland was secured to deliver the works. For the Barrhead scheme, this involves the procurement, construction and commissioning of a new Feeder Station at Elderslie (south-west of Paisley), switchgear modifications at Eglinton Street Feeder Station, a new TSC at Busby Junction and SCADA system enhancements at Cathcart ECR (Electrical Control Room) to supervise and control the new 25kV assets.

The cost of the advance works was £43.5M.

A six-week blockade of the Barrhead line commenced on 24 June 2023 to permit the electrification and upgrade works to take place. During the closure 14,400 metres of OLE were erected and 130 metres of track was laid in. Other infrastructure work included the raising of bridge parapets at Kennishead and Priesthill & Darnley stations, together with a significant amount of work undertaken at Barrhead station. Here, a platform extension was necessary to accommodate the longer four-car EMUs.

The Glasgow–Barrhead line reopened on Friday 4 August 2023, following this £63.3M upgrade.

The £96.3M funding for the Busby Junction to East Kilbride route section was announced by the Scottish government on 6 September 2023 as part of the rolling programme of electrification, which forms a key element of its transport decarbonisation plans. The scheme included the doubling of 6.4km (4 miles) of the route between Busby and East Kilbride.

The 22.4 track-km East Kilbride Electrification between Busby Junction and East Kilbride will enable electric services to serve the route from December 2025. Planned route clearance activities are to address a number of overline structures that would foul the OLE, mostly carrying public highways. Further elements of this enhancement project include a 1.4 km extension of the Hairmyres Loop, the relocation of Hairmyres station, platform extensions at Giffnock and East Kilbride, a new station building at East Kilbride plus step-free access improvements at all existing stations.

Borders Railway Discontinuous Electrification

This scheme demonstrates Transport Scotland's commitment to a phased implementation to rail decarbonisation. Hybrid EMU stock is to be procured to allow electrification to be extended in areas where route clearance challenges require more time to develop affordable permanent solutions. The hybrid EMUs will draw power from the OLE through the pantograph(s) on electrified sections to

Borders Railway

both supply the traction motors and charge the units traction batteries. These batteries will, in turn, deliver stored energy to run the trains on non-electrified lengths.

The sections to be electrified in the first phase are listed below:

Newcraighall (connecting with existing electrified network) to a point
    south of Kings Gate Junction (between Shawfair and Eskbank Stations)
Tweedbank to Bowshank Tunnel (but may need to be extended to
    Galabank Junction north of Stow Station).

A new Feeder Station (FS) is to be installed at Tweedbank, designed to cater for future extensions of the OLE northwards to Shawfair Station. Initially, Tweedbank FS will supply the electrified southern section and allow the hybrid EMU stock to re-charge their traction batteries before returning to Edinburgh. A new TSC at Newcraighall/Millerhill will be installed to provide for electrical separation of the Borders Line from the existing electrified network in the Portobello/Millerhill area.

A separate project is to provide a new FS at Portobello Junction to replace the existing Portobello FS commissioned in 1990 as part of the ECML electrification scheme. The new FS will have two feeding circuits that will increase the electrical capability of the Edinburgh area rail network, including the north end of the Borders Line. This project is fully authorised and in the process of implementation at the time of writing.

Fife Decarbonisation Phases 1 and 2, including the Levenmouth Branch

This scheme is to allow hybrid EMU stock to serve the existing Fife Circle and the Thornton North Junction to Ladybank Junction section plus the Levenmouth Branch reopening project. The route section of ECML North from Haymarket West Junction to Dalmeny Station is also being electrified. The details of the individual route sections that will be electrified follow below:

- Haymarket West Junction to Dalmeny
- Kinghorn (north of the station) to Ladybank
- Thornton North Junction to Levenmouth Station
- Thornton South Junction/Thornton North Junction to a point between Cardenden and Lochgelly stations.

A new FS at Thornton North Junction is being constructed that will be capable of powering all the Fife Circle from Inverkeithing, Charlestown Junction to Longannet, the Levenmouth branch and eventually to Newburgh and Tay Bridge South.

To cope with the additional electrical loadings on the west side of Edinburgh a new FS is being constructed at Currie on the Carstairs to Haymarket East Junction section of the WCML. This involves expansion of the existing TSC at Haymarket and alterations to the feeding arrangements in the Newbridge TSC area of control on the E&G line.

The £116.6M Levenmouth Branch reopening project comprised the reconstruction of a double-track railway approximately 10km (6 route miles) long, with two new stations at Cameron Bridge and Leven. The last BR passenger

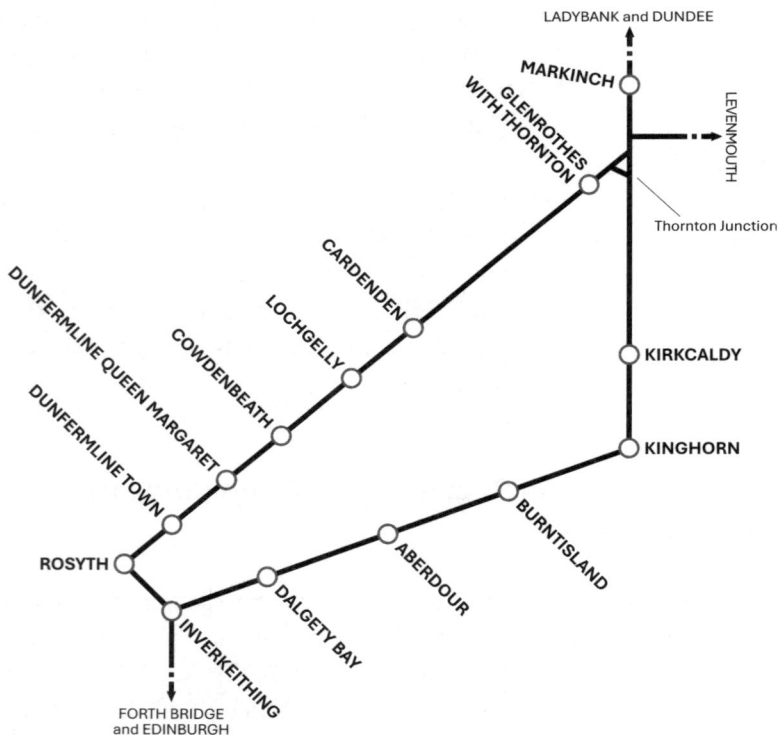

Fife Circle

services operated in 1969, with freight services then declining until these ceased in 2001. Work started in September 2020 with the reconstruction completed in the spring of 2024. The new passenger service to Edinburgh commenced at the timetable change on 2 June 2024. The project included installing the OLE bases, although electrification works will not be completed until the new hybrid EMU rolling stock is delivered. The procurement of the rolling stock is to be funded by Transport Scotland in conjunction with ScotRail. At the opening it was stated that the new electric trains to and from Edinburgh will replace the preliminary DMU services 'at the earliest opportunity'.

Decarbonisation – Traction Power Phase 1

During the development of the Scottish electrification schemes, subsequently delivered in CP5, it became clear from the traction power modelling work that was undertaken with various freight and passenger timetables that additional electrical capacity would be required to sustain the existing network as well as supporting the additional electrified routes. The outputs from the modelling combined with the experience of managing the 25kV OLE distribution system in Scotland informed further development work. This identified that six new Feeder Stations (FS) would be required to allow the expanded network planned for implementation during CP6 and early CP7 to function in a safe and reliable manner. This piece of work also factored in the planned increase in cross-border electric-powered passenger and freight services, which included planned rolling stock changes where operators are upgrading from diesel to electric traction.

The provision of six new Feeder Stations also requires alterations at existing 25kV distribution system sites together with several new TSC locations to manage the extended network and the enhanced existing network. The new FS sites to be commissioned by the end of 2025 are as follows:

- Elderslie
- Newton
- Currie
- Portobello
- Tweedbank
- Thornton.

The works outlined above have been carefully aligned with future phases of the rail decarbonisation strategy to support the electrification north from Dunblane through Perth and Dundee to Aberdeen together with the planned extension of electrification north of Perth to Inverness.

During the development phase for traction power a dialogue was established with the National Grid Company and the TNO (Scottish Power). This proved invaluable, as it highlighted parts of the Grid infrastructure that were not suitable for a railway supply point or would require infrastructure upgrades to meet the railway peak loading requirements.

The traction power programme for Phase 1 also involved ECI (early contractor involvement) to produce outline designs, cost plans and procurement strategies. This allowed Network Rail to cost and subsequently implement a bulk purchase of the distribution switchgear and modular enclosures for the new FS and TSC sites as well as additional equipment required at existing sites. Like the Grid and TNO electrical assets those required on the Network Rail side have long lead times, which supports the need for early engagement and finalising of the electrical distribution layout and design.

The process established with the National Grid Company and Scottish Power is also being used to discuss the electrification proposals for parts of Ayrshire and the South-West of Scotland. These cover the routes outlined below:

- Barrhead to Kilmarnock
- Barassie to Kilmarnock
- Kilmarnock to Gretna Junction
- Mauchline Junction to Newton on Ayr Junction
- Ayr (Belmont) to Girvan.

These are covered by later stages of the Scottish rail decarbonisation strategy.*

---

* As this book went to press, Audit Scotland announced that there are insufficient funds for Scotland's planned £26Bn investment programme and so the $32M Borders railway decarbonisation programme has been paused. In addition, it is likely that a revised Rail Services Decarbonisation Action Plan (DAP), to be publishes in 2025, will slip the decarbonisation target to 2045.

Chapter 14
London Overground

From the 1960s, many of the inner-suburban services around London deteriorated with falling passenger numbers and an associated lack of investment. Some of the routes were old and were operated in isolation, often with disjointed service patterns reflecting their historic lineage.

Towards the end of the 1970s, the Greater London Council took steps to reinvigorate the North London Line and this route was incorporated into the London Underground map, marked up as 'British Rail'. On the West London line, a limited Clapham Junction – Kensington Olympia service via West Brompton had operated for some time (originally from during the Second World War for Post Office employees only) before a more comprehensive passenger service was introduced in 1994 by Network SouthEast. A further positive development occurred on 1 June 1999 when West Brompton Station was opened by the Minister of Transport, Glenda Jackson. However, there was no one organisation focused on improving and integrating the heavy rail inner suburban routes that skirted the capital.

The seeds for the reinvigoration of these inner suburban lines were sown in the early 1980s, although it was to be a long time before they came to fruition. In 1983, the Department of Transport consulted on a proposal to set up a London Regional Transport (LRT) Authority for London for the control of public transport services in the Greater London area. This authority came into being on 29 June 1984 and, on 1 April 1985 the organisation inaugurated London Buses Limited and London Underground Limited as wholly owned subsidiaries of LRT.

Following the privatisation of British Railways, the Waterloo and City line was transferred to London Underground on 1 April 1994. In 1997, the DC electric services of the Euston to Watford line together with the North and West London lines were brought together as part of the National Express Franchise and branded as Silverlink. The LRT remained in control of public transport until control passed to the newly formed Greater London Authority on 2 July 2000, when that organisation's agency, Transport for London (TfL), was formed and took over. Subsequently, in 2006, the Secretary of State for Transport, Alistair Darling, agreed to transfer the control of the metro services operated by Silverlink to TfL. At this time, the outer London Silverlink services were taken over by the London Midland Franchise.

TfL held a vision of developing a new, modern rail network around London that interconnected with the capital's other heavy rail and tube services. They therefore sought tenders in February 2006 for the Silverlink metro services under the provisional title 'North London Railway' as a first stage of achieving that objective. The proposed new services, which were rebranded as London Overground in September 2006, were to be operated as a concession rather than a franchise. At last, there was to be a single entity controlling the Underground system and the key inner suburban heavy rail services. The first operator of the 'North London Railway' services was an alliance between Laing Rail and the MTR Corporation with a contract to run the services until 2016.

The Mayor of London, Ken Livingstone, launched the new London Overground services at Hampstead Heath Station on 11 November 2007, the day on which TfL took over responsibility for the 'North London Railway' operations. Included in the initial London Overground portfolio was the Richmond to Stratford line, the West London line and the Gospel Oak to Barking line together with the former London Underground line between New Cross/New Cross Gate and Shoreditch. For the latter, there were plans to upgrade this route as a heavy rail service with extensions at either end.

This was an important milestone for the capital's public transport system. The date 11 November 2007 marked the start of a process to sweep away the historical operating restrictions of these inner suburban routes and, together with a number of other 'Cinderella lines' around London, forge them into an integrated network of services. The new organisation initially took over 55 stations and initiated a comprehensive station refurbishment programme to progressively provide 'access for all'.

Passenger service quality was vastly improved with the simultaneous introduction of 62 new trains and the Oyster card system. The rolling stock and station branding incorporated a very bold orange stripe, and a similar stripe was used to denote Overground services on the Tube map.

Further Network Rail and London Underground routes have since been absorbed into the London Overground network. From 2016, operations have been run by Arriva Rail London, and this concession is responsible for operating the following services at the time of writing:

- Euston to Watford Junction
- Richmond to Stratford via Gospel Oak
- Gospel Oak to Barking
- Willesden to Clapham Junction via Shepherd's Bush
- Highbury & Islington to West Croydon via Dalston Junction (with extensions to Crystal Palace, New Cross and Clapham Junction)
- Liverpool Street to Enfield Town/Cheshunt/Chingford
- Romford to Upminster.

A brief description of the routes and the electrification details follows.

Euston to Watford

The Euston to Watford line parallels the London and Birmingham Railway main line, which originally opened in 1838. The electric service between Euston and Watford Junction, together with the Croxley Green branch, was completed in the last year of the existence of the LNWR in 1922. The Rickmansworth branch was subsequently electrified by the LMS in 1927.

Passenger services ceased on the Rickmansworth branch in 1952, and on the Croxley Green branch in 1996, although sections of the latter may be brought back into use under the Croxley Link scheme. However, this project is currently stalled due to funding and other difficulties. Further details of the history of the Euston to Watford suburban line and its electrification in the twentieth century are described in Chapters 2 and 9.

The network originally utilised the 630V fourth rail DC system. Subsequently the system was upgraded to 660V, with the fourth rail bonded to the running rails, enabling the operation of both standard British Rail third-rail EMUs and London Underground tube stock. Latterly, the system has been upgraded again to permit the use of London Overground Class 710 Aventra EMUs.

As already described, with the privatisation of British Railways in 1997, the Euston to Watford service became a component of the National Express Franchise, branded as Silverlink. The service was one of the first to become part of the London Overground network in November 2007. At the time of writing, the services generally operate with a twenty-minute service frequency.

Richmond to Stratford

As stated in Chapter 2, the North London line between Richmond and North Woolwich (later cut back to Stratford) was an amalgamation of a number of separate lines constructed between 1846 and 1869. The line was electrified by the LNWR at the start of the First World War from Broad Street through to Richmond.

As previously noted, this was one of the three lines launched as London Overground on 11 November 2007. Class 313 EMUs operated the route until 2010, when these were replaced by Class 378 Capitalstar four-car EMUs operated by London Overground. The new service attracted many new customers so that these units were subsequently strengthened to become five-car formations.

The line's trains moved to new high-level platforms 1 and 2 at Stratford from the low level on 15 April 2009, as the previous low-level platforms were required for the Docklands Light Railway Stratford International service.

Gospel Oak to Barking

The original core of the route was between Upper Holloway to Woodgrange Park, which was established in a piecemeal manner in the late 1800s. The history of the construction and the various service alterations that have taken place over the years is described in Chapter 12. The current Gospel Oak to Barking route was initiated in the 1980s with a two trains per hour service. At the launch of the London Overground Network in November 2007, the Gospel Oak to Barking element of the 'North London Railway' concession became part of the new network.

Transport for London was keen to see the route electrified but electrification was ruled out in 2008 on grounds of both cost and infrastructure difficulties, including unstable ground and the number of restricted infrastructure overhead clearances.

In October 2009, Network Rail published its Route Utilisation Strategy, which indicated a benefit to cost ratio of 2.4:1 for the line. However, there was much to-ing and fro-ing between the Mayor of London, Boris Johnson, and Justine Greening, the Secretary of State for Transport, about who should pay for the proposed investment.

In June 2013 it was eventually announced that £115M (£157M in 2025) of funding was to be made available for the electrification as a component of a number of rail infrastructure upgrades included in the government's 2013 spending round.

The work involved substantial civil engineering works that were beset by a number of problems concerning the design of the enhancements. Therefore, by the time work started, inflation and cost escalation had increased the cost of the upgrade to £133M (£179 in 2025) of which the 'electrification' portion was approximately 30 per cent of the total.

Network Rail awarded the contract to electrify the route in September 2015 to J Murphy & Sons. This utilised the NR Series 2 Overhead Line Equipment range for this scheme, which included a link to the East Coast Main Line at Harringay.

The works also involved platform lengthening, track lowering, bridge reconstructions and extensive retaining wall strengthening. Elsewhere, tight clearances under and above the track required a number of sections of the line to be slab-tracked. Weekend closures were taken from South Tottenham to Barking from June to September 2016, together with a full line block from October 2016 to February 2017.

Commissioning was originally planned for June 2017, but due to the various design and trackwork issues, compounded by the late delivery of materials, completion of the work was further delayed.

The infrastructure works were eventually completed in May 2018, including the commissioning of the electrification. However, late deliveries of the new rolling stock prevented the introduction of the new electric trains during that year. The first two Class 710 EMUs entered service on 23 May 2019 and operated a two-trains per hour service until sufficient new units were delivered to enable the full timetable to be implemented in June of that year. However, it was August before the last new 710 unit was supplied. TfL offered a month's free travel to compensate passengers for the inconvenience caused by the reduced service levels.

A 4km (2.5 miles) electrified extension to Barking Riverside was constructed from the Tilbury Loop line (between Barking and Dagenham Dock stations) to serve the huge new Riverside housing development. The new line was commissioned in 2021, six months ahead of programme. Located in the new town square, the Barking Riverside terminus station opened in May 2022. The extended line together with the brand new station, having interchanges with the Fenchurch Street and District and Hammersmith and City lines at Barking, have significantly improved public transport facilities in the area.

West London Line

As noted in Chapters 2 and 9, the northern section of the West London Line had been electrified by the LNWR in 1914/15. Further third-rail electrification took place in 1993/4 to enable Eurostar trains to run from their depot at North Pole to the then terminal at Waterloo.

In order to permit the proposed Regional Eurostar trains to run from the both the West and East Coast main lines and by-pass London en route to Paris and Brussels (and vice versa), the 25kV AC system from Willesden was extended southwards by a kilometre to Mitre Bridge Junction. Unfortunately the proposed Regional Eurostar services were never implemented.

However, this extension also enabled, after many years, an electric suburban services to be reinstated between Willesden Junction and Clapham Junction, calling at Kensington Olympia. A further station was reopened on 1 June 1999 (the original having closed in 1940), when West Brompton Station was launched by the Minister of Transport, Glenda Jackson.

Associated with the huge Westfield Shopping Centre Development, a new station, Shepherd's Bush, together with an integrated bus interchange, was built by the developer as part of the Section 106 development contribution. Station building work commenced in 2006, but the opening was delayed until 29 September 2008 due to issues associated with platform width. Also inaugurated at the same time was a surface-level interchange with the Central Line station of the same name. The Central Line station had also been upgraded as part of the development scheme.

As it had been originally planned for the new Shepherd's Bush station to open during 2007 it would have been served by Silverlink trains. Accordingly, at contract handover, the station was still branded with Silverlink signage. However, by the time the station opened, this rail franchise had terminated, as the service had passed on to London Overground in late 2007, so the Silverlink signage was replaced.

From the inauguration of London Overground services in November 2007, a four trains per hour service was provided between Clapham Junction and Willesden Junction, with alternate trains running through to Stratford via the North London Line. Southern services between Milton Keynes, Watford and Clapham also formerly used the route, calling at Shepherd's Bush station, but this service has now ceased.

A further station at Imperial Wharf, next to Chelsea Harbour, opened on Sunday 27 September 2009. This station is adjacent to a brownfield site that was transformed into a luxury riverside apartment development.

The West London Line completed the western section of the Overground's orbital rail route on 9 December 2012, when this section met up with the East London Line Extension (ELLX), which had reached Clapham Junction in 2009.

Usage of the services calling at Shepherd's Bush Station increased rapidly, so in April 2015 Network Rail funded the extension of the platforms to accommodate longer, eight-car trains as part of a project to lengthen platforms on the West London Line. In addition, a new entrance and footbridge were provided at the northern end of the station funded as part of the further development of the Westfield Shopping Centre.

Dalston and West Croydon (and Extensions)

The East London Railway Company was opened in 1869, reusing Marc Brunel's Thames Tunnel, which had been built for horse-drawn carriages. On 31 March 1913 the route was electrified with the assistance of the Metropolitan Railway. In 1933, the line became part of the London Underground network, operating between New Cross/New Cross Gate and Shoreditch (later cut back to Whitechapel). The Mayor of London, Ken Livingstone, announced that the East London Line would be upgraded and extended on 12 October 2004 as part of a major Capital Investment Programme. This project was at that time under the control of the Strategic Rail Authority.

Operations remained much the same until 22 December 2007 when services were temporarily suspended for the route to be upgraded. The line was extensively rebuilt over the next few years, reopening as part of the London Overground network in April 2010. This included a northern extension of services to Dalston Junction, and also to West Croydon in the south. The line was re-electrified on the 750V DC third-rail system.

A key associated objective was to help regenerate the deprived areas of London through which the line ran. Subsequently, the project had passed from the Strategic Rail Authority to TfL. The proposed new service was described as a 'metro-style (National Rail) train service' and was later referred to as the East London Line Extension (ELLX).

A total of €660M funding was secured from the European Investment Bank for the initial phase to upgrade the East London line and extend it to Dalston Junction. A further £64M was provided by the Department of Transport and £11M by TfL to connect the route to the South London line as Phase 2. Both phases of the ELLX project were to be delivered within TfL's five-year investment programmes. The scheme was eventually to cost over £1Bn (£1.5Bn at 2025 prices). It involved the extensive rebuilding and re-electrification (at 750V DC) of this former London Underground line, together with extensions to Dalston Junction in the north and Surrey Quays in the south plus the aforementioned link with the South London Line.

Boris Johnson, the Mayor of London, officially opened the core section of route between Dalston and Surrey Quays on 27 April 2010. The initial daytime service comprised eight trains per hour between Dalston Junction and Surrey Quays, four of which continued via New Cross Gate and the other four to New Cross.

The extension to the north diverges off the former route before Shoreditch station (which was closed and replaced in a new location), crossing Shoreditch High Street by means of a bridge, and traversing the former Bishopsgate Goods Yard before running on to Kingsland Viaduct and to Dalston Junction. This last section of the route formerly formed part of the North London Line to Broad Street Station (closed in 1986). New stations were constructed at Dalston Junction, Haggerston and Hoxton together with the replacement station at Shoreditch.

The service was further projected southwards to New Cross Gate, Crystal Palace and West Croydon by means of Network Rail infrastructure. These stations were previously managed by the Southern franchise until transferred to Transport for London. These sections of route were upgraded and re-signalled by Network Rail prior to the commencement of the new Overground services on 23 May 2010.

CRYSTAL PALACE NEW CROSS GATE BROCKLEY SHOREDITCH HIGH STREET DALSTON JUNCTION HAGGERSTON

WEST CROYDON NORWOOD JUNCTION ANERLEY PENGE WEST SYDENHAM FOREST HILL HONOR OAK PARK NEW CROSS SURREY QUAYS CANADA WATER ROTHERHITHE WAPPING SHADWELL WHITECHAPEL HOXTON

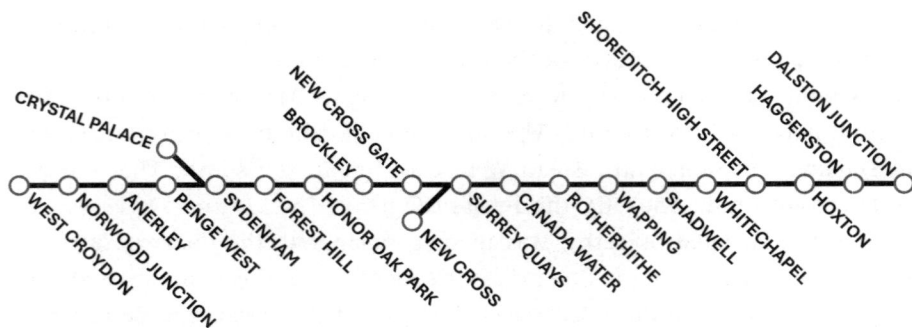

East London Line Extension (ELLX)

By utilising the track bed of the former No. 2 lines, the route was further extended from Dalston Western Junction at the north end to Highbury & Islington, running parallel to the No. 1 lines on 28 February 2011. On 9 December 2012, a link with the South London Line to Clapham Junction was inaugurated. From Surrey Quays trains run on this extension via Queens Road, Peckham, Denmark Hill and Wandsworth Road, branching off at Factory Junction to Clapham Junction rather than running to Victoria Station, the former terminal for this route.

Thus, in 2012, the ELLX Overground project, together with the North London, West London and South London routes, completed an orbital railway around inner London. It should be noted that, for service reliability reasons, among other things, there are no plans for services to run through from the South London Line to the West London Line. However, trains from both routes generally share the same platform at Clapham Junction. This is the former Platform 2, which was divided into two staggered platforms: Platform 1 for Overground services to/from Willesden Junction and Stratford, and a 'new' Platform 2 for services to/from Highbury & Islington.

ELLX Overground services were operated by Class 378/1 Capitalstar EMUs from the start of each phase, serving 30 stations in all.

Liverpool Street to Chingford, Enfield Town and Cheshunt, and Romford to Upminster

On 31 May 2015, the Enfield Town, Cheshunt and Chingford services together with the Romford to Upminster branch became part of the London Overground system (see Chapter 9). These 25kV AC services were transferred from the Abellio Greater Anglia Franchise.

Initially, when these became London Overground lines, the rolling stock that operated all of these services were Class 315 and 317 EMUs. However, from October 2020, these units were progressively replaced by Class 710 EMUs. The new rolling stock also operates the Romford to Upminster line, which has a maximum line speed of only 50kph (30mph).

These new trains are a big improvement on the previous units, having more room, live information screens, USB charging points, free Wi-fi and additional wheelchair spaces.

The train service frequencies are generally two trains per hour on the Enfield Town and Cheshunt lines, but which together provide a 15-minute service as far as Edmonton Green; the service level is somewhat greater in the peak hours. The Chingford line has a 15-minute service throughout the day, whilst the Romford to Upminster shuttle service runs at 30-minute intervals.

Chapter 15
South Wales Metro

The Cardiff Railway Initiative

The Cardiff Railway was one of the 25 Train Operating Companies (TOCs) set up by the British Railways Board in April 1994 in preparation for the government's proposed privatisation of the national network. Co-author John Buxton had been appointed as the designate Managing Director in August 1993, and from April 1994, he spearheaded the arrangements to hive off the Valley Lines from the Western Region and set up an independent, stand-alone TOC.

John and his senior management team soon got to grips with improving customer service, punctuality and safety on this busy but capacity-constrained network, the infrastructure having been heavily rationalised in the previous 25 years. However, between 1985 and 1992 a number of services were restored, and in his previous role as the Divisional Civil Engineer (South Wales), John had overseen the reinstatement works for the reopening of the Aberdare Line in 1988.

As TOC Managing Director, John had a vision for the Cardiff Valley Lines future. He envisaged revamping the network to make the TOC more accessible, with tram-trains penetrating the centre of Cardiff and the Bay off the existing 'heavy-rail' network together with new on-street-running infrastructure by means of 'line of sight' operation.

Looking beyond the break-up of the railway at privatisation, John's aspiration was for the tram-train network to be separated and re-formed as a vertically integrated passenger transport network controlled by a passenger transport executive (PTE). The aim was to make the railway more accessible and integrated with other transport modes and thus to better serve the needs of the Valleys and Cardiff area. This would include expanding the bus feeder services that connected with the Valley Lines network. In addition, John was very keen to improve on-train facilities for cyclists to further encourage a modal shift from private cars.

The new Cardiff Railway Company quickly established good relationships with the local authorities, the Cardiff Bay Development Corporation (CBDC) and the regions' politicians and local stakeholders, including Sustrans.

John championed the idea of renaming the Bute Road Station as 'Cardiff Bay'. John applied to have this station renamed, and organised a ceremony on the official 'renaming' day when a special train was run from Aberdare to the newly renamed 'Cardiff Bay' Station. This special service emphasised the key role that the railway could play in linking the heads of the Cardiff Valleys to the Bay area where new employment opportunities were being created. Pete Waterman officiated at the renaming ceremony which generated a lot of positive publicity, and which helped to get the CBDC thinking about the potential for improved rail links to the Bay.

John became an active member of several working groups, who were focused on improving transportation in the area. Transportation consultants Oscar Faber TPA were already progressing a public transportation strategy for the Cardiff area, which included an option to convert part of the network to light rail operation. John took the idea further by promoting the concept of tram-trains in the form of the remarkable Karlsruhe Stadtbahn tram-train system in Germany. This local electrified network combined the 'on-street' system in the city of Karlsruhe with shared running on the heavy rail lines in the surrounding countryside. John's idea of such a network in the Cardiff area generated a lot of interest within the groups. As a result, in 1996 an 'official' study tour was made to Karlsruhe Stadtbahn, sponsored by the Cardiff Bay Development Corporation and South and Mid Glamorgan Councils.

Outline plans were then drawn up for a new tramway branching off the heavy rail network link at Cardiff Queen Street, running along Queen Street and on to an interchange beneath Cardiff Central Station before continuing to Cardiff Bay. This new tramway would replace the heavy rail line between Queen Street and Bute Road stations, enabling the railway embankment, that had for so long divided the communities along the length of the 2km long line, to be demolished. The new tramway would run alongside the new boulevard that the CBDC intended to construct, with the tram lines extended into the Bay area.

Removing the old railway embankment would also present a much better 'gateway' for both the railways and proposed highway to Cardiff Bay.

Bi-mode electro-diesels were postulated to operate the services but, in the long term, the full electrification of the Valley Lines was envisaged. Taken together with the reopening of 'freight only' and closed lines, plus a feeder bus network, the aspiration was to give Cardiff and its environs a world-class transportation system of which to be proud.

The Managing Director of Powell Duffryn Engineering, a local firm with a long history, based at Cathays in Cardiff, was keen to design and build the tram-trains, having previously built a tram with innovative features for the Blackpool system. The company offered to construct a prototype tram-train as a precursor to – hopefully – winning the order for a fleet of similar vehicles. Had this come about, Powell Duffryn would have taken on and up-skilled a significant number of local people, increasing the economic activity in the capital and, in addition, becoming the UK's only company manufacturing tram-based systems.

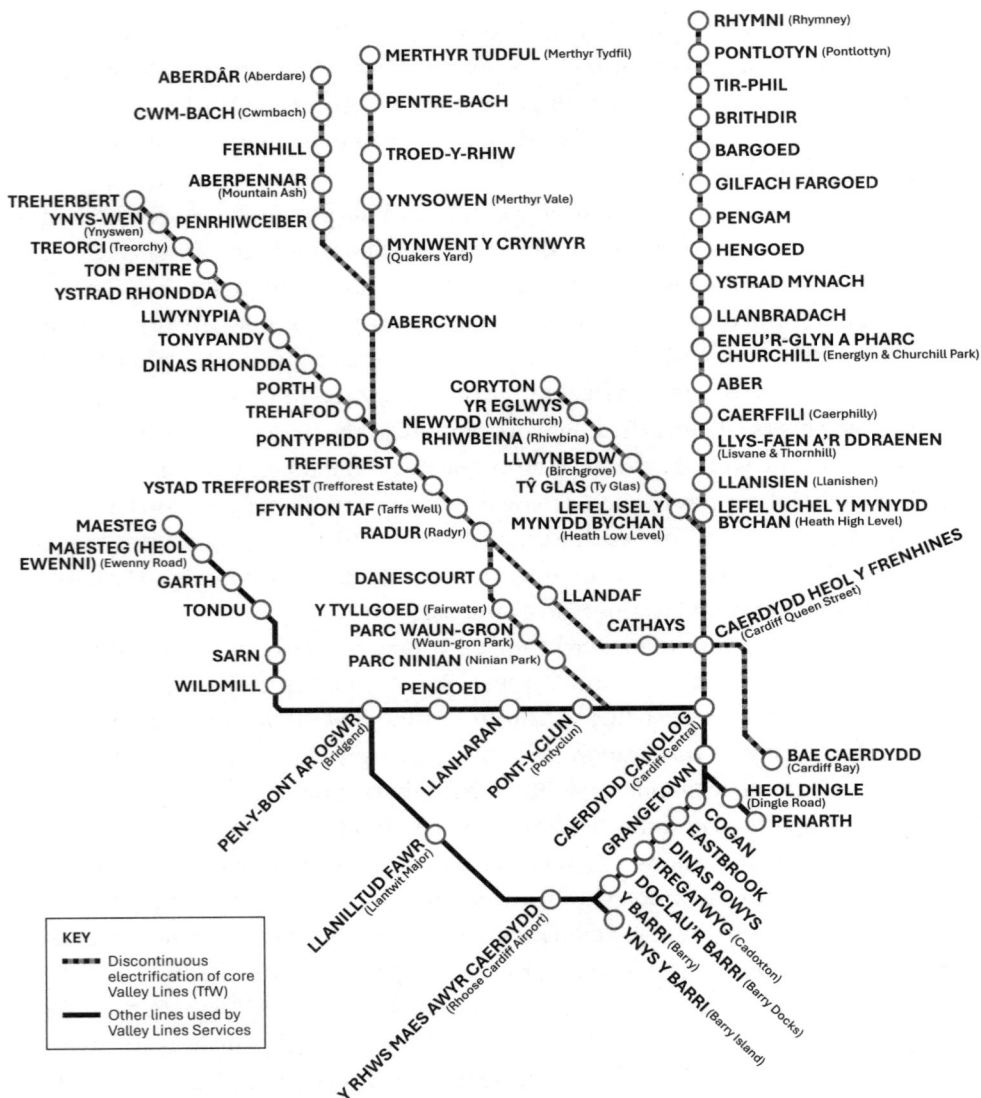

Valley Lines (Llinellau'r Cymoedd)

Just as everything looked to be falling into place, national politics intervened, when, in 1997, it was announced that the local authorities were to be reorganised and the CBDC would be wound up by the end of 1999. These unexpected announcements effectively floored the scheme as the development of the system would obviously take more than two years just to undertake the development tasks, let alone the construction work. In addition, it was also announced that Cardiff Railway would be privatised during 1997.

17 Class 91 locomotive No. 91031 on a northbound service at Doncaster running beneath Mk 3 headspans on 25 June 1995. *D Heath collection*

18 Class 80 AZUMA unit heading south at Doncaster under BR Mk 3b sagged simple OLE. *C J Marsden archive*

19 Drayton Park 750V DC/25kV AC changeover. Note the neutral section. *G Keenor collection*

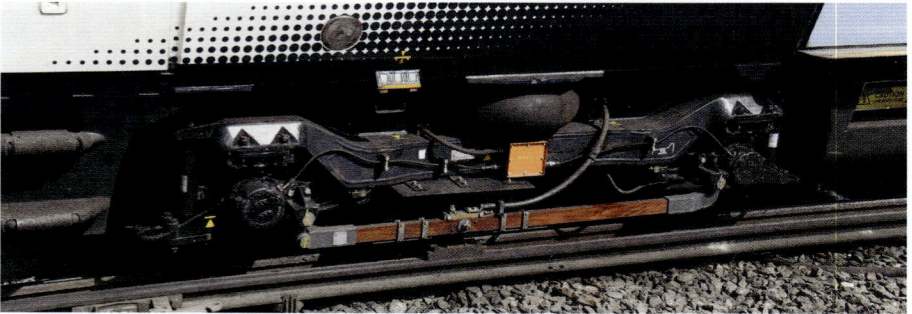

20 Third-rail 750V DC pick-up arrangement beneath a Class 465 Networker EMU. The 'pick-up' shoe slides upon the head of the conductor rail and may move vertically within set limits to take account of undulations in the conductor rail. The metal conductor rail is supported on ceramic insulators. *C J Marsden archive*

21 Class 374 Eurostar set passes over the Medway Viaduct under high-speed sagged simple OLE of French design. Note auto transform feeders attached to each mast.
C J Marsden archive

22 BR Mk 1 portal at Watford Junction. *G Keenor collection*

23 A BR Mk 2 25kV AC portal at Bishopton (between Paisley and Greenock).
G Keenor collection

24 This Great Eastern line 1948 GE anchor portal was initially converted in 1962 to 6.25kV AC and converted again in 1978 to 25kV AC. In 2014 it was rewired with GEFF (Great Eastern Furrer+Frey) equipment. *G Keenor collection*

25 Class 360 'Desiro' EMU beneath an original GE portal with a replacement GEFF (Great Eastern Furrer+Frey) portal behind awaiting the transfer of the OLE. *G Keenor collection*

26 BR Mk 3 25kV AC two track cantilever at Enfield Lock illustrating this structure's usefulness where there is insufficient space on one side of the tracks. *G Keenor collection*

27 25kV AC rigid overhead line at St Pancras International. Note the frequent supports needed for this type of OLE. *G Keenor collection*

28 A novel, albeit over-engineered Series 3 arrangement, utilising a short boom twin track cantilever structure because the support structure foundation is at a significantly greater distance from the running rails than normal. *J Buxton*

29 A self-supporting tension anchor at a wiring overlap, supprisingly with no back stay on the structure. *J Buxton*

30 Mast planted in a concrete foundation (left); an augured pile foundation (centre); driven steel tube pile foundation (right). *G Keenor collection*

31 Installation of the first mast for the Cambridge to King's Lynn electrification 30th September 1989. *C J Marsden archive*

32 Special engineering train with mixing equipment for concreting-in OLE mast foundations. *C J Marsden archive*

33 1950s engineering train for OLE registering and wiring. Note lack of safety barriers.
C J Marsden archive

34 Two images of the HOPS wiring units used on the GWEP. These units comprise power packs and access platforms together with drum carriers and tensioning units which permit the registration and wiring of the OLE including tensioning. The masts and portal structures shown had been erected by other types of HOPS unit. Note the white safety rails that permit Adjacent Line Open (ALO) working. *G Keenor collection*

The CBDC decided that within the restricted time that they had available, the construction of a two-lane highway to the Bay could just about be accomplished. John tried to get the CBDC to make passive provision for the tramway, but this would have required much additional work and, as the 'clock was ticking', this was declined, so any possibility of a tramway in the future was also scuppered.

In addition, Powell Duffryn lost any chance of developing and manufacturing their tram-train product and thus the opportunity to become the only home-based tram manufacturer. With a falling order book, and no prospect of new tram orders, Powell Duffryn closed its Cathays Factory. In addition, the local south Cardiff community remained cut in two by the retention of the existing railway embankment.

A rather disillusioned John departed the newly privatised Cardiff Railway in February 1997. His vision of an integrated transportation system, with an electrified metro at its core, had been shattered by the politically inspired fragmentation of both the railways and local government, coupled with the demise of the CBDC. With him went any hope for an integrated multi-modal transportation system for Cardiff in the 'noughties'. While sad to leave South Wales, John instead assisted in the development of the high-speed Channel Tunnel Rail Link together with a number of successful Metro projects including Dublin, Doha, Toronto and Riyadh.

In place of his positive vision for expansion and development were a succession of downbeat rail franchise contracts that were based on the premise of 'no growth'. Despite the 'nay-sayers' prognostications, ridership did, however, increase, as John had predicted, and services became very over-crowded, especially during the rush-hours. Often, people were physically unable to board the trains and were left at the platforms. Unsurprisingly, the railway came in for much vociferous criticism.

Unfortunately, there were to be 15 wasted years before the idea of a similar network was again promoted as the Cardiff Metro.

The Current South Wales Metro Scheme

Professor Mark Barry of Cardiff University produced a report in February 2011 proposing a regional metro system to connect Newport, Cardiff and the South Wales Valleys. The report, published jointly by the Institute of Welsh Affairs and the Cardiff Business Partnership, envisaged a regional metro system combining electrified heavy rail and light rail systems at a cost of circa £2.5Bn.

With support from the Welsh government, Professor Barry developed a more comprehensive plan in 2013, to include a cross-rail scheme for Cardiff including light-rail operations between a number of the Valley Lines and Cardiff Bay. The initiative was further developed to encompass bus corridors, improved integration between bus and rail, station upgrades, park-and-ride schemes and provision of walking routes and cycleways.

Subsequently, in November 2015, the South Wales Metro (Metro De Cymru) was launched by First Minister Carwyn Jones as a region-wide integrated heavy rail, light rail and bus-transportation system. This was followed by a publication entitled *Rolling Out Our Metro*.

The Welsh Assembly government's intention was for this revitalised network to increase the journeys made by public transport, walking and cycling from the then current 32 per cent, to 45 per cent by 2040. This is one of several integrated strategies, which, taken together, were aimed at achieving a sustainable, low carbon future for Wales.

In June 2018, a £5Bn investment in the Metro was announced, which included five new stations, an on-street extension of the Cardiff Bay line, and a link between the Ebbw Vale branch and Newport, together with a fleet of new conventional trains and tram-trains.

Procurement of the Metro was embodied within the invitation to tender for the Wales and Borders Franchise, with the tendering process managed by Transport for Wales. In May 2018, the franchise was awarded to KeolisAmey Wales. Their plan envisaged partial electrification of the Cardiff Valley Lines and operations by means of multi-modal OLE electric, battery and diesel trains.

Within fifteen days, KeolisAmey placed its first rolling stock deal for five Vivarail three-car hybrid (battery/diesel) trains, to be known as Class 230. These were intended for the Wrexham to Bidston, Crewe to Chester and Conway Valley lines. The train's mode of propulsion is controlled through Global Positioning Systems that cut out the diesel engines in stations and sensitive environmental areas, automatically switching to battery power. They came into use in 2023, and have produced a significant increase in passenger numbers on the Wrexham to Bidston route.

In August 2018, further orders were placed with Stadler for rolling stock, including seven three-car and seventeen four-car tri-mode Class 756 FLIRT (25kV/Battery/Diesel) units and thirty-six Class 398 CityLink Tram-Trains. The latter units are similar to the Class 399 units used in the Sheffield area but have higher floors to enable them to utilise the existing platforms on the network.

Also in 2018, the Welsh Assembly government announced a package costing £738M, including £150M from the UK government together with £125M pledged from the European Union.

Transport for Wales took control of the Core Valley Lines (CVL) and the ownership of the infrastructure from Network Rail in March 2020. These lines include the Treherbert, Aberdare, Merthyr Tydfil, Rhymney, Coryton and Radyr lines but exclude the Great Western Main Line through Cardiff Central together with the Penarth, Barry and Vale of Glamorgan routes.

However, by this time, it was considered that the franchise held by Keolis-Amey was unsustainable and so the agreement was terminated, with Keolis-Amey retaining the responsibility for upgrading the infrastructure.

The plans for the Cardiff Valley Lines envisaged operations as follows:

- Class 756 FLIRT tri-mode units – these are to run partly on 25kV OLE and partly on battery power on the Rhymney and Coryton lines and battery and diesel power on the Penarth, Barry and Vale of Glamorgan services.
- Class 398 CityLink Tram-Trains – again to run on 25kV OLE and battery on the Treherbert, Aberdare and Merthyr Tydfil routes together with the proposed on-street extensions in the Cardiff Bay area and later within the Cardiff area. The Class 398s are to be provided with passive provision to charge from 750V DC power supplies and can operate on 'line-of-sight' operating arrangements.

Bi-Mode and Tri-Mode Operations

The project encompasses the UK's first large-scale 'intermittent electrification' scheme, which incorporated catenary-free sections (CFS) and permanently earthed sections. For the latter, used primarily where the OLE runs under bridges and structures with insufficient electrical clearance, balises are provided. These devices automatically open the 25kV vacuum circuit breaker (VCB) on the trains, and are placed on the approach to the structure, enabling the train to coast without power under the earthed OLE 'dead section' through the restricted clearance, picking up the 25kV power again on the other side. Re-powering of the train then occurs two seconds after the VCB detects the live OLE, when the VCB closes, enabling the unit to take current once again.

For areas where there is insufficient clearance for the OLE wire for the pantograph to remain up or where long stretches of restricted clearance exist (for example, tunnels), CFS is used. A balise to lower the pantograph is located at least 13 seconds running time on the approach to the obstruction. Then, 10 seconds before the CFS, the OLE is terminated at a height above the reach of the pantograph. Thus, if the balise has failed to lower the pantograph, it rises to its maximum height, at which point the pantograph drop-down valve operates, as a 'fail-safe' measure, to bring the pantograph down before it can strike the structure. Another balise is provided immediately beyond the CFS to raise the pantograph to contact the OLE once more.

Power supply to feed the OLE is taken from a new substation at Tonteg, near Treforest Estate Station and from Network Rail's present substation, west of Cardiff Central Station.

Extensive modelling of power and battery performance indicated that the ideal proportion of conventional OLE to be provided should be for 60 per cent of the network. Propulsion for the remaining extent would be by battery power. This 60/40 split also provides the most cost-effective balance, optimising vehicle performance and battery life against infrastructure capital costs.

Depot Strategy

Initial thoughts were to base all the Class 756 FLIRT units and the new diesel units (Classes 197 and 231) at the Cardiff Canton High Level Depot, which would be upgraded to suit. The works were to include the installation of OLE and battery-charging facilities.

There is, however, insufficient room at this site to accommodate the Class 398 Tram-Trains and, with their unique configuration, it was considered that a brand new bespoke depot was essential. The ability of tram-trains to operate on 22m radius curves meant that a far smaller site would be required compared to that needed for conventional units.

Accordingly, in the early days of the project, a 10-acre industrial brownfield site was therefore earmarked for the depot adjacent to Taffs Well Station. Despite the compact nature of the site, there proved to be sufficient space to incorporate the network's signalling control centre and the SCADA, too (to be known as the Core Valley Lines Integrated Control Centre – CVLICC).

The £150M depot was completed in 2024, and has three 100m long maintenance pits, one having retractable OLE. For overnight charging there are ten stabling sidings electrified with OLE, but these are to be the only external OLE lines in the depot, as all tram–train movements will be battery powered. A total of 400 train crew and more than 50 other staff will be based at the new depot.

As the project progressed, it became apparent that insufficient allowance had been made for the servicing and stabling of the Class 756 FLIRT units. In 2020, TfW therefore initially acquired the Pullman Works and Sidings located at Low Level adjacent to their Cardiff Canton Depot. However, this site was busy with train refurbishments and bogie overhaul contract work, so little additional storage and stabling facilities were available.

In 2021, TfW approached John Buxton, owner of Cambrian Transport Ltd, the operator of the private Barry Railway, to ascertain if it would be possible to acquire the Barry Depot and the associated Heritage railway. Cambrian Transport had a long lease from the Vale of Glamorgan Council, who had acquired the Depot, running lines and sidings in 1997. The site included a five-road shed and seven other sidings together with a main line connection. A deal was quickly struck between TfW, the Vale of Glamorgan Council and Cambrian Transport and the site was acquired in May 2022. Following initial refurbishments, the stabling of Class 150 DMUs commenced in July 2022 and the first FLIRT units arrived in September of that year. Eventually, only Class 756 FLIRT units are to be allocated to the depot.

As an aside, John Buxton's Barry Railway Heritage railway operations were moved to another smaller shed and two sidings at Plymouth Road, Barry Island. Although this meant that the rolling stock numbers had to be reduced, operations could continue utilising some of the TfW infrastructure, subject to adherence to a mutually agreed Operations Agreement.

Infrastructure Works

KeolisAmey were quick off the mark in awarding early contracts for the enhancements needed for the South Wales Metro. In June 2019, Balfour Beatty were awarded the track, power and fibre broadband packages, Siemens Mobility the signals and telecoms and Alun Griffiths the civil engineering works. To maximise the efficiency of project delivery, these organisations formed an alliance together with TfW and Amey. The alliance constructed a material storage and distribution centre at Treforest, which was commissioned in early 2020.

While the infrastructure work is quite extensive, the intermittent nature of the project has obviated the need for a large amount of the usual electrification works. There is, however, a significant cost involved in delivering the line capacity upgrades, including improvements to running times. This requires the provision of six new loop lines, four loop extensions and approximately 30 kilometres of line speed enhancements. In connection with this, 13 underbridges are to be widened to take double track. Compared to a conventional electrification project, the work needed to provide adequate OLE clearances is relatively modest: just two overbridges need to be reconstructed. However, at more than 50 structures, parapets need to be raised. In addition, 20 track lowers are also required.

Other civil engineering and building work includes improved facilities at 68 stations, plus platform alterations to achieve level boarding at 110 stations, together with a number of platform extensions and eight new platforms. This includes a new platform at Cardiff Bay Station. In addition, 12 new station footbridges are also to be constructed. Cardiff Queen Street junction is to be remodelled with 17 new turnouts and, in addition, two new stations are to be constructed at Gabalfa and Butetown.

With regard to signalling and telecoms, 50 new signals and more than 300 axle counter sections are to be provided, together with nearly 100 new signalling location cabinets. The present telecom and data system is to be replaced with a new 10Gbit network.

Overall, the scheme required 170 single track kilometres (STK) of the Valley Lines to be electrified requiring circa 3,000 masts. There are to be 60 permanently earthed sections and 30 CFS locations including Cardiff Queen Street, Cardiff Central Station and Caerphilly Tunnel. The intermittent electrification has saved in the order of £150M capital cost. However, this must be set against the higher first costs for the trains plus the additional operations and maintenance costs for the heavier and more complex rolling stock. This includes the expensive battery replacement costs required every seven or so years for the bi-mode and tri-mode units.

Commencement of Operations

During 2022, infrastructure work proceeded more slowly than planned due to restrictions imposed by the Covid pandemic. Over the next twelve months, however, some of the programme slippage was recovered by utilising long blockades to undertake the work more efficiently. By these means, wiring of the railway line from Pontypridd to Aberdare and Merthyr Tydfil to Abercynon was completed during 2023.

Following a long blockade to install the OLE together with the upgrading of the track and signalling system, the Treherbert line was energised and tested during May 2024. Electric services began operating in the summer of 2024. Energisation of the Rhymney and Coryton lines is planned to follow before the end of 2024.

The Cardiff Bay branch is to be commissioned, as a 'line of sight' light rail route, in December 2024. Through tram-trains will run to Cardiff Bay from Treherbert, Aberdare and Merthyr Tydfil. Passive provision will be made to extend the line as a tramway beyond the station to the Bay, and another on-street spur is to be constructed from the new Butetown station to Cardiff Central (possibly beyond) via Callaghan Square.

Stadler trains should be operating all Cardiff Valley Services by May 2025, when the new timetables are introduced, doubling previous service frequencies.

On completion of the project, the new metro will be a cleaner and greener railway than hitherto. It is estimated that the new bi-mode and tri-mode trains will reduce CO_2 emissions across the Cardiff area rail network by some 25 per cent.

Chapter 16

Electrification – The Economics and Politics

It is interesting to compare and contrast the attitudes to state support/intervention of the railways in mainland Europe and Britain. The individual stances explain the different approach to railway engineering and electrification adopted in the UK compared to that on the continent.

From the early days, the European governments took a guiding hand in the evolution of their railway systems, in both the planning of their respective networks and the provision of funding. The development of railways in the UK on the other hand was uncoordinated, funded entirely with private capital, orchestrated by promoters and driven primarily by the profit motive. As a result, there was no strategic national plan for the building of railways in Britain. Instead, a piecemeal, free market approach resulted in the construction of many independent, intertwined networks, which resulted in much needless duplication of railway services, many totally uneconomic lines and thus, in overall terms, a weaker commercial system.

In the twentieth century, this lack of co-ordination caused severe organisational issues within the railways as they struggled to effectively marshal resources and reduce costs to compete with rapidly growing road competition.

On mainland Europe, generally a more logical approach to the development of the railway systems had been adopted. With a higher degree of state control, the railway network was built to serve communities and industry with little if any duplication. Also, the railway organisations that evolved were simple but strong structures providing a much firmer base on which to grow and enhance the infrastructure and the services that operated on it. In the twentieth century, this resulted in a more strategic approach to railway electrification than in the UK. In addition, as there was less duplication and fewer uneconomic lines, there was a lesser requirement to prune the network. Consequently, the anguish of rail closures endured after the Second World War in the UK was not experienced to anything like the same extent on the continent.

The classic, historical example of the shortcoming of the nineteenth-century UK system pertinent to railway electrification in the twentieth century, was the closure of the Woodhead route just thirty years after the overhead wires had been energised (see Chapter 4). With a more considered approach, fewer routes across the Pennines could not only have better served the needs of industry and local communities but also, subsequently, avoided the embarrassing closure of an electrified main line route.

From the conception of the steel wheel running on steel rails at the height of the Industrial Revolution, a century of bold, intensive railway development and inexorable expansion followed. This 'railway age' exuded a sense of audacious, self-possessed confidence and permanence. Even the track on which the trains ran was referred to as the 'permanent way'.

Perhaps because of this impudence, there has long been a diffident relationship between railway organisations and legislators. Things didn't get off to a very promising start: at the opening ceremony of the world's first inter-city passenger line, the Liverpool and Manchester Railway, William Huskisson, the President of the Board of Trade, decided against advice to alight from the train when it stopped in order for the locomotive to take on water, and a train approaching on the adjacent line, hauled by the famous Rocket locomotive, ran over his leg, shattering it. He subsequently died of his injuries later that day. It was not to be the first time that a politician's folly impacted adversely on both the railway and the individual's own wellbeing!

Nevertheless, with the successful opening of the Liverpool and Manchester line, a railway mania swept the country in the following decades. The railways, with their technological advantages, took most of the medium to long distance traffic from the canals and the roads to become the dominating force in the transport industry during the second half of the nineteenth century.

Successive governments therefore introduced statutory restrictions and obligations on the railway companies, the most onerous of which were twofold – the setting of fixed mileage-based charging rates and the requirement to act as a 'common carrier'. These burdensome obligations foisted upon the railway companies the extremely restrictive requirement to carry all traffic offered at a fixed rate, whether it was commercially viable or not. These measures dented not only the industry's profitability, but also its confidence.

The railways' erstwhile sense of invincibility was further challenged by the growth of electric urban tram systems in major conurbations during the early twentieth century. However, this time the affected companies countered the competition, albeit belatedly, by undertaking a number of early heavy rail electrification schemes, particularly in the London area. Later, the calamitous effects of the deprivations of the First World War, and its aftermath, were to shatter both the industry's self-assurance and stability to the core. Furthermore, government control during the war also eroded the railway companies' sense of independence.

Another consequence of the war was the termination of the limited but ongoing efforts at electrification, particularly the works south of the Thames by the London and South Western Railway and the London, Brighton and South Coast Railway. The North Eastern Railway's aspirations to commence the electrification of the East Coast Main Line were also still-born.

Prior to the war, the government had showed little interest in railway electrification. The limited number of schemes that did go ahead were instigated by the railway companies themselves. The driving forces were generally competition and operating cost reductions, but the inclination to experiment also played a part. These include electrification (albeit to differing standards) of the routes from:

- Liverpool to Southport
- Manchester to Bury
- Lancaster to Morecambe and Heysham
- London to Watford, and
- Broad Street to Richmond.

Often passed over in the annals of railway history, perhaps because it was subsequently entirely dismantled, was the significance of the programme of electrification initiated by Sir Vincent Raven on the North Eastern Railway (see Chapter 3). This was the first time that proposals for main line electrification were seriously considered. The North Eastern went on both to electrify passenger services around Newcastle and to introduce the first coal-carrying electric freight service between Shildon and Newport. Plans were afoot to electrify the main line between Newcastle and York for both passenger and freight operations, but the scheme was unfortunately cut short by the onset of the First World War.

At the end of hostilities, the availability of surplus army lorries and a plentiful supply of cheap petrol triggered a huge rise in competition from road transport. As the railway companies were obliged by law to publish their rates, it was easy for the new road-based hauliers not only to undercut railway prices, but also to cherry-pick the most profitable traffic. Unfortunately, this scenario was repeated twenty-five years later after the Second World War, and it continued into the period of the nationalised network.

Unsurprisingly, the rapid haemorrhaging of traffic to road after the First World War dented the railway industry's financial health. It soon became clear that many of the smaller railway companies would either need the support of their stronger neighbours or, more radically, to be combined into a single railway system.

The government was therefore faced with two options: amalgamate the railways into one organisation or group the companies into a small number of concerns so that, in effect, the 'strong could support the weak' by cross-subsidisation.

This posed a real dilemma for the Liberal government at the time. Amalgamation was in theory the best option but would result in the formation of by far the largest company in the land, if not the world. Who would own this monolith? How could it be controlled to ensure that it met the national interest rather than just the shareholders' benefit? Should it be nationalised? What about competition? However, with clear vision, a firm plan and strong project management, these issues could have been overcome.

On the other hand, saddling the viable elements of the nation's railway with the many unremunerative branch lines and duplicate secondary routes was also not a particularly attractive way forward. How could these new companies remain profitable in order to finance essential investment and enhancements? What would happen if one or more of the companies became insolvent? How could vital rail services be maintained in such circumstances? Who would bail out such bankrupt companies?

Despite all the shortcomings, the second option was considered to be politically less painful, quicker to implement and much less controversial. In addition, the fact that competition between the new companies would still exist, albeit to a limited degree, also made this option politically more attractive.

The government therefore took the easier and quicker way out, deciding to retain an element of competition within the railways system. The 1921 Railways Act was enacted to group the companies into what became known as the 'Big Four'.* In so doing, the conception of a general management of the railways was effectively rejected, leaving these four concerns to earnestly try to extract business from one another while the real competition, primarily in the form of road transport, was growing at an exponential rate.

During the changing circumstances of the time, it was evident that the railways were often not the most suitable, nor the most economic means of meeting the country's transportation needs. It should have been quite clear to the nation that the railways hitherto dominant position in the arena of transportation was ebbing away. Unfortunately, in the early years after the Great War, both political and public perceptions of the role that railways should play lagged behind this notion. Regrettably, these sentiments persisted through much of the twentieth century to the detriment of the railways, and, ultimately, the country's best interests.

With the benefit of hindsight, it can be seen that the flawed 1921 'Grouping' legislation was just the first of a series of political interventions over the last hundred years that, unfortunately, generally failed to deliver the desired

---

* The London, Midland & Scottish Railway (LMS), the London & North Eastern Railway (LNER), The Great Western Railway (GWR) and the Southern Railway (SR)

outcomes. Thus, at the 'Grouping', which actually was implemented in 1923, the four newly formed companies, handicapped by inflexible pricing policies and the post-war doldrums, struggled to stay solvent. Every effort was therefore made by the companies to avoid capital expenditure on infrastructure. This was particularly so on the LNER, where revenues in the depressed, industrialised areas of the North-East had diminished substantially. To their credit though, the company boards did authorise sufficient funds to maintain adequate infra-structure standards, even though they often failed to meet their obligations to shareholders. During this financially depressed period, maintenance could so easily have been allowed to deteriorate, tipping the railways into a spiral of decline that would have been difficult, if not impossible, to escape. Subse-quently, prospects for any electrification, even in the longer term, would then have been extremely bleak.

Some opportunities did, however, present themselves in the difficult financial period of the 1920s and 1930s. In particular, huge technological advances made in rail transportation improved efficiency and brought new and faster passenger services. Revolutionary new forms of signalling were introduced, and steam locomotive design reached the pinnacle of its development. The latter was, in part, due to the fact that it was far cheaper for the cash-strapped railway companies to develop their existing traction technology than to invest in main line electrification, which was patently unaffordable on a significant scale in the prevailing circumstances. However, had funds been available, this approach would have been a much better investment for the industry.

The political view of railway electrification in the 1920s can best be described as being 'ambivalent'. On the positive side, the government did encourage the railways to focus on standardising electrification systems by setting up the Kennedy and Pringle Committees. (The electrification of the LMS/LNER joint line between Manchester and Altrincham being the first scheme to follow one of the recommended standards, albeit not the preferred one.) On the other hand, no political interest, let alone financial support, was forthcoming to help the railway companies invest in electrification.

As a result, in the 1920s the LMS did no more than completing the LNWR's Euston to Watford lines network with the energisation of the Rickmansworth branch. Only the Southern, spurred on by the need both to reduce operating costs and face down external competition, undertook any further significant electrification works.

Ten years after the Grouping Act, the government set up the Weir Committee in 1931 to report on 'the economic and other aspects of the electrification of the railway systems in Great Britain with particular reference to main-line working' (see Chapter 5). The committee's report was very positive and pointed out the modest expenditure required by electrification compared to the spending on roads! The prognosis for electrification looked good. However, the Railway Companies Association, which represented the 'Big Four', advised the government that they would want to know what government assistance would be available before showing any interest in the prospect of main line electrification. Initially, the government was not responsive, and this was particularly disappointing to the LNER's chief electrical engineer, H.W.H. Richards, who had given significant assistance to the Weir Committee in the preparation of the Weir Report.

One company, the Southern Railway, did, however, continue to significantly extend its electrified network in the 1930s. This was partly due to the competition in London from other forms of transport, but also because the company took advantage of the belated support offered by government funds, which were made available to the railways during this difficult economic period, under the Railways (Agreement Act 1935). It should be noted, however, that the government support was primarily made available to reflate the economy, rather than to initiate a nationwide electrification initiative.

The Southern Railway's General Manager, Sir Herbert Walker, spearheaded the company's rolling programme of electrification, and under his competent leadership and the guidance of the company's Electrical Engineer, A. Raworth, remarkable progress was made with electrification schemes south of the Thames. Extreme economy was the order of the day – the 'new' electric passenger units were usually low-cost rebuilds of steam-hauled stock! The modest first cost of the third-rail system, requiring only minimal infrastructure improvements, meant that long stretches of railway could be electrified in relatively short timescales. While the technical aspects of the Southern Railway's third-rail system may be questionable, the overall benefit of this extremely cost-effective investment to the London and the South-East region cannot itself be called into question.

Similarly, under the Railways (Agreement) Act 1935, the LNER Chief Electrical Engineer, H.W.H. Richards, was at last able to lay down plans for the Woodhead and the Liverpool Street to Shenfield schemes. However, the delay in the receipt of financial assistance thwarted any significant physical progress, before all works on railway electrification schemes were terminated by the Second World War. Thereafter, comparatively little progress was made for more than a decade as, even after hostilities had ended, the nationalisation process and the severe financial constraints of the time stymied progress.

Having considered the politics and economics of the railway from the early days of electrification to the end of the Second World War, it is interesting to postulate how railway activities may have developed had the option to amalgamate the railways been adopted instead of the 'Grouping'. The haphazard way in which railway schemes were promoted and sanctioned in the nineteenth century resulted in the construction of many unviable lines and a plethora of duplicated routes. Even when railways dominated the UK transportation scene before the First World War, many of these railways were totally uneconomic.

Had amalgamation gone ahead in the early 1920s there would have been an opportunity to weed out and close the truly unviable branch lines. In addition, the duplicate secondary routes could have been rationalised with little impact on the transport service available to customers. A thorough revision of main line passenger operations, breaking through the discontinuities caused by the old railway company's boundaries, could have streamlined services while offering passengers faster journey times and more regular services. As a result, fare box income would have increased. In addition, these more effective and efficient working practices would have required significantly less resources, so reducing costs.

The government would undoubtedly have been prevailed upon by this powerful organisation speaking with 'one voice' to reduce the railways statutory obligations and regulation. A thorough review of freight activities would then have allowed the railways to carry the traffic most suitable for rail at better rates, while also permitting a rationalisation of marshalling yards and freight depots, reducing transit times. Thus, freight revenues would have also increased while a more efficient system would have significantly reduced operational costs.

The need to transition to cleaner and more efficient traction than the steam locomotive should also have been foreseen. Steps could have been taken to identify and implement, over time, the most efficient and effective way of operating the different elements of the network and a long-term transition plan developed. A key element of the programme should have included a rolling programme of electrification of core and suburban routes.

This transition would thus not have been anything like the wasteful and hurried implementation of the 1950s Modernisation Plan, but instead there would have been an evolutionary, steady, progression during the 1920s and 1930s into the modern era. In concept, the plan would have been similar in its inception to the aspirations of the contemporary, 'up and coming' railway engineer, Robert Riddles, who later attempted to put such a plan in place in the newly formed British Railways following nationalisation. Of course, the Second World War would have intervened, but the revitalised, integrated railway would have been better placed to serve the country's war needs, and a sense of continuity would have prevailed through the conflict and into the post-war years.

With good management foresight, this alternative scenario would have resulted in an effective and measured realignment of activities over a reasonable timescale to transform the railways into a more stable, sustainable and financially viable system. It has to be acknowledged that strong opposition was encountered to line closures that actually occurred in the 1950s. Thus, the suggested changes would have met an even higher level of resistance. However, a more carefully thought-through plan, pursued in a measured way, but with vigour, would have achieved this aim. Ironically, had this approach been taken, more routes would have been retained compared to the over-zealous 1960s Beeching Plan, rushed in a decade later as the railway's financial position became ever more acute.

Post-Second World War to the Modernisation Plan

Towards the end of the Second World War the railway companies began framing their post-war strategies. So, with the envisaged 'Brave New World' following the war, why was only one of the Big Four Railway Companies seriously contemplating future electrification of its network? This was primarily because only the Southern Railway foresaw the revenue-enhancing potential of providing the public with frequent, fast and clean electric services. Many years later, in British Railways' time, this was to be termed the 'sparks effect'. The other three companies primarily saw electrification as a way of improving operating efficiency and thus reducing costs, which, of course, the Southern Railway also took into account. Therefore, apart from the Southern Railway, only the LNER included electrification in their post-war aspirations. These were, unsurprisingly, the revival of the Liverpool Street to Shenfield Scheme and the Woodhead New Works Project, both of which had been planned before the war.

Following the 1945 General Election, the incoming Labour government had announced that the railways were to be acquired 'for the nation' by means of a Transport Act, which was enacted in 1947. This caused much uncertainty and, as a result, many experienced senior managers chose to leave the industry. For those who stayed much of their time was focused on organisational matters in preparation for nationalisation.

In the early post-war years, the railways were extremely run-down but could be kept working, albeit at a low level of efficiency. Funding was extremely tight with efforts focused on rehabilitating the system rather than the restoration of pre-war main line passenger timings or the development and modernisation of the infrastructure.

One casualty of the political uncertainty of the time was the re-establishment of the Southern Railway's third-rail electrification plans that had been cut short by the war. A hiatus of more than a decade ensued before any further physical progress was made.

The implementation of the nationalisation of the transport sector on 1 January 1948 brought into being an organisation known as the British Transport Commission (BTC). Under this huge organisation, agencies, termed 'Executives', were established to discharge functions 'delegated' to them by the Commission. One such Executive was established to run the newly unified railway system – the Railway Executive. Unfortunately, as so often is the case when major political legislation is enacted, the loose definition of the word 'delegation' in the act caused many issues between the Railway Executive and the Commission.

Within the Executive, six geographical 'Regions' were established to run the day-to-day operations, headed up by chief regional officers. The Regions generally followed the old Big Four Company boundaries with two exceptions: – the LNER area was divided into two (forming the Eastern and North Eastern Regions), and the new Scottish Region took over territory from both the LMS and LNER.

Regrettably, the Commission also set up all the Executives by 'form' of transport rather than by 'function'. Furthermore, by perpetuating a regional railway structure rather than setting up new route/network profit centres, the opportunity to streamline operations and run these organisations on effective commercial lines was squandered.

Inexplicably, the Members of the Executive were appointed by the Minister of Transport, rather than by the Commission, so the latter organisation was not even able to ensure that the Members selected were whole-heartedly in favour of integration.

Taken together, these misjudgements gave the Executives the scope not only to increase their powers by default, but it also gave them a strong motivation to look inwards. As a result, the focus was often more about sorting out their own internal affairs rather than looking outwards, to play their part towards the goal of integrating with other forms of transport. Thus, the government's objective of optimising the use of the nation's roads, railways and waterways was lost even before the new arrangements were put in place.

In this light, the Railway Executive looked upon the Commission as a supervisory body, akin to the pre-Nationalisation Company Boards. They considered that this body should take the lead on matters of transport integration and not much else, leaving the 'self-reliant' Railway Executive, for example, to establish a unified railway system. The Commission saw its role quite differently. This overarching organisation would take the lead on strategic and policy matters ranging from initiating infrastructure enhancement projects to designing future customer charging schemes. It wished to work with the Railway Executive and share in the major decision making. As a result, it was inevitable that relations between the Commission and the Railway Executive were strained from the beginning.

With the monumental task faced by the Railway Executive to forge the four companies into one, it is perhaps understandable that the Members of this organisation took the decisions that they did. They certainly approached their assignment with great energy, deciding from the start on a functional form of organisation. This was sensible in the circumstances that prevailed at the time as this was the fastest way to reduce costs and restore operating performance. This centralised approach was, however, also seen as vexatious by the Regions who felt that they should have more say. So, relationships here too were often thorny.

The appointment for the Member responsible for mechanical engineering, electrical engineering, road motor engineering and scientific research was R.A. Riddles. He had a railway career dating back to 1909 and had formerly been vice-president of the LMS. Prior to this he had been the mechanical engineer for Scotland for the LMS and before that, the principal assistant to LMS Chief Mechanical Engineer. During the war, as Deputy Director General, Royal Engineer Equipment at the Ministry of Supply, his peerless engineering and managerial skills enabled him to successfully deliver a wide range of equipment for the war effort, ranging from D-Day Mulberry Harbours to 'austerity' steam freight locomotives.

Having shown an interest in the benefits of electrification in his early days when apprenticed to the London & North Western Railway at Crewe, it is clear that Riddles considered electric traction to be the ideal form of railway motive power. However, with the economic circumstances prevailing at the time of nationalisation, he considered that steam traction would be the main form of traction for the medium future, so he went on to develop the range of BR Standard types. In these early years of the Executive, he did, however, play a key role in the revival of the aforementioned Great Eastern suburban and Woodhead electrification schemes. He also initiated the conversion of the Lancaster to Morecambe and Heysham branch into a 50 cycle AC test bed, which proved very successful, and fundamentally changed the way that main line electrification was to be taken forward in the future.

Riddles is often referred to as a man of steam, but it can be seen that, with his broad span of expertise, he was no more dedicated to steam locomotives than to other facets of mechanical engineering. However, Riddles was extremely firm in his view that steam should be the principal form of traction on main and suburban lines until these routes were electrified. As far as diesels were concerned, he determined that diesel-electric traction could take over shunting duties and diesel railcars could be deployed on branch lines and secondary routes. He did not see main line diesel traction as a way forward and outlined the reasons for which in a paper, those being:

- Investment funds were severely limited.
- There would be high costs – five steam locomotives could be purchased for the cost of one diesel locomotive of equivalent power.
- New costly depot infrastructure would be required.
- Oil supplies were subject to strategic issues and foreign currency restrictions.
- Steam locomotives were far more reliable than diesel locomotives.
- Twin diesel locomotives were required to equal the power of the most powerful steam locomotive, and this would require some trains to be reduced in length by one carriage to fit within permissible station limits.

There was another reason that Riddles was against the introduction of main line diesel locomotives. Although he never formally recorded it, his deeply held concern was that the introduction of diesel traction, even as an intermediate step, would inevitably defer, if not kill, the rolling programme of electrification that he had in mind.

However, as noted in Chapter 6, Riddles's fears were realised when C.M. Cock and a number of senior BTC management started to consider a programme based on diesel-electric traction, thus avoiding the infrastructure costs associated with electrification. As a result, many in the industry came to the view that diesels would be needed, not only to operate off future electrified core routes, but also to fill the gap prior to electrification being completed, or indeed as a long-term alternative.

There followed a period of intense controversy. Riddles disagreed strongly with Cock on rapid dieselisation and the general direction that senior BTC management were starting to take. He 'stuck to his guns', believing that a rolling programme of electrification of all core routes and suburban conurbations should be initiated, to form over time a comprehensive and coherent network. As noted earlier, his strategy considered that the operation of secondary routes and branch lines would be served by the new diesel railcars then being authorised. Under his plan, the elimination of steam traction would be phased over a longer period, but he believed that his proposal was the most robust from a practical and economic point of view.

However, a sense of frustration was growing within many levels of the industry to the Riddles perceived obsession with steam, particularly with the construction of a 'needless' new range of steam locomotives. In addition, there was a realisation that under Riddles's plan, modernisation for many parts of the network would be a long time coming. From 1953, with the movement towards a decentralised organisation and the winding up of the Railway Executive, a vehement reaction to the Riddles strategy took hold and the charge towards a major dieselisation programme began. This, together with the politically inspired reorganisation of the railways taking place at that time, caused Riddles to decide to resign from his post.

In many ways this was a great pity, but his was an unwillingness to recognise that diesel locomotive technology would advance over time and that, in the longer term, with improving national social and economic conditions, steam would not be able to meet the demands of a modern railway. Had he compromised and agreed a diesel pilot scheme early on, the later reaction by senior officers who favoured partial dieselisation, against the Steam Lobby's rigid policy, may not have been so vehement. In addition, it would have formed a base for the development of a plan for a more measured and gradual transition from steam. The precipitous reaction to this 'straight steam to electrification' policy resulted in a looming 'dash for diesels' programme from the mid-1950s with its many unsatisfactory consequences. Ironically, the Steam Lobby's adherence to its rigid, unyielding approach brought about the earlier demise of steam traction than would otherwise have been the case!

It is not widely known but towards the end of the Railway Executive's existence a more balanced long-term 'Development' plan for the railways was drafted. The study was headed up by J.L. Harrison, Chief Officer (Administration) in April 1953. Tempering to a degree Riddles's 'anti main-line diesel' views, this document put forward a coherent proposal for a modernisation strategy taking cognisance of both viewpoints. Envisaged in Harrison's £500M (£12.5Bn at 2025 prices) programme was an evolutionary transition into the modern era over a period of up to twenty years.

While less ambitious than the subsequent 1956 Modernisation Plan, there is little doubt that it would have been more cost effective and less disruptive in its implementation. A rolling programme of electrification was to be the major item in the programme, costed at £160M (£4.8Bn). The proposed main line schemes included the West Coast Main Line from London to Glasgow, taking in Birmingham, Manchester and Liverpool. Also, London to Manchester via the Midland Route and London to Bristol and South Wales.

Major expenditure was also to be spent on track and signalling, stations, plus the rationalisation of depots and marshalling yards. Significant funds were also to be allocated for the introduction of a large diesel railcar fleet and for the fitting of continuous brakes to freight rolling stock. The building of diesel shunters would continue, and initially, a modest pilot programme for the acquisition of main line diesels would be implemented – dieselisation of a secondary main line such as the Great Central would have been an ideal proving ground.

Passenger service improvements would then be delivered by the electrification programme and phased dieselisation.

With the subsequent abolition of the Railway Executive, this programme was put 'on the shelf' pending the emergence of the new order following the protracted major organisational changes that were to follow. With the simultaneous departure of Riddles from the scene, and the increasing interest in diesel motive power in the minds of other key railway officers, it was clear that these proposals would never be 'taken down' and considered further.

This was probably the post-war industry's most significant lost opportunity, for there is little doubt that had this plan been followed through, operating performance would not have declined to the low levels experienced in the 1960s and 1970s as a result of the inherent unreliability of much of the main line diesel locomotive fleet hurriedly acquired under the later Modernisation Plan.

Reverting to the relationship between the Commission and Railway Executive, as noted earlier, with the organisational arrangements and the characters involved, relations were never going to be close. By 1951, relationships were starting to improve between the Executive and the Commission, athough the Regions still considered that decision making was too centralised. However, largely due to the drive by the Railway Executive, the condition of the infrastructure was recovering, line speeds were being restored to former levels and the railway was running more efficiently. For example, net ton miles per engine hour had increased by 30 per cent. However, passengers were more aware of the fact that, despite these improvements, pre-war express passenger service timings still had not been restored. While things were moving in the right direction, the railways still had more to do, but the many politicians and the public at large were getting impatient. Support for the social changes and state control introduced in the late 1940s was waning.

The General Election of October 1951 brought a landslide victory for Conservatives, who went on to form a government that totally rejected the 1947 Act. The new administration set about eliminating the idea of integration and reintroducing 'healthy rivalry between areas' in a newly decentralised railway. The organisation that what they had most firmly in their sights was the Railway Executive.

The Transport Act of 1953 embodied these 'competition' principles and received Royal Assent in May 1953. The Railway Executive was abolished and replaced by six area authorities on 1 October of that year. These new area bodies reported to the British Transport Commission, which was retained under the leadership of Sir Brian Robertson. The railways were then run under interim arrangements until the new organisation was implemented in January 1955. Thus, the railways had endured another three-year period of flux, as the system had previously done in the 1940s, for little, if any benefit. Unfortunately, the opportunity to remove the rates and charges structure enshrined in the 1921 Act was only partially enacted, which was unfortunate as, to a large extent, this was the main reason why British Railways fell into a spiraling state of financial decline from 1952 onwards. This unhappy state of affairs was to seriously occupy the government and British Railways for nearly forty years thereafter.

The time spent re-organising the railway in the 1940s and 1950s was over-shadowing the need to improve efficiency and enhance the railway system. Progress was being made, but not on a scale to deliver a modern railway within a reasonable timescale. Fortunately, some electrification work following the successful Liverpool Street to Shenfield Scheme had been initiated by the Railway Executive during these difficult times. Electrification of the London, Tilbury and Southend lines was put in hand as was the Kent Coast scheme on the Southern. However, with the huge distraction of the reorganisation, little thought had been given to a rolling programme of electrification during this period by the either the railway authorities or the government.

Astonishingly, before the new railway organisation had time to settle in, and before the role of railways in the envisaged competitive environment had been determined, the 1955 Railway Modernisation Plan was announced. With such fundamental changes arising from the 1953 Act, an assessment of the future role and need for railway services should clearly have been carried out before initiating a large-scale investment into what was by then an ailing industry.

In any event, the £1240M (later inflated to £1660M – £51Bn at 2025 prices) Modernisation Plan went ahead alongside the Commission's worsening financial position. Past under investment, the decline in heavy industries' output and government intervention to prevent railway price increases had all contributed to the railway's unsatisfactory financial position. It was hoped that the further relaxation of the regulatory structure, the perceived benefits of the Modernisation Plan and the granting of new commercial freedoms would take the railways out of deficit by 1961/2. This, however, was a vain hope. It should have been obvious at the time that these initiatives could not bring any significant financial benefits for many years. The Commission recognised this but were so desperate to seize the investment funds being offered to them to improve their declining railway, they did not demur.

In exchange for the granting of the Modernisation Plan funding, the Com-mission, as already noted, had to forgo the price increase that they had proposed for 1956. The government's view was that the Commission had been given suffi-cient commercial freedom to enable it to comprehensively reshape the industry and rapidly achieve profitability. The press and public expectation seemed to be that the railways would become profitable from the very first year. This was totally unrealistic in the short term; the implementation was expected by the

Commission to take up to five years with the full financial benefits coming through in the 1970s.

Set against these record levels of capital funding, the revenue budget was in decline during the second half of the 1950s. The Commission considered that the only way to balance the operations and maintenance budget was to significantly reduce expenditure. Efforts to reshape rail freight activities, however, got off to a slow start, as the process relied upon pricing freedoms that still had not been fully granted. In addition, the devolution of authority to the Regions fragmented what needed to be a national strategy.

It therefore seemed at the time that the only option left to the Commission for rapid cost reduction was to initiate a programme of large-scale withdrawal of unremunerative passenger services. A closure programme was drawn up, but it was soon found that the necessary consultative process with users and staff involved much management effort and a considerable amount of time. In addition, contracts had to be negotiated with bus companies to provide alternative services, which required further management effort. Ironically, the early financial results following the introduction of new diesel railcar sets on branch and secondary lines seemed, for a time, to indicate that many rail services might, after all, become viable. This was because passenger usage was generally found to increase with the deployment of the new trains, while operational and maintenance costs were reduced – the more attractive diesel units costing 60 per cent less than steam-hauled services. In virtually all cases however, the new trains merely reduced the losses rather than eliminating them. Nevertheless, optimism often prevailed over hard evidence, which further stalled the closure programme. Consequently, the Commission's financial performance continued to worsen. The lack of action to rationalise both the network and unremunerative services was a major misjudgments. The government should take some of the blame too, for effectively regulating passenger fares and retaining the regulatory restrictions on freight.

With hindsight, it is clear that the commercial reshaping of the railways, together with the removal of the regulatory restrictions, should have taken place before the physical modernisation of the system. The actual reality was, unfortunately, that a significant proportion of the new infrastructure, trains and equipment provided under the Modernisation Plan was aimed at traffic that was in decline or of uncertain value. Consequently, much of the investment funds were wasted. Clearly, resources could have been put to better use had they been applied to a rail network that had been judiciously trimmed to operate a reduced range of viable services.

A key feature of the Modernisation Plan was the electrification of the East Coast Main Line between London King's Cross to Leeds and Doncaster.

The scheme was to utilise the 25kV, 50-cycle AC system, which, the plan had announced, would be the standard for all future overhead electrification schemes. The plan effectively took forward and extended the proposal for the electrification of the East Coast route that had been previously considered by the Railway Executive.

Electrification of the West Coast Main Line was not at first to be included, but the Chief Officer, Electrical Engineering, S.B. Warder, persuaded the Commission to include it by giving his assurance that sufficient resources were available to proceed with two main line schemes. Stanley Warder was considered by many colleagues to be a 'born-optimist', and, unsurprisingly, it subsequently became evident that neither the Commission nor the railway industry suppliers could support both schemes. With the greater business potential of the London to the West Midlands, North-West and Glasgow corridor, the decision was made to concentrate on the West Coast route and postpone the East Coast project.

At this time, the West Coast Main Line was carrying a significant proportion of traffic that was of questionable value. It was therefore unfortunate that many track miles were electrified, only to be later de-wired under subsequent traffic rationalisation policies.

The Modernisation Plan also included a number of other routes that were actually being taken forward or were under active consideration for 25kV electrification as follows:

- Shenfield to Chelmsford and Southend Victoria (circa 48 route km or 30 route miles) and under construction
- London, Tilbury and Southend Central (circa 100 route km or 63 route miles)
- Moorgate and King's Cross to Hitchin and Letchworth/Royston, including the Hertford Loop (circa 99 route km or 62 route miles)
- Glasgow Suburban Lines (circa 144 route km or 90 miles).

Four hundred km (two hundred and fifty miles) of third-rail DC schemes were also included in the plan, including the Kent Coast scheme, which had been given outline approval way back in 1952!

In addition to the West Coast and (later deferred) East Coast Main lines already mentioned, extension of electrification of the lines from London Liverpool Street to Ipswich and the Clacton, Harwich and Felixstowe branches were also to be considered. In the event, electrification beyond Colchester did not proceed.

The Commission wisely set up an Electrification Committee to advise on the overall programme to be included in the Plan. The members of this Committee were as follows:

- Chairman: S.B. Warder (Chief Officer, Electrical Engineering, BTC)
- R.C. Bond (Chief Officer, Mechanical Engineering, BTC)
- C.W. King (Chief Civil Engineer, BTC)
- J.H. Fraser (Chief Signal and Telecommunications Engineer, BTC)
- R.F. Harvey (Chief Officer, Operating and Motive Power, BTC)
- J.R. Pike (Chief Commercial Officer, BTC)
- A.E. Robson (Chief Officer, Carriage and Wagon Engineering, BTC).

The Committee published its prioritised recommendations for the project in March 1957. This envisaged an ambitious programme spanning over three timescales as follows:

Modernisation Plans and possible additions to 1970 3145 route km
                                                           (1965 route miles)

The period 1970–80	3132 route km (1957 route miles)
After 1980	1293 route km (808 route miles)
Total route mileage	7570 route km (4730 route miles)

The Committee emphasised the need for a rolling programme of work, incorporating associated non-electrification works and enhancements, to ensure that skilled staff were retained and progress was maintained in a cost-effective manner.

The BTC endorsed the committee's report and itemised additional routes that might be included for 25kV electrification by 1990, namely:

- extension of the East Coast Route from Leeds and York to Newcastle, Edinburgh, Dundee and Aberdeen
- Weaver Junction to Carlisle, Glasgow, Perth and Kinnaber Junction (for Aberdeen)
- Glasgow to Edinburgh
- East Anglian lines east of Ipswich and Bishop's Stortford
- principal routes in Lincolnshire, South Yorkshire and Nottinghamshire
- principal lines in the West Riding and North-East
- the Midland Main Line from London St Pancras to Leeds and Manchester
- London Waterloo to Southampton, Bournemouth and Weymouth.

Interestingly, no Western Region lines were included, probably in part because this Region, more than any other, retained much of its fiercely independent (Great Western) culture. At the time, this Region's traction ambitions were focused firmly on a diesel-hydraulic future.

The Modernisation Plan was re-appraised in July 1959 with the publication of the White Paper 'A Reappraisal of the Plan for the Modernisation and

Re-equipment of British Railways'. This document took stock of the progress made to that point in time and looked forward for the next five years and beyond. Besides the switch from the East Coast Main Line to the West Coast route, the paper noted the 1956 decision to extend the Great Eastern 50 cycles/ sec AC electrification to Chelmsford and Southend. Also inserted was the electrification from Bishop's Stortford to Cambridge to be completed by 1964, while deferring conversion of the route to Ipswich and Harwich. As it turned out, the electrification did not proceed between Bishop's Stortford to Cambridge, nor Colchester to Norwich for more than a quarter of a century. The only other electrification work undertaken on the Great Eastern Lines in the latter half of the 1960s was on the Lea Valley line between Clapton Junction and Cheshunt, which came into use in 1969.

Contemporaneously with these schemes, electrification of suburban lines in the Scotland area were also going forward. Sir Robert Inglis, the LNER Divisional Manager in Scotland, first outlined these proposals in 1946. His ideas covered both LNER and LMS territory in the Glasgow area. With the impending nationalisation looming, no progress was made in the 1940s but the plans were explored again in 1953. Over the next two years, there were two 1500V DC schemes, which were authorised as the Stage 1 element of the plan, namely:

- North Clyde Lines comprising Glasgow Queen Street and Helensburgh, Balloch, Milngavie, Bridgeton, Springburn and Airdrie
- South Clyde lines between Glasgow Central and Neilston and Motherwell (via Kings Park) and the Cathcart Circle.

Stage 2 was to follow, namely:

- Glasgow Central to Gourock and Wemyss Bay.

In 1956, the schemes were changed to the 25kV AC system. The North Clyde line was brought into use first with the introduction of passenger services on Sunday 5 September 1960. In these early years there were many failures of the units' mercury arc rectifiers until these were replaced by solid state types.

The Stage 1 lines south of the Clyde followed, without the major failures experienced north of the river. In 1965, work commenced on Stage 2, which was completed in 1967. Interestingly, the two electrified routes were not connected until the electrification of the Argyle line was completed in 1979.

Returning to the West Coast Route, the West Coast scheme was authorised in stages, the first section being Crewe to Manchester, which was given the 'go ahead' in October 1955. This scheme was considered to be a main line pilot 'proving ground' for the 25kV AC system. It was subsequently determined that it would be prudent to undertake the electrification of the Styal Line, a diversionary route for the Crewe to Manchester route, which generally operated with a low traffic density. The Styal line was commissioned in May 1958 and used to test a variety of fixed equipment, locomotives and electric multiple units. The former Western Region Gas Turbine locomotive, No. 18100, was converted into a 25kV electric traction unit and used for testing and driver training on the newly electrified line.

Electrification of the Crewe to Manchester route was completed to time in September 1960. In parallel with progress on this route, planning on the other sections of the West Coast route continued and, in fact, following a Commission directive in October 1958, the entire programme was accelerated and given priority over all other electrification schemes in the country. While there was no effect on schemes that had already commenced, some schemes that had not started were deferred. In the event, the planned speeding up did not materialise, as the Commission's finances worsened.

In summary, the elements of the Modernisation Plan that ran into difficulties were not all of the BTC's making, but the Commission must bear a sizeable responsibility for its misjudged decision to speed up the conversion from steam to diesel traction, when the priority should have been to reduce the size of the network and take out unviable activities.

Faced with another railway financial crisis, the government decided to intervene again with ideas of a further reorganisation and appointed a committee, chaired by Sir Ivan Stedeford (Group Managing Director and Chairman of Tube Investments), to undertake a review of British Transport, including the railway's financial performance. The remit included a review of the West Coast electrification, and the Committee evidently considered that the investment couldn't be proved. Severe differences in opinions erupted between the Chairman and another 'heavyweight business' member, Dr Richard Beeching. These disagreements may have been the reason that the report wasn't published until much later. On 15 March 1961, Dr Beeching, the Technical Director of ICI, was given five years' leave by his company, and was appointed as the designate Chairman of British Railways Board. This new body was to take over from the British Transport Commission following the abolishment of the latter organisation under the 1962 Transport Act.

Fortunately, the West Coast electrification project was not cancelled. However, following the appointment of the H.C. Johnson as General Manager of the London Midland Region in 1962, a full review of the electrification was undertaken to determine whether all the elements of the scheme should be progressed. This effectively added a year to the timescale of the project. Fortunately, though, all but a few minor sections having low traffic density were allowed to go forward.

The project was opened, stage by stage, between Manchester, Liverpool and London Euston in April 1966 and to Birmingham in March 1967. In his paper, read to the Chartered Institute of Transport in January 1968,* entitled 'Main Line Electrification – A First Appraisal', H.C. Johnson noted that in the first four weeks following electrification, journeys had increased by 59 per cent on the London to Manchester route and 60 per cent up on London to Liverpool services. In the 24 weeks since the London to Birmingham electric service had been introduced, journeys had increased by 24 per cent.

From the experience gained from this very first electrification of a major trunk route, two key lessons were learned. First, future electrification projects should more thoroughly assess the opportunities for rationalising the infrastructure. Secondly, higher speeds, resulting in shorter journey times increase patronage, so opportunities to improve track alignment and enhance the infrastructure to improve line-speeds should be fully considered in the development of projects, along with the operating requirements and the commercial prospects.

The BTC had accepted in principle in May 1957 that the West Coast electrification should be extended from Weaver Junction (where the Liverpool line diverges) to Gretna and across the border to Glasgow. This was justified on the basis of the reduced journey times significantly increasing patronage, plus the anticipated traffic that would be diverted to the West Coast line should the Settle and Carlisle line be closed.

Detailed planning eventually commenced in 1966 and the £55M (£1Bn at 2025 prices) scheme was authorised in February 1970, with work commencing in 1971. Learning from the earlier stages, a thorough rationalisation of the route was undertaken. Full through electric services between London Euston and Glasgow Central with a five-hour journey time commenced in 1974. The scheme was delivered on time and within 10 per cent of budget.

As noted earlier, the Modernisation Plan included 400km (250 miles) of low voltage DC electrification on the Southern Region. The work was split into two stages – Stage 1 included the conversion of the following routes:

- Gillingham to Margate and Ramsgate
- Faversham to Dover
- The Sheerness Branch.

Stage 2 works included:

- Sevenoaks to Ashford, Dover and Ramsgate
- Maidstone East to Ashford, Canterbury West and Minster
- Maidstone West to Paddock Wood.

---

* Volume 32, Number 8, January 1968.

The Kent works included extensive infrastructure work, including quadrupling tracks. Stage 1 services were initiated in June 1959, and Stage 2 in June 1962.

The next line to be electrified was the London Waterloo to Bournemouth route. This was the Southern Region's first foray into 100mph running, requiring electrification from the limit of the existing electrified lines at Sturt Lane, near Farnborough, to Southampton and Bournemouth. In the early planning stages, consideration was given to:

- retaining the low-voltage third-rail system over the whole route
- adopting the 1500V DC overhead system from Sturt Lane to Bournemouth
- instigating the 25kV overhead system between Sturt Lane and Bournemouth
- converting the existing third rail lines between London Waterloo and Sturt Lane to the 25kV AC overhead system and continuing this system to Bournemouth.

It was estimated that for both dual third-rail/DC overhead and third-rail/AC overhead arrangements, the cost would be 20 per cent higher than a straightforward third-rail DC system. There were also concerns about the reliability of the 'change-over' arrangements. The cost of converting the existing route between London Waterloo and Sturt Lane was, as expected, exorbitant, and so it was decided to retain the third-rail system throughout.

There was a compelling operational case to extend the electrification from Bournemouth to Weymouth, but this could not be justified financially. Therefore, an innovative solution was adopted whereby high-powered four-car EMUs propelled one or two unpowered units from Waterloo to Bournemouth. On arrival, these units were detached, and non-motored units were worked to Weymouth by a Class 33 locomotive operating as a 'push–pull' train.

The 1962 Transport Act and the Beeching Years

During 1960, the government determined that yet another reorganisation to decentralise the railway by creating autonomous regional boards and pruning the network would solve the industry's problems. This was even though contemporary studies of area financing had not predicted very favourable outcomes. Nevertheless, the government pressed ahead with the legislation, passing the Transport Act 1962, and implementing the new organisation a year later – the British Railways Board (BRB) came into existence on 1 September 1963. In the run up to the introduction of the new organisation, Dr Beeching held the reins of the British Transport Commission before taking up the Chairmanship of the BRB.

British Railways was separated from all other pre-BTC activities, and as a result was a much slimmer corporate body, albeit with a far from ideal structure. The one good feature of the legislation was the removal, at last, of all remaining regulatory controls over rail freight, removing the 'common carrier' requirement and allowing the railway to set its own charges.

The general mood in the country was that the railways couldn't go on as they were, and, having broken away from the restrictions of the past, the task was to make the railways profitable again. There was, at this time, a general acceptance that the system would need to be reduced in size. However, as rail closures became more unpopular as time went on, this support waned, even though many of those complaining hardly, if ever, used rail services.

At board level, there would be a return to a functional organisation, which, it will be remembered, was abolished under the previous government-inspired organisational change in 1953 because it was seen as a significant weakness!

Dr Beeching brought in 'new blood' from outside the industry and set about trying to make the railways profitable. He published the famous 'Reshaping of British Railways' Report just three months after the British Railways Board had come into being. Today, his legacy seems to be solely about railway closures, and while he was responsible for reducing the network route mileage by a third, his proposals to improve inter-city passenger services and introduce efficient modern methods of freight working, including block and liner trains, are largely lost sight of.

Many rail managers of the time had similar views to Dr Beeching, recognising that it was better to invest in a smaller flourishing network than a larger, failing system. As part of the analysis, Dr Beeching and his team scrutinised all current electrification schemes. While critical of many of the original figures used for justification, he took account of the investment already made in these schemes, and thus he did not cause the progress of any of the schemes that were underway to be impaired. In fact, two new, albeit modest, electrification projects were initiated, following a detailed financial analysis, in the four years that he led the railway.

It should be recorded that in undertaking the extensive analyses, without the benefit of the information and computerised systems available today, he and his team pretty accurately identified the strengths and weaknesses of the railway system at that time. His proposals to modernise and improve the efficiency of the inter-city passenger and bulk freight operations would subsequently make it easier to justify the electrification of the core routes carrying these traffics.

The tragedy of the years following Beeching was that a study was not under-taken to identify and reserve the railway rights of way where reinstatement of closed routes might be justified in the future. This aspect was not even consid-ered at the time even though it was quite evident that road congestion was not only clogging up many cities but was also further increasing at an exponential rate.

By Dr Beeching's own admission, he never solved the relationship between the BRB headquarters and the Regions. This was, however, not because he was not determined, nor clever enough. The structure was flawed from the very beginning, and it was not until 1982, with the introduction, under the then Chairman Bob Reid (1), of the Business Sectors, that the issue was starting to be properly addressed.

The 1970s

As far as electrification was concerned, the 1970s started off well with the Glasgow extension of the West Coast Main Line and the Great Northern Suburban electrification projects progressing. In addition, the Conservative government asked BR to make a proposal for a significant electrification project, which resulted in the authorisation of the St Pancras to Bedford (Bedpan) scheme going forward.

While the actual implementation went well, the introduction of the new electric services was severely delayed due to a dispute with the ASLEF Trades Unions concerning the introduction of 'Driver Only' (DOO) operation. This called into question the wisdom of investing in an organisation with such difficult labour relations.

Prior to these difficulties, in November 1977 the government agreed to review with the Board the general case for mainline electrification and, in May 1978, it was agreed that the joint review would be undertaken by the Depart-ment of Transport and the BRB (see Chapter 7).

Recapping, the review, undertaken in two stages and concluded in 1981, rec-ommended electrification of a number of key routes. All electrification options that were evaluated, apart from the smallest, gave an internal rate of return of 11 per cent. It was stated that the best course of action, providing railway finances were not constrained, would be the largest and fastest programme. This latter programme would take twenty to thirty years, and, on completion, the report estimated that more than 80 per cent of passenger and 70 per cent of freight traffic would then be electrically hauled. It was reckoned that this extensive programme would give a net present value of £305M (£1.25BN at 2025 prices).

The report proposed the setting up of three or four electrification construction teams to undertake a programme of work extending into the future. The group noted that the alternative of 'ad hoc approvals' of individual schemes could not accrue the benefits of a rolling programme.

Calls for a rolling programme of electrification have regularly been made for more than a century and it appeared that the industry was on the cusp of at last achieving this 'castle in the air'. Significantly, while the strongest voice for electrification had always come from the industry, it looked for a while as though the politicians and civil servants were on board too!

The Thatcher Years

The joint review was published early in the Thatcher government's first term and recommended electrification of a number of key routes. As noted earlier, of all electrification options that were evaluated the report stated that the best course of action, providing railway finances were not constrained, would be the largest and fastest programme. Mrs Thatcher had discounted as too difficult privatisation as a way forward for the industry because of the continued need for subsidy, so considered that the best course of action was to make the industry more cost effective, including selective investment where a payback could be foreseen.

The joint report forecast a payback of up to 11 per cent, and, for a while, it looked as though the government would agree to a rolling programme. Unfortunately, the report was being considered against a background of the ASLEF DOO dispute, the closure of the Woodhead electrified route and the country's poor economic circumstances. Unsurprisingly, therefore, the recommendation for a rolling programme was not taken forward. However, the East Coast Main Line electrification scheme was authorised and this was followed by a series of cost-effective schemes by the new Business Sectors during the decade (see Chapter 9).

Margaret Thatcher is not generally seen to be a friend of the railways, but ironically, under her government in the 1980s, British Rail managed to electrify more route miles during the decade of her premiership than any other administration before or since.

Sectorisation

The concept was largely developed by one of Bob Reid's board members, John Welsby, who continued to develop his ideas through the 1980s and early 90s.

In April 1992, he was to implement BR's full business 'Sectorisation' where, at last, 'Accountability' went hand in hand with 'Responsibility'. During this period, a plethora of proposals for electrification, supported by robust business plans, side-stepped the usual political blockages, to enable Sector management to deliver the successful programme mentioned earlier.

On taking up the role of chairman, John Welsby went on to introduce the 'Organising for Quality' initiative, changing not only the industry's organisation but also its culture and customer service. The new arrangements were viewed by most rail managers, including your authors, as the best UK railway organisation of all time. Terry Gourvish summed up the new organisation perfectly when he noted in his book* that 'it offered Britain the best prospects of a more stream-lined, customer-oriented, empowered organisation in an integrated form'.

Privatisation

In April 1992, John Major's Conservative government was re-elected with the pre-election promise that BR was to be privatised. Another reorganisation was on the way with the prospect, yet again, of the resurrected but failed concept of more competition within the industry. Privatisation commenced in 1993 and was rushed in before the 1997 General Election. Privatisation resulted in the introduction of short-term franchising of the passenger services, and this virtually killed off electrification for a decade.

Apart from a couple of modest schemes progressed by the Scottish government, it was 2007 before the DfT made the first rumblings that a programme of electrification may go ahead in the rest of the UK. However, when the then new Transport Secretary, Ruth Kelly, introduced the Labour government's White Paper, 'Delivering a Sustainable Railway' (the thirty-year plan for the future of the railways), electrification was effectively again left on the sidelines (it being declared that such schemes would only be authorised on a 'case by case' basis and subject to impossibly short payback periods). Instead, the government appeared to engage in another dalliance with alternative forms of traction energy, especially hydrogen fuel cells. The fact that the properties of hydrogen would limit its use to passenger services on secondary routes and branch lines was either not recognised or ignored.

On the plus side, Crossrail was included under the Plan with work commencing in 2009. Also in that year, Lord Adonis, the Secretary of State for Transport, first proposed the construction of High Speed 2 (HS2).

---

* Terry Gourvish, Oxford University Press

Lord Adonis went on to announce the electrification of the £162M (£255M at 2025 prices) Liverpool to Manchester line (via Chat Moss) together with the £874M (£1.35Bn at 2025 prices) Great Western route (GWEP) from London to Bristol and South Wales. These two schemes signified the start of the first major UK rail electrification projects in the country for two decades. The latter project was heralded as the first electrification scheme in Wales with journey time savings of 20 minutes from London Paddington to Swansea. It was estimated at the time that the pay-back for the scheme would be 40 years.

Following the General Election, a plethora of electrification schemes were authorised, driven in part by the political desire to achieve 'net zero'. While many schemes were delivered in England and Wales over the next decade, there were many cost-overruns, most dramatically highlighted by GWEP. This caused the UK government to lose confidence in the ability of the industry to deliver electrification in a cost-effective manner. Accordingly, the scope and specification of a number of troubled schemes were subsequently curtailed or cancelled altogether. Schemes in Scotland, authorised by the devolved Government on a rolling programme basis, fared rather better (see below).

In February 2017, the House of Commons Transport Committee deemed the UK's rail franchising model to be 'no longer fit for purpose'. The MPs' report noted that franchising had not increased competition in the way the government had hoped in the 1990s. The report stated that 'many metrics of performance are plateauing, and the passenger is not receiving value for money'. Recounting the history of rail reorganisations since 1921, it is perhaps not surprising that the privatisation model chosen did not work.

The hailed increase in passenger numbers had more to do with the positive UK economic growth and increasing road congestion. That is not say that many of the private operators did not create further growth through improved marketing techniques, harnessing the power of the internet, but their services were operating on a dysfunctional, fragmented system. During the first decade of privatisation, infrastructure costs increased three-fold and electrification progress was stalled. Those of us in the industry in the 1990s and 2000s can vouch for the fact that privatisation of passenger services was flawed from the start. When questions about shortcoming of the system were raised, these were inevitably met with a retort along the lines ....'ah! but look at the passenger growth!'

It was disheartening to witness the politicians trading in, on a whim, the best railway organisation of all time for a hugely flawed structure in the 1990's. Privatisation would undoubtedly have had a better chance of success had Bob Reid 1's Sectors been brought to the market.

On 21 September 2020 the government finally accepted the inevitable and permanently abolished the rail franchising policy.

Scotland Post-Devolution: Electrification Strategy 2005 to 2025

The almost continuous progress of railway electrification in Scotland since the turn of the millennium contrasts starkly with attempts to wire up the network in England and Wales. Since the responsibility for railways was devolved to Holyrood in 2005, the Scottish Parliament has consistently supported the progressive electrification of the network. Politicians on all sides have been convinced of the economic and social benefits that a rolling programme would achieve, and a consistent policy has been maintained. More recently, Scotland has risen to the sustainability challenge and the government has developed a comprehensive decarbonisation strategy for rail, which was published in July 2020.

The politicians' faith in the industry has been repaid with the delivery of a cost-effective 25kV electrification programme. Although some scheme costs were coming in above budget (for example, EGIP), a concerted collaborative effort by key stakeholders north of the border helped to drive down costs and deliver projects within programme timeframes.

The managing director of Scotland's Railway is in charge of the operator and Network Rail, Scotland under an alliance agreement. Since this change was implemented, there have been significant improvements in operations and maintenance performance in addition to the more effective implementation of renewals and new works.

Single track kilometre costs have generally been delivered at £1.5M or less, comparing well with those achieved on the continent. These rates are one-third of the Great Western Main Line electrification figures.

Electrification Efficiency

An internal report entitled 'Enabling Efficient Electrification in Scotland' was prepared by Network Rail in 2019, which summarised the lessons learned from these projects. This report provided a blueprint for the future, outlining how the efficiency and effectiveness of electrification schemes might be further improved.

Consequently, it is not difficult to see why things that are done differently north of the border have delivered such a cohesive, successful and forward-looking network.

Over the last ten years in Scotland, approximately £1Bn has been invested in electrifying circa 500 track kilometres, including the associated infrastructure enhancements. Prior to the Covid-19 pandemic, the 'sparks effect' held true with passenger numbers carried on electrified services increasing by 23 per cent from 2018 to 2020.

By early 2020 just over 40 per cent of Scotland's route miles were electrified and more than three-quarters of passenger journeys were made on electric trains. A total of 45 per cent of freight haulage was by electric traction.

Electrification works continued during the pandemic, albeit at a slower rate. Generally, costs were held under reasonable control despite the restrictions that had been imposed due to Covid-19.

Sir Peter Hendy's 2015 Report to the Secretary of State for Transport on the Replanning of Network Rail's Investment Programme

By 2012 costs on the Liverpool to Manchester electrification scheme were escalating due primarily to the difficult ground conditions across the famous Chat Moss. In the same year confidence in Network Rail's ability to deliver electrification projects to budget and programme was still high however, and the electrification of the Midland Main Line from Bedford to Derby/Nottingham and Sheffield was authorised. Originally budgeted at a cost of £800M, by 2014 it was clear that the final cost would be significantly greater. During this same year, the new electric trains on the Liverpool to Manchester line were also scheduled to start running, but these services were deferred until 2015.

Similarly, in 2014 it became apparent that GWEP costs had doubled, and the project was deemed to be a year behind schedule. A variety of reasons were given for the cost over-runs and delays. In addition, it was noted that while the new High Output System (HOPS) equipment had entered service, it had not achieved anything like the anticipated production levels.

Thus, the costs and programmes of all three schemes (Liverpool to Manchester, Great Western, and Midland Main Line) were all over-running. In addition, Network Rail had entered the Control Period 5 (CP5) Capital Investment Programme (2014–19), with the most ambitious schedule of works ever attempted, probably since the original construction of the railways. It soon became apparent that there were neither the physical resources nor the skilled labour to fulfil all the planned works, which were budgeted at nearly £40Bn.

Not surprisingly, the government became very concerned about the overall situation on the railways. As a result, in June 2015, the Secretary of State for Transport, Patrick McLoughlin, ordered a review. Sir Peter Hendy, who had been newly appointed as the Chairman of Network Rail, was asked, as a priority, to undertake a full appraisal of the CP5 Capital Investment Programme and to report back in the autumn.

Sir Peter's review acknowledged that the cost and timescale on a small number of significant enhancement projects had, for a variety of reasons, increased beyond the expectations of the CP5 budget. He concluded that there had been over-optimism on costs and timescale and that the planning processes, both within and outside Network Rail, had been inadequate. In addition, there had been changes in scope during project development and delivery, together with a recognition that a number of large projects 'should have been managed on a holistic basis rather than piecemeal'.

Concerning electrification projects, Sir Peter pointed out that the railways had not carried out any significant electrification projects for two decades and, as a result, there was limited information to generate reliable cost estimates. There were also scope definition and planning deficiencies. Alluding to the fact that one of the main reasons for the review was the unsatisfactory situation with the GWEP project, the report specifically highlighted the fact that early cost estimates for electrification schemes had been inadequate. In addition, scope creep had also inflated costs, albeit some changes were outside Network Rail's control, for example, resulting from revised interoperability (TSI) requirements.

The fact that this overall situation was partly as a consequence of the many skilled and experienced people who were encouraged, and sometimes pressed, to leave the railways at privatisation, was not specifically mentioned. The report indicated that 80 per cent of the cost over-runs were due to the electrification schemes, with a further 10 per cent primarily due to the overspending on the Thameslink Project. Most of the remainder was due to projects spilling over from CP4, plus scope changes required by the DfT.

The report went on to describe how effective control of these errant projects was to be restored. The costs and programmes were being scrutinised and revised accordingly. For the worst performing scheme, GWEP, approximately 30 per cent of the work was still in the development stage and this was to be re-evaluated. More effective controls were to be put in place and site productivity enhanced. The report considered that the wiring to Cardiff would be completed within CP5 at a predicted cost of £2.8Bn (in 2012–13 prices), which included a 20 per cent contingency. The electrification beyond Cardiff would be deferred, being 'expected to be completed in CP6'.

The review highlighted the fact that the CP5 huge programme had stretched the railway's internal and contracted resources to the maximum. In particular, signalling resource, essential as a necessary precursor to electrification, was particularly stretched. It was recognised that the industry needed to recruit many more apprentices and engineers to complete the ongoing and future works.

Singling out electrification design parameters, the review acknowledged that it may not be economically or practically viable to meet normal European standards on electrical clearances, which had been rigidly adhered to on Great Western. It was recommended that, in such cases, a risk assessment should be undertaken and, if acceptable, a timely exemption applied for and obtained from the relevant standards authority.

In summary the report pointed out that Network Rail was, at the date of publication, already comprehensively revising its management of enhancement programmes, including ensuring that the organisation implemented more rigorous sponsorship and change control processes. In addition, it noted that a more effective working arrangement with the government was needed. A key requirement in the future would be to ensure that projects would be better developed before funding commitments were put in place and work commences.

Chapter 17
Achieving Sustainability, Decarbonisation and Affordability

The House of Commons passed an amendment to the Climate Change Act on 24 June 2019, which committed the UK to achieve net zero carbon emissions. In response, Network Rail published its Traction Decarbonisation Network Strategy (TDNS) in July 2020. Based on the Rail Industry Decarbonisation Task-force's work, the plan indicated how carbon emissions could be minimised to help meet the UK's 'net zero' target by 2050. The Scottish Government's Rail Services Decarbonisation Action Plan was also launched in July 2020, aimed at decarbonising the railways north of the border by 2035.

These plans were in line with the UK's commitment to meet the 1994 United Nations Framework Convention on Climate Change (UNFCC). Annual Conference of the Parties (COP) meetings have been held since the signing of the Treaty, to consider the progress being made to reduce emissions. The most significant conference was undoubtedly COP21, when the '2015 Paris Accord' was agreed to take action to avoid global temperatures exceeding 2°C.

COP26 was scheduled to be held in Glasgow in 2020, but the Covid-19 pandemic delayed the conference until October 2021. It was hoped that world leaders would agree in Glasgow to limit global temperatures to 1.5°C. After much debate many countries agreed to further limit their emissions of CO_2, although it did not prove possible to achieve unanimity.

Various rail organisations were represented and the home country, Scotland, made great play about its plan to decarbonise the railway network north of the border by 2035. Vivarail showcased its Class 230 battery train, with its innovative fast-charging system. It performed well during the thirteen-day duration of the conference, running between Glasgow and Kilmarnock. Two hydrogen trains were also on display, one being a converted Class 310 Hydroflex unit supported by Porterbrook, and the other a converted Class 314 led by Arcola Energy and showcased by Angel Trains.

With emissions of an estimated 97 metric tonnes of carbon dioxide per annum in 2020, transport is the largest emitter of CO_2 in the UK, accounting for 27 per cent of total emissions. At the time of the conference, Network Rail's TDNS proposal for further main line electrification, including the utilisation of battery and hydrogen power for secondary lines, foresaw the passenger railway achieving 'zero carbon' status by 2050. Achieving modal shift from road and air to rail could also make a significant reduction in overall CO_2 emissions, further assisting the government to meet its legally bound target. However, it should be noted that fully decarbonising rail freight is challenging, even with 'in-fill' electrification. Nonetheless, providing additional strategic rail freight interchanges could, with modal shift, still result in a much larger proportion of UK freight traffic moving more sustainably by rail, regardless of the form of traction.

Progress in Scotland

A more positive approach to sustainability and decarbonisation has been made in Scotland compared to the rest of the UK. In addition, the electrification costs per STK have generally outturned below the rates for the rest of the UK. A review of the Scottish proposals follows:

The Scottish Government's Rail Services Decarbonisation Action Plan was launched by Transport for Scotland on 28 July 2020, and this set out the strategy to decarbonise its railway earlier than the rest of the UK. This plan is a key element of the government's Scottish National Transport Strategy for a sustainable, inclusive, safe and accessible transport system. As previously noted, Network Rail's TDNS for the UK was also published in July 2020.

Both documents conclude that rail decarbonisation requires the electrification of a significant proportion of the routes that are presently not electrified. Battery or hydrogen traction would be deployed on the remaining lines.

It is interesting to note the fundamental difference between this Scottish Plan and Network Rail's TDNS. The Scottish Government's Rail Services Decarbonisation Action Plan is an instruction to the Scottish rail industry while the Network Rail TDNS document was merely a recommendation from the UK rail industry to the UK government.

Transport accounts for 37 per cent of all emissions in Scotland, with buses and rail transport accounting for 3.2 per cent and 1.2 per cent respectively. Although rail has a low emissions share, action to decarbonise is still a priority as zero transport emissions would not be possible without modal shift to get people and goods off the roads. Freight transport decarbonisation demands completely revised distribution system requiring a substantial shift to rail transport, which, in turn, enhances the case for rail electrification.

It is envisaged that by 2045 all routes in Scotland would be electrified, with the exception of the following:

- the Far North Line beyond Tain
- the Kyle of Lochalsh line
- the West Highland line
- the Stranraer line beyond Girvan.

For these routes, it is envisaged that alternative traction technologies in the form of battery or hydrogen power would be deployed.

The document sets out a programme for decarbonisation including transition plans, for example partial electrification and bi-mode operation, and alternative traction technologies. It also covers the usual associated infrastructure enhancements together with other necessary alterations, for example, provisions for the safe and effective battery charging and storage and supply of hydrogen to trains. The range of battery traction and hydrogen is currently considered to be between 55 and 400 miles respectively, but developments in both technologies should extend the range significantly. The innovative Vivarail fast charging system, since being taken over by First Group, should also be a game changer.

It is interesting to compare the unit costs per mile[*] for EMU, BEMU and hydrogen trains that ScotRail have made public. These are:

- EMU   £1.32
- BEMU   £1.62
- Hydrogen £2.46

EMUs are clearly the most cost effective and with their more basic configuration, likely to be the most reliable of the three.

While no new diesel trains will be procured, the plan recognises that diesel trains will continue to make an important contribution to transport decarbonisation as their emissions are lower than those of road transport. Although all the DMU fleets currently in use are due to reach the end of their lives between 2025 and 2035, it is envisaged that modifications (for example, use of fuel additives and/or replacing fluid flywheel transmissions with hydro-mechanical types) may be made to these trains to further reduce emissions.

The electrification plan is very challenging, requiring 1,800 STKs to be wired, with eighteen feeder stations and up to 560 bridge modifications. On average, 130 STKs per annum of the Scottish network will need to be wired to achieve the target.

Electrification work is currently proceeding on the routes between Muirhouse South Junction and Barrhead, including the branch from Busby Junction to East Kilbride. Wiring should then follow to Kilmarnock.

The following routes, amounting to more than 900 STKs, also feature firmly in the plan:

- Glasgow to Anniesland via Maryhill
- the Borders line from Edinburgh to Tweedbank

---

[*] Note that these figures do not include capital cost.

- the Levenmouth branch
- partial wiring in Fife as a precursor to a fully electrified route
- Dunblane to Perth.

Subsequently, it is envisaged that, by 2035, electrification of the routes between Glasgow and Carlisle via Gretna and the main lines to Aberdeen and Inverness will have been completed, so that all the railways across the Scotland's central belt will be fully wired.

It is also planned to electrify the Aberdeen to Inverness route by 2035, with the use of bi-mode traction as an interim step. Following on from this, a cross-Inverness extension of electrification to Tain, on the Far North line, is also envisaged.

The permanent solution for the Far North, West Highland and Kyle lines are considered to be in the form of alternative traction technology, with the three routes considered as a package taking cognisance of their individual requirements. Bi-mode battery or hydrogen technology is also envisaged as a transitional remedy for the Ayr to Girvan line, and as a permanent solution between there and Stranraer.

Electrifying the route between Carlisle and Glasgow via Gretna will provide the necessary strategic capacity for electric haulage of rail freight, and a diversionary route when the West Coast main line is closed. Through freight services are envisaged to both Aberdeenshire and Inverness-shire. Line capacity for an additional four trains per day in each direction has been identified. The plan highlights the fact that the use of electric traction is key for rail freight to decarbonise.

There is much to commend the Scottish plan. The basis is simple:

- Work collaboratively with relevant rail parties and stakeholders.
- Identify the quantum in single track kilometres.
- Adopt and adapt best practice.
- Determine designs appropriate for the network.
- Derive a sustainable design and delivery work-bank.
- Assess the annual single track kilometres that can be practically delivered.
- Derive a rolling programme.
- Develop appropriate contracts with clear specifications and scope.
- Work collaboratively with competent contractors.
- Drive continued and sustained reductions in unit costs.
- Measure key contractual KPIs and wider outcomes, for example, modal shift, impact on climate.
- Identify lessons learned on each project and feed into future schemes.

There is no doubt, from the experience gained in Scotland, that a rolling programme is vital for the delivery of efficient, sustainable and cost-effective railway electrification.

The Scottish Rail Services Decarbonisation Action Plan provides an exceptional political, economic, environmental and transport vision of the future. Furthermore, it is gratifying to see the level of collaboration that exists between Transport Scotland, Network Rail and ScotRail that is turning the plan into reality. Building on the recent past, these organisations, together with key stakeholders, are undoubtedly making Scotland's railway greener while improving their efficiency and economics. It's clearly a win-win-win!

Railway Industry Association (RIA) Electrification Proposals and the Traction Decarbonisation Network Strategy (TDNS)

This section covers the attempts by the UK rail industry to influence the government to back a rolling programme of overhead electrification by:

- Confirming the industry's commitment to 'net zero'
- Demonstrating how the industry can deliver cost effective electrification

The industry's initiatives were in response to the government's request to 'provide a vision for how it will decarbonise'. This question was proposed in February 2018 by the Rail Minister, Jo Johnson. In particular, he asked what would be required to remove all diesel-only trains from the National Network by 2040.

The industry's response was to initiate a study by the newly set-up Rail Industry Decarbonisation Taskforce. This body published its report in July 2019.

Things moved up a gear for the prospects of rail electrification when the House of Commons unanimously passed an amendment to the Climate Change Act on 24 June 2019, which committed the UK to net zero carbon emissions by 2050. This followed a report by the Committee for Climate Change that predicted that Net-Zero was possible, although demanding. In response to this new legislation, Network Rail developed the Traction Decarbonisation Network Strategy (TDNS).

However, with senior politicians still nervous about restarting a major electrification programme and the government's paymasters still considering that the cost of electrification per STK was still too high, only existing authorised schemes were being progressed after 2017. In the meantime, the RIA went on to develop two reports, the 'Cost Challenge Initiative' (published in March 2019) and the 'Why Electrification?' study (launched in April 2021) to try to get the need for a rolling electrification programme back on the agenda. Unfortunately, it became clear that Network Rail's TDNS proposals began to wither, so the RIA published a new strategy in April 2024 indicating how a lower cost, higher performing net zero railway might be achieved by 2050.

For the record, a brief outline of the TDNS and RIA reports concludes this section.

Traction Decarbonisation Network Strategy (TDNS)

The UK government's support for decarbonisation has strengthened in the decade following 2010, and a bold step was taken in June 2019 when a legislative commitment was made to achieve net zero carbon emissions by 2050. Following on from this the Department for Transport (DfT) requested that the rail industry explore whether all diesel-only trains could be eliminated from the network by 2040 in England and Wales.

The Scottish Government's Rail Services Decarbonisation Action Plan set an even tighter target to decarbonise domestic passenger rail services by 2035.

As the vast majority of rail carbon emissions emanate from diesel traction, Network Rail believe that it is logical to consider extending electrification of the main network and deploying other renewable energy sources to power trains on lightly used lines.

The Rail Industry Decarbonisation Taskforce identified three possible traction technologies that are sufficiently mature to replace diesel traction, namely, battery, electric, and hydrogen. Each of these technologies has different technical capabilities, which mean that not all are suitable for all types of rail services.

Working collaboratively with rail industry partners, Network Rail developed the TDNS. This report, which built on the Rail Industry Decarbonisation Taskforce's work, indicates the different ways that carbon emissions could be minimised by means of overhead electrification, battery or hydrogen fuel cells. The TDNS study reviewed every section of non-electrified route in the country to find the optimum solution for each line.

Of the 15,400 STKs that are not electrified in Great Britain, Network Rail considered that they would need to provide infrastructure for the three options as follows:

- overhead electrification – between 11,700 and 13,040 STKs
- battery – between 400 and 800 STKs
- hydrogen – between 900 and 1300 STKs.

For the remaining 260km where no clear decision was made a likely technology was identified within the TDNS recommendations.

The report recommends reinstating the cutbacks instigated in 2017 to the Midland Main Line and Great Western Electrification projects. In addition, the inclusion of the following routes are 'firmly' recommended:

- Taunton to Derby via Bristol and Birmingham
- Severn Tunnel Junction to Gloucester
- Penzance to Newbury
- Salisbury to Guildford
- Guildford to Reading
- Reading to Redhill
- Didcot to Worcester via Oxford
- Oxford to Coventry via Aynho Junction
- Oxford to Bedford
- London Marylebone to Birmingham
- Tilbury Branch
- Felixstowe to Birmingham via Ipswich
- TransPennine Routes
- Settle and Carlisle
- Carlisle to Newcastle and on to Sunderland.

Unfortunately, the TDNS was, to all intents and purposes, terminated in September 2022 – the government's 'Comprehensive Spending Review' effectively making the programme undeliverable.

In its place a five-phase Electrification Proposal was to be developed but, at the time of writing, only the 900 STK of electrification proposed in Phase 1 was committed. The other four phases were, unsurprisingly, subject to the usual political, economic, financial and programme uncertainties. Subsequently, the TDNS was quietly dropped and, consequently, the original 2050 decarbonisation timescales will be missed – the 'best case' assumption appears now to be 2065–70.

RIA Electrification Cost Challenge Report March 2019

The RIA Cost Challenge Report, published on 19 March 2019, was produced as a rejoinder to the perception, particularly in political circles, that the management of electrification projects by the railway industry was poor and that delivery was inefficient, over-budget and generally late. Thus, railway electrification, in the minds of many key decision makers and stakeholders, was deemed unaffordable and off the agenda.

These opinions arose primarily as a result of the serious cost and timescale overruns on the Great Western Main Line 25kV Electrification Project (GWEP), although some other electrification projects were also in some difficulty at this time. As a result, to reduce the cost over-runs, a number of schemes were cut back or cancelled, but further consequential increased costs arose in some cases because of the need to provide bi-mode rolling stock rather than straight electric units.

The RIA report contended that the cost and time over-runs were due to an unrealistic programme of work plus an unpreparedness in the use of novel technologies, which resulted in poor productivity. The historical 'start–stop', 'feast or famine' approach to railway electrification was also denounced.

The report summarised UK railway electrification progress since 2007, including reviewing the Great Western Electrification Project in some detail to ascertain the reasons for its failures. The lessons learned were drawn out and highlighted.

The paper then goes on to demonstrate how electrification could be delivered for 33–50 per cent below the cost of recent schemes, using evidence garnered from successful schemes delivered in the UK and internationally.

The report appealed for the railway industry to join together, to reassure politicians and stakeholders that mistakes wouldn't be repeated and urge the government to renew its commitment to electrification. A ten-year rolling programme was called for to build capability thereby reducing long-term costs.

The report endorsed a contemporary report by the Rail Industry Decarbonisation Taskforce, which identified three options to replace diesel, namely:

- battery
- overhead electrification
- hydrogen fuel cell.

The study clearly identified rail electrification as the first choice in a hierarchy of options to decarbonise the network by 2040, also highlighting this solution as the optimal solution for intensively used routes.

Emerging technologies (battery and hydrogen) were cited as options for lightly used lines.

Railway Industry Association's 'Why Electrification?' Report, April 2021

This report, published on 22 April 2021, exhorted the government to a rolling programme of electrification immediately, in order to achieve the administration's own legal commitment to achieve net zero by 2050.

Building on Network Rail's TDNS, the document explains why electrification is essential to decarbonise the railway, emphasising that it is both a future-proof technology and a good investment.

The publication of the report was accompanied by the release of an open letter to the Transport Secretary, Grant Shapps, signed by 18 businesses, calling on the government to begin an immediate programme of electrification.

Railway Industry Association's Report 'A New Strategy for Delivering a Lower Cost, Higher Performing Net Zero Railway by 2050'

In March 2024, the government confirmed that, under current plans, the proportion of electrified routes would increase to 51 per cent of the rail network. Welcoming this commitment, on 11 April 2024, the RIA published its latest ideas on how the railway can become a lower cost, higher performing net zero industry over the next 25 years. These proposals took cognisance of the RIA-commissioned report by Steer, which indicated that passenger numbers would grow between 37 per cent and 97 per cent by 2050, depending on future rail policy.

Building on current government infrastructure plans, the RIA proposed that electrification of a further 15 per cent of routes together with the implementation of alternative energy solutions on secondary lines, would ensure the decarbonisation of all passenger services and 95 per cent of freight services by 2050.

The RIA's system blueprint estimates that most of the remaining third of the network could be decarbonised by utilising battery trains and indicates how, by making logical rolling stock procurement decisions, 'a co-ordinated track and train approach could deliver improved outcomes for passenger and freight users, taxpayers and the supply chain'.

A network map incorporating the RIA proposals is shown in Plate 1.

Chapter 18
Prospects for Electrification Projects Going Forward

There is no doubt that a UK electrified railway network would be a worthwhile investment, delivering a more cost-effective and higher performance system. For the passenger railway, historically it has been proven that the increased business generated as each new route is commissioned improves the viability of the network as a whole.

Electrification also benefits the freight railway, as electric locomotives are typically twice as powerful as diesels, and so are capable of hauling heavier trains at higher speeds, thus increasing network capacity. Furthermore, besides being cheaper to operate, the procurement cost of electric trains is lower than diesel trains, thus reducing the cost of replacing the existing ageing UK diesel fleet.

Electrification is therefore clearly beneficial in its own right.

Decarbonisation is a relatively new priority for rail, but it should not be taken as the driving force for electrification. However, a rolling programme of electrification could play a major role in achieving the UK government's commitment to achieve 'net zero' carbon emissions by 2050 in two ways.

First, as a large proportion of rail services are currently powered by diesel trains, electrifying the network is the key to reducing the railway's carbon emissions.

Secondly, electric trains can make a much bigger contribution than just eliminating the emissions from the trains. This is because electrification also involves upgrading the infrastructure, and, taken together, these enhancements make rail travel more attractive by improvements in speed, connectivity, reliability and capacity. This phenomenon, known as the 'sparks effect', delivers rail services that are better used, with a good proportion of the new business coming from other modes of transport.

Anecdotally, an electric network would deliver a significant degree of modal shift, enabling rail to 'punch above its weight' from an environmental point of view. Thus, the imperative to decarbonisation our world significantly increases the value of a comprehensive electrification strategy.

The authors therefore consider that it is essential to initiate a rolling programme of Overhead Line Electrification of the UK railway's core routes, together with a small number of 'in-fill' third-rail schemes. It is the only practical technology for main line passenger and freight operations.

While battery and hydrogen solutions can also play a part, it should be recognised that these technologies are unlikely to be suitable for most heavy rail operations and should not detract from the need to fully electrify the core routes. Battery and hydrogen technologies are only ever likely to be suitable for light passenger trains for operations on branch lines and possibly, as an interim step, on secondary routes.

The threat to railway electrification from decarbonisation of other forms of transport needs to be recognised. Significantly, steps are already being taken to decarbonise road transport in the UK. Electric cars, battery- and hydrogen-powered buses and vans are already becoming more widely used. Plans are afoot to trial motorway electrification for the trunk haulage of heavy goods vehicles (HGVs). All these initiatives are in direct competition with rail, so it is important that railways secure their place at an early date for 'net zero' funding before other transport sectors abstract it. In this scenario, a tainted railway system may be seen as a less favoured means of travel.

Having completed this dissertation on the history of UK electrification in just about all its aspects, including anecdotal personal information from many of those involved in the schemes described, how can the necessary funding be secured? What are the lessons learned to be acted upon in the future to meet this challenge in a cost-effective manner?

There are three key requirements in the authors' view, namely:

- winning back political confidence on a national and local scale
- driving down the first cost of electrification, and
- initiating a rolling programme to deliver a firm national strategy, incorporating robust project planning.

Political Confidence

The key to achieving the necessary funding for electrification is the necessity of securing political confidence. Unfortunately, in the light of the difficulties experienced in some of the major electrification schemes in the past, the railway often finds itself on the back foot.

Not every decision maker in the UK government is entirely convinced that a rolling programme of electrification is necessary. While there is a ground-swell understanding that decarbonising is a 'must', some senior politicians still hold on to the hope that an alternative solution, such as hydrogen power, would render further wiring of the rail network unnecessary, thus saving a considerable expenditure by the Treasury, despite the evidence brought forward to the contrary.

The ability of the industry to effectively cost electrification projects and deliver them to time and budget is also often called into question by politicians. The example of GWEP is often referred to, even though there have been a considerable number of successful schemes delivered in more recent years, particularly in Scotland.

However, some within the industry seemed under the illusion that the government's requirement to meet its legally bound decarbonisation goals, or, indeed, to help deliver its 'levelling up' agenda, would somehow improve Network Rail's bargaining power, particularly concerning acceptance of past electrification costings. This, however, was always extremely unlikely, and this line was not pursued by the Network Rail board.

In fact, perhaps in view of recent history, there appeared to be a nervousness and a degree of risk aversion at a senior level in Network Rail concerning the development of the Programme Business Case for that organisation's 2020 TDNS. This exercise recognised that electrification of 13,000 STKs would be required to deliver a carbon neutral railway. However, two years after the publication of the study, the key issue as to how an acceptable cost per STK could be delivered, remained unresolved. Therefore, schemes were simply not being put forward because the Programme Business Case costings could not be met. As a result, the DfT's confidence that TDNS could be delivered at an affordable rate waned. Consequently, with the added financial pressures caused by the financial crisis that swept across the UK in the autumn of 2022, the government quietly allowed the dates for decarbonising the railway to slip.

Getting back to fundamentals and to conclude this section on a more positive note, it is important that the industry continues to extol the benefits that electrification brings to the areas it serves. While the 'sparks effect' benefits are now understood by those within and close to the industry, these benefits need to be promoted to a wider audience. It is also important that the resulting social and economic benefits to the local areas served by the route are taken account of, including the initial employment opportunities for local people during construction, involving training and upskilling as well as the longer-term benefits. Historically, it has been proven that every electrification scheme provides an immediate boost and an ongoing benefit to the local economy.

These 'external' benefits should clearly be fully identified, costed and incorporated into future electrification proposals to help make the case for the proposed investment.

Driving Down Costs

To get an electrification strategy translated into a rolling programme, the primary focus must be to drive down costs.

Historically, the 'scheme by scheme' approach in England has invariably resulted in electrification projects being long in gestation, high in start-up and completion costs, with timescales inflated by having resources and materials procured on a 'one-off' basis. Procuring components for each individual project all add expense compared to the lower unit costs and overheads that an efficient ongoing rolling programme would deliver. Consequently, overheads in the UK are estimated to be several times those of the rest of Europe where electrification is generally delivered as a continuous process.

In addition, from a staffing point of view, the current 'start–stop' approach also often means that teams are disbanded at the end of each project and valuable experience is therefore lost. To grow a strong base of engineering talent, the supply chain needs a reliable flow of schemes. This also enables the lessons learned by management, staff, and technical perspectives to be carried forwards from job to job. Assisted by regular training modules, it is easier and cheaper to develop a dedicated workforce maintaining a high level of competence, compared to the recruitment/redundancy cycles that have so often been experienced in the past.

Lessons from past projects, both positive and negative, must be clearly learned and acted on.

If the clock is turned back to the electrification of the East Coast Main Line, the cost per STK at 2025 prices came in £660 per STK. This was one of the most successful major railway projects ever. When this has been compared over the last decade to the spiralling cost of GWEP, the riposte was that the ECML had been delivered on the cheap and that this was the root cause of regular 'de-wirements' in operation. This is, however, strongly refuted by those involved with the ECML electrification, for the works met all the contemporary approved specifications.

In addition, one of your authors, as the project director for the scheme, can vouch that the main cause of 'de-wirements', particularly in Railtrack days, was due to the lack of the requisite maintenance. This was not surprising, as many parts of the OLE system were initially omitted from Railtrack's outsourced maintenance contract. At the end of the day, the ECML scheme was delivered to the timescale and budget that was acceptable to the government. Future schemes need to emulate this.

Both authors of this book attempted to proffer advice to reduce costs on GWEP, but very few of our recommendations were taken up. There is no doubt, from this first-hand experience, that significant savings could have been made but at the time of these interjections the project was already heading for a significant cost over-run. The primary causes were due to an unrealistic programme, under-estimating, over-specifying, high-risk new technology, bad contract arrangements and the resulting poor productivity.

The GWEP cost therefore outturned at £2.8M per STK, approximately three times the original budgeted rate.

So how can the cost per STK be reduced and what, in 2025, was being done to address this key issue? To understand the background, it is pertinent to review the costing initiatives that have been undertaken over the last fifteen years or so.

Looking back to 2007/8, a study undertaken for the DfT by the Rail Safety and Standards Board estimated electrification cost from £500,000 to £650,000 per STK, equating to £780,000 to £1,020,000 per STK at 2025 prices.

The Railway Industry Association's 2019 Electrification Cost Challenge study indicated that 'a well delivered "simpler" electrification project should cost £750,000 to £1M per STK, with more complex' schemes being delivered for under £1.5M per STK. This body's report did much to demonstrate how the industry had learned from past experiences, and it indicated that the way to achieving acceptable STK costings was by implementing a continuous production process.

In addition, the more positive rail electrification experience achieved in Scotland since 2005 by means of an ongoing rolling programme is helpful in showing the way forward for the rest of the UK. North of the border, the Scottish government has been committed to a rolling programme, which, while limited at first, has grown to become a key part of the industry's decarbonisation action plan and Scotland's national transport strategy.

The cost of electrification in Scotland has generally outturned at £1.5M to £2M per STK. Although this compares well with the £2.5 to £3k per STK for England and Wales, it is recognised that the rate in Scotland, as well as for the rest of the UK, needs to be reduced substantially.

Taking cognisance of the foregoing and the Scottish experience, there are therefore currently three impediments that need to be surmounted in order to secure the Westminster government's backing for a UK-wide rolling programme:

- reducing the cost of electrification to £1 to £1.5M per STK
- raising competence by means of effective and comprehensive training arrangements
- securing trust and confidence that the industry can deliver such a programme in a timely and efficient manner.

To meet this challenge, Network Rail and the DfT set up the National Electrification Efficiency Panel (NEEP) in 2021 (see Appendix 7), on which the ORR also is represented. The panel was set up jointly chaired by Professor Andrew McNaughton (former Chief Engineer, Network Rail and HS2 technical director) and Peter Dearman (Senior Electrification Engineer).

Therefore, on first reading, the figures associated with the most recently authorised electrification scheme (Wigan East to Lostock Junction, near Bolton) do not look encouraging. The cost of £78M for 10 kilometres of double-track railway equates to £3.9M per STK. While it must be borne in mind that this is a complex project that includes station upgrades and platform lengthening, together with a lot of civil engineering work, the actual electrification costs per STK are similar to the outturn costs of the Great Western Electrification Programme (GWEP).

Another major costing factor within the Wigan East to Lostock Hall electrification that needs to be highlighted is the risk associated with old mine workings in the area. These risks have often been underestimated in the past. It should be remembered that this was one of the main reasons that delayed the completion of the North-West Electrification Programme, causing the costs to over-run.

Notwithstanding these issues however, electrical clearance modification work is still a major cost factor and, in the authors' opinions, should be at the focus of efforts to reduce the cost per STK going forward. The former British Rail standard laid down the following electrical clearances:

- 'normal' clearance between catenary and structure – 270mm
- 'reduced' clearance – 200mm
- 'special reduced' clearance – 150mm.

In BR days, engineers were generally allowed to exercise their engineering judgement to decide on the optimum solution regarding the design clearance, taking cognisance of the site conditions and the costing of the potential options.

Unfortunately, Network Rail subsequently imposed a far too rigid decision-making regime, requiring a minimum clearance of 270mm unless an individual risk assessment was undertaken. These risk assessments are often undertaken by consultants, who, being very conscious of maintaining their reputation, usually opt for 'normal' clearance.

Given this scenario, an individual safety case for each bridge is often seen as a risk or penalty in time and cost or both. It also encourages estimators to simply assume that bridges will be lifted or rebuilt when costing projects, rather than going for reduced clearances. A simple solution to this issue would be to have a generic safety case that can be applied to bridges with similar clearances issues to give guidance, thus allowing engineers closer to the 'coal face' to make more cost-effective decisions.

This conundrum affects both modifications to bridge and track lowers. For the latter, Network Rail lays down a 100mm allowance for track lifts to allow for future tamping. This is far too generous in the authors' opinion. A change to the tamping regime by adopting 'Sprinter Tamping' would significantly reduce the inevitable (although not desired) compounding of track lifts over time so that this 100mm (4 in) allowance could easily be halved to 50mm (2 in) -- see Appendix 7.

Typically, bridge reconstructions or 'track lowers' come in at £2M to £4M pounds per site. To get these costs down, more should be done to roll out the application of surge arrestors and insulating coatings at bridge sites to minimise the necessary clearance work, as was done at the Cardiff Intersection Bridge (see Chapter 12 and Appendix 3).

Perhaps the main thrust in costing electrification ought to be to consider the potential bridge solutions first. This is the view of the National Electrification Efficiency Panel (NEEP) (see Appendix 7) who are concerned that clearances at bridges are being considered at the end of the design process rather than driving it. Whether out in the countryside or in built-up areas, bridge reconstruction is expensive. This is particularly so in the latter case where services, such as gas, electricity, water or sewage may also be carried within the structure. In addition, usually the gradient of approach roads must be maintained, which can also prove costly. Thus, bridge reconstruction should generally be the option of last resort.

While bridge costings are highlighted above, it is essential that the industry does not lose sight of the need to drive down costs and innovate across the piece. For example, it needs to be recognised that Network Rail's GRIP process is an inefficient means of managing electrification works. The process gives the Route Asset Managers (RAM) a big say in the solution, and while these managers have the responsibility for the technical aspects, under GRIP, they have no accountability for the costs. Almost inevitably therefore, their decision errs on the side of caution regardless of the cost, which their (usually) non-technical superiors find it uneasy to challenge. Therefore, GRIP unfortunately tends to inflate costs rather than control them, and, furthermore, increases timescales involving much unnecessary bureaucracy. Consequently, it has been said that UK electrification requires more paperwork per single track kilometre than any other country in the world!

Accordingly, it was belatedly recognised by Network Rail that the eight-stage linear GRIP process was too inflexible. Network Rail have therefore recently developed a new more streamlined process: PACE (Project Acceleration in a Controlled Environment). This is being utilised on a trial basis, and it is hoped will streamline project processes.

National Strategy and Project Planning

While it is beyond the scope of this book to identify an electrification plan for the future, the authors consider that the Network Rail Decarbonisation proposal represented a good first step towards the development of a strategy, although, as noted earlier, the scope for future electrification has been pared back and timescales extended. In 2024, the Railway Industry Association (RIA) proposed a more modest option for electrification focused on core passenger routes and key heavy freight links, showing how, with the integration of battery-powered routes, a more cost-effective decarbonisation programme could be delivered (see Plate 1). Whatever the final outcome, proposals need to be firmed up and costed, with the scope, specification and management plans comprehensively fixed and documented.

Despite the difficulties experienced in some of the major electrification schemes in the past, there have been many successes (for example, ECML and HS1) and, in a wider sense, the infrastructure for the UK Olympics. It is interesting to note that the latter project was headed up by Sir John Armitt and included many other ex-HS1 personnel.

These schemes all had effective Project Management structures, which encompassed an appropriate balance of competent, skilled personnel who possessed the knowledge, experience and passion to deliver.

Taking cognisance of work done by the NIC, together with the experience of those who have delivered successful projects, when planning a railway electrification project, the following key points should be carefully considered:

- Identify and prioritise a long-term plan within an adaptive framework.
- Scope and specify each project to maximise the benefits that electrification can deliver.
- Ensure costings are realistic and robust at an early stage.
- Scrutinise the economics to ensure that the payback and the 'spin-off' benefits are robust.
- Kindle a positive political will at all levels.
- Implement firm project controls.
- Instigate a robust safety management system.
- Warrant a competent technical approach.
- Choose an appropriate type of contract.
- Manage contractors in a firm but fair manner.
- Carefully select KPIs (measure twice and feedback!)

In the light of the difficulties experienced in some of the major electrification schemes in recent years, the challenge faced by the teams with the responsibility for developing the new generation of electrification projects should not be underestimated. They will have a huge, multi-faceted task on their hands. However, with a clear strategy, a firm project plan, a skilled workforce and a positive attitude, there is no reason why a national rolling electrification programme cannot succeed.

Great British Railways (GBR)

Although a 'transition team' had been at work since 2021, at the time of the General Election in 2024 it was still unclear what form the new Great British Railways organisation would take, or whether it would bear the name originally proposed. In fact, as this text was being drafted, there was some doubt as to whether the elected Labour government would proceed with the setting up of the organisational structure envisaged in the Williams Shapps plan.

It is understood that currently the plan is to devolve electrification projects to the Regions. However, the authors firmly believe that this is a mistake. A national electrification rolling programme must be directed and controlled from the centre to avoid a repeat of the failure of Network Rail's High Output Track Renewals Train when it was prematurely regionalised. The Regions, in running the day-to-day railway, are too busy and have too many distractions to focus on a national project. The electrification work is likely to be the first thing to be deferred or stood down during a perturbation or a resource shortage. Also, this uncoordinated approach could cause the contracting market to overheat if each Region places a high level of orders at the same time. This contributed to the delays and cost over-runs on GWEP.

It is therefore hoped that Great British Railways (or whatever name is chosen for the new organisation) will set up an Electrification Directorate to manage and control the rolling programme of electrification. This organisation would be charged with integrating the Infrastructure Enhancements and Renewals and Maintenance teams to best effect. This body should also interface with the National Grid to ensure that the necessary power supply upgrades are delivered to meet the required programme timescales. In addition, it should review and set appropriate electrification standards, including agreeing safety standards within and without the industry. Headed up by a senior, experienced engineer with project management experience, the Electrification Directorate should implement a delivery methodology that will furnish the nation with an enhanced, cost-effective main line electric railway system.

With the key elements of the industry brought back into a unitary organisation in Great British Railways, together with the government's support for

Northern Powerhouse Rail and whatever elements are taken forward from the Integrated Rail Plan, there is an opportunity to take the industry forward on a more stable path so that it can deliver the transportation service that the country needs for the future. The structure of the new national railway organisation needs to be carefully considered – the government and the industry must get it right this time! Cognisance needs to be taken of the frequent reorganisations that have been foisted on the industry to implement flawed politically ideologies over the last century. The organisation must be capable of adapting to the societal, commercial and environment changes that the future will bring.

For its part, the industry must reach out to provide an effective, efficient, attractive and environmentally responsible service for the nation as a whole. Electrification of the core routes is recognised as a key component to achieve this aim, so the industry must reinvigorate initiatives to reduce the cost of electrification by means of innovation and more efficient use of resources. It is imperative that electrification is made affordable! From experience on the continent and in Scotland, it is clear that a continuous production process is key, both to drive down costs and achieve the rate of progress required.

In addition, it is important for the leaders of the industry to work closely with the government to focus on keeping national and local politicians abreast of developments – elected officials make better choices when they are well-informed.

In writing this book, giving a concise history of UK electrification, the authors' earnest hope is that, going forward, the industry, stakeholders and political decision makers will take cognisance of the lessons that should be learned from the past. While there have been many successful schemes, there is much to learn from the history as there have also been many flawed decisions made both within the industry and at various political levels.

As the authors see it, the foregoing is the only way that a prudent Great British Railways (or whatever name the new integrated rail organisation takes), a risk-averse DfT and a sceptical government can be persuaded to authorise a rolling programme of electrification, with the confidence that each scheme will be delivered to time at an acceptable price.

As the time of writing, even with the watering down of the initial ambitious rail proposals since 2008, the industry stands on the cusp of a potential transformation – a fervent hope is that the opportunity for a rolling programme of electrification does not slip from view once again. In the authors' opinion, the Railway Industry Association's April 2024 strategy would form a good plan to take forward. Note especially that electrifying only 50 miles of short section track 'infills' on otherwise completely electrified end-to-end freight routes would be an easy 'win', economically, operationally, and environmentally.

Electrification OLE Information (courtesy of G. Keenor)

OLE TYPE	Designer	Introduced	Voltage	Location(s)	Notes
SCS-02	BICC	1956	1500V DC	Shenfield-Witham, Shenfield - Southend	Converted to 6.25KV AC1960-2; 25KV AC1979
GE/MSW	BICC	1949 1952	1500V DC 1500V DC	Liverpool St – Shenfield, Christian Street Jn (near Fenchurch St) – Gas Factory Jn (near Bow) Manchester – Sheffield via Wath (curtailed to Ardwick – Hadfield 1981)	Converted to 6.25KV AC1960; 25KV AC1978-87 Converted to 25kV AC 1984
BR Mk 1	BICC	1957	25kV AC	Euston-Weaver Jn - Liverpool; Colwich – Manchester, Crewe – Manchester; Rugby – Stafford via Birmingham; Roade – Rugby via Northampton; Glasgow Suburban Stage 1; Christian St Jn – Shoeburyness; Barking – Pitsea via Grays;	
UKMS	NR		25kV AC	Chelmsford – Colchester; Colchester – Clacton & Walton; Bethnal Green – Bishop's Stortford, Hertford East, Enfield and Chingford Locations in East Anglia	Mk 1 Upgrade
BR Mk 2	BICC	1966 1968	25kV AC 25kv AC	Manchester – Styal Glasgow – Wemyss Bay; Glasgow – Gourock (Glasgow Suburban Stage 2) Clapton – Cheshunt (Lea Valley)	Now removed
BR Mk 3A	BR / BICC	1972	25kV AC	Weaver Jn – Glasgow; King's Cross – Royston & Hertford Loop, Witham – Braintree	
BR Mk 3B	BR 1996	1978	25kV AC	St Pancras – Bedford; Hitchen – Edinburgh, Doncaster – Leeds; Leeds – Ilkley, Skipton & Bradford; Carstairs – Edinburgh; Bishops Stortford – King's Lynn; Royston & Stanstead; Colchester – Norwich & Harwich; Romford – Upminster; Wickford – Southminster; Paisley – Ayr & Largs; West London Line, Paddington – Heathrow	Modified system Stockley Jn – airport Terminals
BR Mk 3C	BR BR / NR	1980 2010	25kV AC 25kV AC	Stockport – Hazel Grove, London Fenchurch St – Christian St Jn Airdrie – Haymarket via Armadale & Bathgate	
BR Mk 4	BR / BB	1992	25kV AC	Not used	Developed for WCML upgrade but not used

Name	Year	Contractor	Voltage	Route	Notes
BR Mk 5	1993	BR / BB	25kV AC	Dolland's Moor Yard	
CT	1994	BB	25kV AC	Cheriton – Frethun (Channel Tunnel)	
SICAT5	2005	Siemens	25kV AC	Haughead – Larkhill; Shields – Gourock	
SNCF Systra	2003	Amec Spie	+/- 25kV AC	HS1 CTRL Fawkenham Jn – St Pancras	
V360	2025	SNCF Reseau	+/- 25kV AC	HS2 Euston – Birmingham	360kph OLE. (Note Type for tunnels TBA)
NR UK1	2003	BB Atkins	25kV AC	Crewe – Kidsgrove	UK1 also used on some sections of WCML upgrade
NR Series 1	2012	Furrer+Frey	25kV AC	GWML Electrification Stockley Jn – Reading – Bristol Parkway – Cardiff; Wootton Bassett – Chippenham – Thingley Jn; Reading - Newbury	
NR Series 2	2012	NR	25kV AC	Manchester Victoria – Edge Hill; Manchester – Preston – Blackpool; Rutherglen – Coatbridge; Cowlairs – Cumbernauld – Greenhill; Greenhill – Larbert; Polmont – Larbert; Stirling – Dunblane – Alloa; Gospel Oak – Barking; Midcalder Jn – Holytown Jn via Shotts; Walsall – Rugeley; Tinsley Viaduct – Rotherham	
NR Series 2T	2014	NR	25kV AC	Glasgow Queen St Tunnel – Newbridge; Larbert – Stirling	
	2017	NR	25kV AC	Glasgow Queen St	
NR UKMS125	2017	NR	25 KV AC	Bedford – Kettering – Corby & Market Harborough	
ROCS	2017	NR	25kV AC	GWML Chipping Sodbury Tunnel between Swindon and Bristol Parkway	Rigid Overhead Conductor Bar (Copper)
ROCS	2019	Furrer+Frey	25kV AC	GWML Severn Tunnel between Bristol Parkway and Newport	Rigid Overhead Conductor Bar (Aluminium)
ROCS	2016	Crossrail/ Atkins	25kV AC	Crossrail Tunnels (Pudding Mill Lane – Paddington)	Rigid Overhead Conductor Bar (Copper)
CR	2008	NR Furrer+ Frey	25kV AC	Crossrail Pudding Mill Lane; Liverpool St - Shenfield	

Appendix 2

Geographical Description of the West Coast Main Line (WCML) Route and Electrification Dates

Route Description

The WCML was not conceived as a single trunk route as an entity; rather it was formed over time by linking up lines constructed by a number of different railway companies. Often, the railway promoters side-stepped opposition to their railway projects by avoiding rural towns and landowners' estates and, in addition, tended to follow natural contours through hilly areas. As a result, the linespeed on many of today's main lines, and in particular, the WCML, are constrained by the curvature of the route.

The WCML also features significant gradients, immediately upon leaving its southern terminus at Euston, through the Chilterns Hills at Tring, the Watford Gap, the Northampton uplands and the Trent Valley. There are particularly steep and long climbs to the summits at Shap in Cumbria and at Beattock in Lanarkshire. The speed of steam worked express passenger trains was constrained by the physical features of the route and freight trains often needed the assistance of banking engines. The line was therefore expensive to operate with extended journey times for passenger services.

The WCML is a network of routes that diverge and rejoin the central core route between London and Glasgow. The section from London to Rugby is the original London to Birmingham railway designed and built by Robert Stephenson and opened in 1837. The line from Birmingham to Wolverhampton and Stafford opened in 1839. It was the original main line until a cut-off was constructed via the Trent Valley in 1847. There is a loop to the south of Rugby that serves Northampton.

From Stafford, the route continues to Crewe and then on to Warrington. At Crewe there is a branch to Manchester via Wilmslow, and 15 miles further north there is a branch from Weaver Junction to Liverpool. Beyond Warrington the route continues northwards, serving Wigan, Preston, Lancaster, Carlisle, Carstairs and Motherwell and terminating in Glasgow. A further branch at Carstairs diverges to Edinburgh.

Manchester is also served by two further loop lines, one from Colwich Junction in the Trent Valley south of Stafford, another from the north of Stafford, both of which go via Stoke-on-Trent and Macclesfield.

West Coast Main Line Electrification Dates

The various sections of the southern part of the route were energised in stages as follows:

Styal Line and Manchester Oxford Road to Piccadilly and Stockport:
Testing from September 1958
Open to traffic: 27 November 1959
Manchester to Crewe and Macclesfield: 12 September 1960
Crewe to Weaver Junction: 1 January 1962
Liverpool to Weaver Junction: 14 January 1962
Crewe to Stafford: 7 January 1963
Stafford to Tamworth: 8 July 1963
Tamworth to Nuneaton: 8 March 1964
Nuneaton to Rugby: 30 November 1964
Rugby to Northampton: 9 July 1965
Mitre Bridge Junction to West London Junction/Sudbury Junction:
27 September 1965
Northampton to Willesden Yard: 27 September 1965
Maiden Lane Junction to Camden Junction: 15 November 1965
Euston to Willesden Yard: 22 November 1965
Rugby to Hanslope Junction: 3 January 1966
Bescot to Portabello Junction/Cranes Street Junction/Bushbury Junction:
18 April 1966
Aston North Junction to Perry Bar and Bescot: 3 October 1966
Coventry to Stetchford North Junction: 3 October 1966
Stetchford North Junction to Aston South Junction: 3 October 1966
Aston South Junction to Aston North Junction: 6 December 1966
Bescot to Bescot Junction and Walsall: 6 December 1966
Birmingham New Street to Wolverhampton: 6 December 1966
Perry Bar, Soho, Proof House and Aston North Junctions: 6 December 1966
Rugby to Coventry: 6 December 1966
Wolverhampton to Stafford: 6 December 1966
Colwich to Stone: 6 March 1967
Norton Bridge to Cheadle Hulme: 6 March 1967
Stetchford North Junction to Birmingham New Street: 6 March 1967
Weaver Junction – Bamfurlong Junction: 19 March 1973
Bamfurlong Junction – Preston: 23 July 1973
Glasgow Central to Muirhouse Central Junction, Cathcart East and West
Junctions: 1 September 1973
Cathcart East Junction to Motherwell Junction: 1 September 1973
Preston to Carlisle: 25 March 1974
Carlisle to Kingmoor Yard: 25 March 1974
Carlisle to Motherwell: 22 April 1974

Appendix 3

Reduced Clearances at the Cardiff Intersection Bridges

Working closely with the RSSB and ORR, Network Rail commissioned and supervised a series of tests at Southampton University to establish the possible minimum electrical clearance between the catenary wires and an overhead metal structure.

Surge voltages occur due to lightning strikes and fault currents. BS EN 60060-1 (high-voltage test techniques) indicates that to simulate a lightning strike, an impulse voltage of an amplitude of 193kV should be used for testing purposes. This voltage was used in the Southampton University tests.

The programme focused on the testing of three mitigating measures, namely:

- an insulated covering (for metal surfaces) developed by GLS Coatings
- utilising insulated under-bridge arms currently in use by Network Rail
- incorporating surge arresters developed by Siemens.

The Siemens equipment diverts the surge current through the surge arrester to earth. Siemens warranted their equipment to arrest a surge voltage down from 200kV to 70kV.

When tested at 193kV under dry conditions without any other mitigation measures, a Siemens surge arrester achieved a minimum clearance of 40mm between a British Rail designed under-bridge arm and a steel earthed plate. This was reduced to 30mm when the new Network Rail insulated under-bridge arm was tested.

The tests were repeated for the Network Rail under-bridge arm under wet conditions and the minimum clearance under these conditions was found to be 70mm.

Tests on the Network Rail arm, when all three of the mitigations were applied under dry conditions, showed that the arm could physically touch the insulated coated metal surface without any collapse of voltage.

Following these successful tests, the Cardiff intersection bridges were coated with the GLS 100R insulating coating and the OLE fitted with insulated under-bridge arms together with Siemens surge arresters on either side of the bridges. The concrete surface of the culvert received some 'patch' repairs and ballast, to a depth of 100mm, laid beneath hardwood timber sleepers. The tracks beneath the bridges were 'glanded' to the track on either side, utilising old bullhead rails laid on their side.

Appendix 4
Electrification Reports and Documents

All available on: www.railwayarchive.co.uk

1912 (6 Dec)	Electrification of the London Suburban System (H Walker Letter to LSWR Board)
1912 (6 Dec)	Report Upon the Electrification of the LSWR Suburban Lines
1920 (12 Jul)	Electrification of Railways Advisory Committee: Interim and Final Reports
1921 (Jan)	Proposed Substitution of Electric for Steam for Suburban, Local and Main Line Passenger and Freight Services (LBSCR)
1927 (23 Jul)	Report of the Railway Electrification Committee (1927)
1928	Requirements for Passenger Lines and Recommendations for Goods Lines of the Minister of Transport in Regard to Railway Construction and Operation
1931	Report of the Committee on Main Line Railway Electrification 1931
1932	Railway (Standardisation of Electrification) Order
1939 (Jul)	Electrification of LNER Suburban Lines – Liverpool Street and Fenchurch Street to Shenfield: Contract with British Insulated Cables Ltd (for LNER)
1944 (May)	Report on Proposed Extensions of Electrification (Southern Railway)
1950	Requirements for Passenger Lines and Recommendations for Goods Lines of the Minister of Transport in Regard to Railway Construction and Operation
1951	Electrification of Railways: Report of a Committee Appointed by the Railway Executive and the London Transport Executive
1956	Modernisation of British Railways: The System of Electrification for British Railways
1960	British Railways Electrification Conference London 1960: Railway Electrification at Industrial Frequency – Proceedings

1966	Railway Electrification on the Overhead System – Requirements for Clearances
1968	Route Improvements Electrification – Weaver Junction to Glasgow
1972	Railway Electrification on the Overhead System – Requirements for Clearances
1977	Railway Construction and Operation Requirements – Structural and Electrical Clearances
1978 (May)	Railway Electrification: A British Railways Board Discussion Paper
1981 (11 Feb)	Review of Main Line Electrification: Final Report (BR)
1988	Railway Electrification: 25kV AC Design on BR (Pamphlet)
2007	Study on Further Electrification of Britain's Railway Network (RSSB)
2007 (23 Oct)	White Paper and the Case for Electrification
2007 (9 Nov)	The Case for Electrification (DfT)
2009 (14 Apr)	Network RUS Working Group 4 Electrification – Network Rail's Remit
2009 (15 May)	Network RUS: Electrification Strategy – Draft for Consultation (Network Rail)
2009 (23 Jul)	Britain's Transport Infrastructure – Rail Electrification (DfT)
2009 (Oct)	Network RUS Electrification (Network Rail)
2010 (25 Nov)	Rail Investment – Secretary of State for Transport's statement confirming the electrification of the Liverpool and Manchester line and the first stage of the GWML
2011 (1 Mar)	Intercity Express and Rail Electrification – Secretary of State for Transport's statement confirming the ordering of Intercity Express Programme (IEP) Trains and the second stage of the GWML
2020 (31 Jul)	Traction Decarbonisation Network Strategy – Interim Programme Business Case (Network Rail)
2021 (Mar)	Wales and Western Regional Traction Decarbonisation Strategy (Network Rail)

Appendix 5

London Tilbury and Southend and Great Eastern Recollections – the Sparks Effect*

Having inherited his passion for railways from his father, Theo Steele became 'one of the most enterprising and popular railway managers either side of privatisation'. He was appointed Passenger Manager, Liverpool Street in 1982 and was subsequently promoted to Assistant General Manager of BR Eastern Region. After spells away from 1997 (Western Region and Channel Tunnel Rail Link), he returned to Liverpool Street as Commercial and Deputy Managing Director of First Great Eastern between 1998 and 2007.

Theo read Geography at Pembroke College, Oxford, graduating in 1970. Herewith are his recollections, in his own words, of a study he undertook in his final year at university regarding passenger growth on the Great Eastern and London, Tilbury & Southend lines between 1955 and 1968.

'The "Sparks Effect" was a term coined after the West Coast electrification of 1966/7. However, a read of my degree Geography Dissertation completed in 1970 unearthed some figures for Season ticket holdings at South-East Essex stations for 1955 (pre-electrification of either route to Southend) and 1968 six years after the LTS route electrification was finally completed – about 2 years later than anticipated. While trains were counted visually there was no centralised LENNON allocation to stations in those days and the best indicators of growth were season ticket holdings which were normalised to annual holdings.

'The figures for stations Billericay and Laindon eastwards are shown for 1955 and 1968 below:

GE Route	1955	1968	LTS Route	1955	1968
Billericay	1,768	4,815	Laindon	584	2,422
Wickford	809	1,005	Pitsea	463	827
Rayleigh	1,432	4,356	Benfleet	868	6,815
Hockley	586	2,693	Leigh	1,985	3,062
Rochford	178	747	Chalkwell	1,078	1,287
Prittlewell	358	304	Westcliff	1,598	1,033

---

* 'Sparks Effect' – where the electrification of passenger rail systems leads to significant increases of both passengers and revenue.

GE Route	1955	1968	LTS Route	1955	1968
Southend Victoria	397	448	Southend Central	1,132	837
			Southend East	687	899
			Thorpe Bay	742	1,167
			Shoeburyness	70	150
Total	5,528	14,368	Total	9,207	18,499

'Not unexpectedly the largest growth is at stations like Billericay, Benfleet or Rayleigh where there was a lot of building around the time of electrification – Benfleet UD had a population of 20k in 1951; 33k in 1961 and 42k in 1971 and Rayleigh UD grew from 10k in 1951 to 30k in 1961. In Southend (Leigh east), the population growth was more modest as the area was largely built up by 1960 and indeed in the centre housing was giving way to a ring road and office blocks – there was a double whammy of a very significant retired population.

'The GE route had about 18 trains a day before electrification, which brought 3 trains per hour off peak and a 5/10 min interval service in the high peak with many trains expanded from 8 to 12 cars by 1962.

'Obviously since these figures Basildon station has been built. My dissertation also notes that, "unusually" both lines covered their costs in the grant submissions submitted in 1969.'

Theo Steele
October 2020

Theo sadly passed away on 20 October 2021, aged 72, while this book was being written.

Appendix 6

Great Western Electrification Project – Monitoring Systems to Ensure Safe Track Lowering Works in Tunnels

Co-author, John Buxton recalls his involvement initiating a safe system of working for track lowering jobs.

In conjunction with my involvement with bridge, structure and tunnel design issues, I played a key role to initiate a scheme to ensure structural and safety assurance during track lowering works in tunnels. Major track lowering was necessary at the Alderton, Box and Patchway Tunnels.

Working with Aecom and their sub-contractor, Senceive, I helped devise a structural monitoring system, utilising sensor devices that were developed specifically for tunnel and bridge works. The digital measuring sensor devices developed (tilt sensors and shape arrays) were extremely sensitive with an accuracy of +/- 0.01mm.

These devices were fixed to the tunnel walls and soffit at locations pre-determined at each structure to be monitored, by an extensive structural analysis. Movements within the tunnel structure during track lowering works, which were often quite small, were thus very accurately measured. These 'real-time' outputs were transmitted via a sophisticated data communication system to an external control centre. This centre was manned 24/7 by competent technicians, who kept a close eye on the monitoring outputs. In addition, the comms system also automatically texted signals to the smart phones of key personnel involved with the works.

This arrangement provided real time structural monitoring, with alarms giving the location of any settlements that exceeded pre-determined tolerances.

Backing up the clever technology was a well-thought-through monitoring procedure. A Monitoring Action Plan defined the responsibilities, protocols and procedures to be implemented depending on the quantum of structural movement detected.

During the execution of the site works, monitoring systems and procedures were overseen by key project personnel organised into a formalised structured management system to ensure the monitoring data was properly analysed. A safety review group (SRG), comprising client and contractors' technicians, met at the end of every shift to consider the monitoring outputs for the period.

In addition, an Engineering Review Panel (ERP) of senior engineers, which I chaired, convened to review the overall situation every 24 hours.

A graded suite of tolerance limits had been determined, together with actions to be taken should any set limit be exceeded. An ad-hoc SRG was immediately called if lower tolerance limits were breached, to consider and implement a pre-determined action plan to arrest any excessive settlements. This usually required the excavation extents to be reduced, For structural movements which exceeded the maximum permitted tolerance, a 'safe stop' of the works was immediately implemented and the ERP would be quickly convened.

When such an event occurred, an emergency action plan would quickly be developed and implemented to stabilise the situation. When the ERP was satisfied that all was in order again, a 'safe start' procedure would be initiated. The system worked very well and generally ensured that the works proceeded without interruption. When any set tolerances were exceeded, prompt action forestalled any emerging difficulties – no situations arose that required the evacuation of any worksites.

The system was first successfully utilised in Alderton Tunnel. Lessons were learned from this experience and a more extensive arrangement was installed in Box Tunnel for the track renewal and lowering works there. Subsequently, the system was adapted for the track lowering works for the extensive and very sensitive structural retaining walls and other bridges and structures through Bath Sydney Gardens.

Following this, the 'fragile' Patchway Tunnels were particularly challenging from a structural point of view. However, the system undoubtedly improved both the efficiency and safety of the works. The technology and the procedures that we developed had much to commend them. As a result, our combined team won two awards, namely:

- NCE Tunnelling and Underground Space Award 2015
- Project of the Year (Infrastructure) South West Built Environment Award 2016.

The methodology is capable of being adopted for the monitoring and management of track lowers within or around any sensitive structures. It particularly provides essential safety and operational assurance in a cost-effective manner for electrification track lowering operations.

Appendix 7
National Electrification Efficiency Panel (NEEP) Initiatives

The National Electrification Efficiency Panel (NEEP) has identified, as a first phase, nine actions to reduce the cost of electrification as follows:[*]

- Bridge parapets – secure widespread adoption of deriving parapet heights from risk assessment rather than blanket application of 1.8m height
- Voltage-controlled clearances – roll out standards for the use of surge arrestors and insulated coatings as applied at Cardiff Intersection Bridges to allow lower clearances to be derived from a suitable risk assessment
- Track vertical allowances – derive economically affordable track lift allowance to protect locations with tight vertical electrical clearance
- Trial hole alternatives – avoid hand-digging trial holes at every OLE foundation with, for example, the use of ground-penetrating radar
- Insulated pantograph horns – adopt this type of pantograph, which are used throughout Europe
- Wire gradients – update design principles to avoid infrastructure interventions in close proximity to level crossings and bridges
- OLE structure spacing – change design rules to reduce the number of structures per mile by optimising spacing
- Rationalise traction distribution principles – reduce the number and scale of electrical substations with designs that use the best modern practice in electrical switchgear and control design
- OLE structure design range – limiting the available range of structure types in the UKMS design range to reduce supply-chain complexity and improving the visual appearance of OLE.

These initiatives are all being progressed, and some have reached the stage where Technical Advice Notes have been issued as a precursor to a 'standards' change.

Future phases are planned to consider training and competence, procurement models, risk ownership, plant and overhead costs – the first and last items being the most urgent.

The action 'bullet' points above are all endorsed by the authors of this book. Some comments on the issues of which your authors have first-hand experience follow to expand on the proposed solutions.

---

[*] *Rail Engineer Magazine*, May - June 2022.

Bridge Parapets

In developing solutions for the Brunellian bridges through the Bath area for the GWEP electrification, it was clear that raising the parapets in the standard manner to the 1.8m minimum height would not be acceptable, particularly through the Sydney Gardens section. Although not implemented due to the subsequent cancellation of the electrification works between Thingley Junction and Bristol Temple Meads, some elegant, 'non-standard' solutions were considered, and derogations would have been sought for their implementation.

Voltage-controlled Clearance

Typically bridge reconstructions or track 'lowerings' come in from £2M to £4M per site and are to be avoided, if at all possible.

For 25kV AC systems, Network Rail require a minimum electrical clearance of 270mm but a reduction to 150mm is possible in many circumstances, providing an individual risk assessment is undertaken. These risk assessments are, in the authors' experience, often carried out by engineers who are often both too risk averse and too conscious of maintaining their reputation. Usually the 'safe' option of 'normal' clearance is chosen, which results, often un-necessarily, in the decision coming down in favour of a bridge reconstruction.

The adoption of a generic safety case, with guidelines, would encourage engineers to make more cost-effective decisions.

In addition, the simple but innovative solution adopted at the Cardiff Intersection Bridge described in Chapter 12 should be rolled out at bridge sites to further minimise any necessary clearance work.

For track-lowering designs, the allowance for track lifts to allow for future tamping also needs to be reviewed. Network Rail currently lays down a 100mm allowance for future 'maintenance' track lifts, and a reduction is proposed by NEEP. A change to the tamping regime with the adoption of 'Sprinter Tamping' would significantly reduce the inevitable (although often undesired) compounding of track lifts over time, enabling this 100mm allowance to be halved to 50mm.

Trial Hole Alternatives

Ground-penetrating radar has been used extensively for investigating the condition of ground conditions in tunnels and should therefore be capable on a wider infrastructure basis of adapting to identify cables, services and boulders that may impede the use of mechanised equipment for OLE mast erection. It is, however, worth mentioning that, ironically, the pits for foundations of the Heathrow Express electrification were all dug by hand on weekdays so that there were virtually no cables cut nor services broken for that project at these sites, compared to the problems later experienced on GWEP.

Wire Gradients

The experience from the GWEP scheme's Steventon Road Bridge has demonstrated that wire gradients can far be steeper than the current standard. These need to be codified.

Other Initiatives that Should be Adopted

Structural Monitoring

The structural monitoring system developed for track lowering in tunnels and in the vicinity of other potentially sensitive structures proved to be very worthwhile on GWEP, achieving a higher level of safety assurance than had been possible in the past. This system should be adopted as a standard procedure in such locations.

Bridges and Tunnels

In the authors' experience the viability of most electrification schemes centre around the risks and costs of the required bridge and tunnel works. For GWEP, rash assumptions were made concerning bridge and tunnel clearance works and, as a result, these works were seriously under-estimated for both cost and time. Had these costing been more thoroughly carried out at an early stage the overall project estimate would have been more realistic, and undoubtedly a range of more cost-effective alternative solutions would have been generated for consideration.

In any event, bridge reconstruction should always be seen as a last resort due to the huge costs of not only reconstructing the structure but also the need to replace utilities that may be carried within the structure. In addition, the works are often required to improve the gradients of approach roads carried by the bridge are also extensive and thus often very expensive, particularly if property acquisition if required.

While bridge and tunnel costings have been specially highlighted, it is essential that the industry does not lose sight of the need to drive down costs and innovate across the piece. In so doing, careful assessment of the risks must also be undertaken. On GWEP, unfortunately, a snap, 'politically inspired' decision to electrify a route that was way down the industry's agenda required a project plan to be put together in haste. Insufficient consideration was given to many aspects of the project, which, together with a range of last-minute good ideas (for example, HOPS), compromised the scheme from the start.

Bibliography

Railway companies', British Transport Commission's and British Railways Boards' annual and other reports, publications and documents

British Railways' Electrification Conference, London, 1960 – 'Railway Electrification at Industrial Frequency' proceedings

Proceedings of:
  The Institution of Civil Engineers
  The Institution of Electrical Engineers
  The Institution of Locomotive Engineers
  The Institution of Mechanical Engineers
  The Institution of Railway Signal Engineers
  The Permanent Way Institution

Euston Main Line Electrification – Technical Conference: October 1966

Modern Railways

Railway Engineer

Rail News

The Railway Gazette

The Railway Magazine

Acts of Parliament, consultative documents and White Papers on railway electrification

Electrification of Railways Advisory Committee: interim and Final Reports: July 1920

Proposed Substitution of Electric for Steam for Suburban, Local and Main Line Passenger and Freight Services, January 1921

Report of the Railway Electrification Committee, 1927: Ministry of Transport

Report of the Committee on Main Line Electrification, 1931: Ministry of Transport

Railway (Standards of Electrification) Order, 1932

The Case for Electrification of the Railways: British Electrical Development Association: 1935

Acts of Parliament, Consultative Documents and White Papers on Railway Electrification

Modernisation and Re-equipment of British Railways: British Transport Commission 1955

Modernisation of British Railways: The System of Electrification for British Railways: British Transport Commission 1956

The British Transport Commission – Proposals for Railways: HMSO 1956

BR: Re-Appraisal of the Plan for the Modernisation and Re-equipment of BR: HMSO July 1959

Railway Accidents – Failures of Multiple-Unit Trains on BR: Final Report: MOT: March 1962

Railway Electrification on the Overhead System – Requirements for Clearances, MOT: 1967

Review of Main Line Electrification – Interim Report: HMSO, 1979

A Lifetime with Locomotives: R.C. Bond 1975

The Birth of British Rail: M.R. Bonavia, 1979

Bob Reid's Railway Revolution: G. Muir, 2021

BR in Transition: O.S. Nock, 1963

Britain's New Railway: O.S. Nock, 1966

Britain's Railways – 1997–2005 Labour's Strategic Experiment: T.R. Gourvish, 2008

British Rail – The Making and Breaking of our Trains: C. Wolmar, 2023

British Rail 1974–1997 – From Integration to Privatisation: T.R. Gourvish, 2002

British Railways 1948–1973 – A Business History: T.R. Gourvish, 2011

Bulleid, Last Giant of Steam: S. Day-Lewis, 1964

DC Electric Trains and Locomotives in the British Isles: R.L. Vickers, 1986

The Docker's Umbrella: P. Bolger, 1992

Electric Euston to Glasgow: O.S. Nock, 1974

Electric Railways 1880–1990: M.C. Duffy, 2003

Electric Trains and Locomotives: B.K. Cooper, 1954

Electric Trains in Britain: B.K. Cooper, 1979

Electric Trains (Their Equipment and Operation), Vols 1 and 2: W.A. Agnew, 1937

Electrifying the East Coast Route: P. Semmens, 1991

History of the Great Western Railway (1930–39): P. Semmens, 1985

History of the Southern Railway (combined Vol 1 and 2): C.F. Dendy Marshall and R.W. Kinder, 1968

The InterCity Story: M. Vincent and C. Green, 1994

The Last Steam Locomotive Engineer: Colonel H.C.B. Rogers, 1970

The LBSCR Elevated Electrification: S. Grant, 2011

London Underground (The Story of the Tube): O. Green, 2020

The Network SouthEast Story: M. Vincent and C. Green, 2015

The Organisation of British Rail: M.R. Bonavia, 1971

Overhead Line Electrification for Railways (6th Edition): G. Keenor, 2021

The Provincial Railways Story: G. Pettit and N. Comfort, 2015

Railway Electrification: Case for a Continuous Programme: Aims of Industry Pressure Group, 1973

The Severn Tunnel (Construction & Difficulties 1872–1887): T.A. Walker, Kingsmead Reprint, 1969

Sir Herbert Walker's Southern Railway: C.F. Klapper, 1973

Southern Electric: J. Glover, 2001

Southern Electric (4th Edition): G.T. Moody, 1968

The Southern Electric Story: Michael H.C. Baker, 1993

Underground (How the Tube Shaped London): D. Bownes, O. Green, S. Mullins, 2012

Abbreviations

4TC	a 4-coach unpowered set
AC	alternating current
ASLEF	Associated Society of Locomotive Engineers and Firemen
ATP	Automatic Train Protection
BAA	British Airports Authority
BEMU	Battery Electric Multiple Unit
Bn	Billion
BR	British Railways or British Rail
BRB	British Railways Board
BTC	British Transport Commission
CB	Commander of the Bath
CBDC	Cardiff Bay Development Corporation
CCTV	closed-circuit television
CEGB	Central Electricity Generating Board
CFS	catenary-free sections
CIÉ	(Córas Iompair Éireann) Irish State Railways
CMG	Commander of the Order of St Michael and St George
CP	control period
CTA	Central Terminal Area, Heathrow Airport
CTRL	Channel Tunnel Rail Link
CVO	Commander of the Victorian Order
DB	Deutsche Bundesbahn (later Deutsche Bahn)
DC	direct current
DfT	Department for Transport*
DMU	diesel multiple unit
DOO	driver only operation
DoT	Department of Transport*
DRS	Direct Rail Services
DVT	driving van trailer
ECML	East Coast Main Line
ECR	electrification control room

ECS	empty coaching stock
EF1	electric loco (freight)***
EFE	electrification fixed equipment
EGIP	Edinburgh Glasgow Improvement Project
ELLX	East London Line Extension
EM1	electric mixed traffic locomotive type 1
EM2	electric mixed traffic locomotive type 2
EMU	electric multiple unit
ES1	electric loco (shunting)***
FRGS	Fellow of the Royal Geographical Society
GBE	Knight Grand Cross of the Most Excellent Order of the British Empire
GEC	General Electric Company
GRIP	Governance for Railway Investment Projects
GWEP	Great Western Electrification Project
GWML	Great Western Main Line
GWR	Great Western Railway
HOPS	high output plant system
hp	horsepower
HQ	headquarters
HS1	High Speed line 1 (London to the Channel Tunnel)
HS2	new high speed line between London and Birmingham
HST	high speed train
Hz	Hertz (cycles per second)
ICI	Imperial Chemical Industries
IEE	The Institution of Electrical Engineers
K	thousand
kmh, kph	kilometres per hour
KPI	key performance indicator
kV	kilovolt
kW	kilowatts
L&Y	Lancashire & Yorkshire Railway
LBSCR	London, Brighton & South Coast Railway
LCR	London and Continental Railways
LENNON	computerised system that collects, processes and distributes ticket sales data

LLP   limited liability partnership

LMD   light maintenance depot

LMR   London Midland Region

LMS/LMSR   London, Midland & Scottish Railway

LNER   London & North Eastern Railway

LNWR   London & North Western Railway**

LOR   Liverpool Overhead Railway

LSWR   London & South Western Railway

LTSR   London, Tilbury & Southend Railway

M   million

Mbit   megabit (1 million bits of data)

MD   managing director

Mk3 or 3b   type of overhead electrification system

MML   Midland Main Line

MP   Member of Parliament

mph   miles per hour

MR   Midland Railway

MTR   MTR Corporation Limited, a Hong Kong based rail operator

MW   megawatts

NATM   New Austrian Tunnelling Method

NB   nota bene – please note

NCE   New Civil Engineer magazine

NEEP   National Electrification Efficiency Panel

NER   North Eastern Railway

NIC   National Infrastructure Commission

NLR   North London Railway

NR   Network Rail

NSE   Network SouthEast

OCS   overhead catenary system

OHLE or OLE   overhead line equipment

OLEC   mandatory courses for working on Network Rail overhead line equipment

ORR   (Government) Office of Rail and Road

OSD   overhead system design

PCM   pulse code modulation system

Pic-Vic (Manchester) Piccadilly and Victoria proposed but never built railway

PSB   power signal box

PTA	Passenger Transport Authority
PTE	Passenger Transport Executive
R&C	Rutherglen and Coatbridge Project
REB	relocatable equipment building
RIA	Railway Industry Association
RSSB	Railway Safety and Standards Board
S&T	signalling and telecommunications
SECR	South Eastern and Chatham Railway
SER	South Eastern Railway
SHES	South Hants and East Sussex
SNCF	French State Railways
SR	Southern Region
SRA	Strategic Rail Authority
STK	Single track kilometres
T&RS	traction and rolling stock
TAWS	Transport and Works Act Scotland (or Order under this Act)
TDNS	Traction Decarbonisation Network Study
TfL	Transport for London
TGV	French high speed train
TI	Traction (Immune) track circuit
TNO	transmission network operator
TRUB	trailer, restaurant buffet
TSC	track sectioning cabin
TSI	technical specifications for interoperability
UD	urban district
UK	United Kingdom
USA	United States of America
V	volts
WCML	West Coast Main Line

---

* These abbreviations varied according to the style that the Secretary of State wished the department to be called.

** From its formation up until the early 1900s, this was the biggest joint stock enterprise in the world.

*** North Eastern Railway sub-classes.

Acknowledgements

The authors wish to acknowledge the help that a large number of colleagues have given them over the period of writing this book and wish to thank:
Clive Atkins, Colin Berry, Laura Booth, Richard Carr, Bob Clarke, David Clarke, Richard Connelly, Alan Cooksey, Gary Decoine, Julia Decoine, Richard Fearn, Hugh Fenwick, Peter Forbes, Chris Green, Terry Hutton, John Jouques, Gary Keenor, Clive Kessell, Rod Lees, Kevin McCallum, Brian Mallon, Colin Marsden, Chris Mew, Frank Paterson, Roger Pease, Garth Ponsonby, Geoff Quinn, Dr Malcolm Reed, Bill Reeve, Kevin Robertson, Alan Ross, Ted Sadler, Alan Sarbut, Andy Savage, David Shirres, Richard Spoors, Michael Staff, the late Theo Steele, Jim Summers, Brian Sweeney, Jim Wheeler, Chris Willey, Julian Worth

The authors are also grateful to the publishing staff for their helpful suggestions regarding the layout, presentation and promotion of the book, in particular: Ryan Gearing and Lauren Tanner at Unicorn; jacket designer, Matt Carr; and book designers, George Newton and Nick Newton.

Special thanks also to our wives, Janet Buxton and Alison Heath, for their encouragement, support and patience during the protracted period we spent drafting this book. John would also like to express gratitude to his son Oliver Buxton and daughter, Selina Ellacott, for their helpful ideas and advice.

In conclusion, Messrs Buxton and Heath acknowledge the kindness of Noel Broadbent in providing the foreword, and Peter Dearman for the introduction.

Index

Page numbers in italics indicate a rail map or other diagram. Page numbers for appendices are indicated by the page number plus the suffix 'Ap' e.g. 297Ap1.